DARK MATTER AND COSMIC WEB STORY

Advanced Series in Astrohysics and Cosmology

ISSN: 1793-1312

Series Editor: Remo Ruffini *(ICRA, Pescara & University of Rome "La Sapienza", Italy)*

Published*

Vol. 2 Galaxies, Quasars and Cosmology
        edited by L Z Fang and R Ruffini

Vol. 3 Quantum Cosmology
        edited by L Z Fang and R Ruffini

Vol. 4 Gerard and Antoinette de Vaucouleurs: A Life for Astronomy
        edited by M Capaccioli et al.

Vol. 5 Accretion: A Collection of Influential Papers
        edited by A Treves, L Maraschi and M Abramowicz

Vol. 6 Lectures on Non-Perturbative Canonical Gravity
        by A Ashtekar, Notes prepared in collaboration with R S Tate

Vol. 7 Relativistic Gravitational Experiments in Space:
        First William Fairbank Meeting
        edited by M Demianski and C W F Everitt

Vol. 8 Hawking on the Big Bang and Black Holes
        by S Hawking

Vol. 9 Accretion Disks in Compact Stellar Systems
        edited by J C Wheeler

Vol. 10 The Chaotic Universe —
        Proceedings of the Second ICRA Network Workshop
        edited by V G Gurzadyan and R Ruffini

Vol. 13 Exploring the Universe: A Festschrift in Honor of Ricardo Giacconi
        edited by H Gursky, R Ruffini and L Stella

Vol. 14 Dark Matter and Cosmic Web Story
        by J Einasto

**To view the complete list of the published volumes in the series, please visit
http://www.worldscientific.com/series/asac*

Advanced Series in Astrophysics and Cosmology – Vol. 14
Series Editors: Fang Li Zhi & Remo Ruffini

DARK MATTER AND COSMIC WEB STORY

Jaan Einasto
Tartu Observatory, Estonia

NEW JERSEY · LONDON · SINGAPORE · BEIJING · SHANGHAI · HONG KONG · TAIPEI · CHENNAI

Published by

World Scientific Publishing Co. Pte. Ltd.
5 Toh Tuck Link, Singapore 596224
USA office: 27 Warren Street, Suite 401-402, Hackensack, NJ 07601
UK office: 57 Shelton Street, Covent Garden, London WC2H 9HE

Library of Congress Cataloging-in-Publication Data
Einasto, Jaan, author.
   Dark matter and cosmic web story / Jaan Einasto.
     pages cm -- (Advanced series in astrophysics and cosmology ; Volume 14)
   Includes bibliographical references and index.
   ISBN 978-9814551045 (hardcover : alk. paper)
   1. Inflationary universe. 2. Cosmology. 3. Dark matter (Astronomy) 4. Astronomers--Estonia--Social conditions--20th century. I. Title. II. Series: Advanced series in astrophysics and cosmology ; Volume 14.
   QB991.I54E36 2013
   523.1'126--dc23
                                                       2013035528

British Library Cataloguing-in-Publication Data
A catalogue record for this book is available from the British Library.

Cover image credit: NASA, ESA, D. Coe (NASA Jet Propulsion Laboratory/California Institute of Technology, and Space Telescope Science Institute), N. Benitez (Institute of Astrophysics of Andalusia, Spain), T. Broadhurst (University of the Basque Country, Spain), and H. Ford (Johns Hopkins University)

Copyright © 2014 by World Scientific Publishing Co. Pte. Ltd.

All rights reserved. This book, or parts thereof, may not be reproduced in any form or by any means, electronic or mechanical, including photocopying, recording or any information storage and retrieval system now known or to be invented, without written permission from the publisher.

For photocopying of material in this volume, please pay a copying fee through the Copyright Clearance Center, Inc., 222 Rosewood Drive, Danvers, MA 01923, USA. In this case permission to photocopy is not required from the publisher.

Typeset by Stallion Press
Email: enquiries@stallionpress.com

Printed in Singapore by Mainland Press Pte Ltd.

To Liia with love and thanks

Preface

The 20th century has been a golden time for astrophysics and cosmology. An excellent discussion of the development of modern astrophysics and cosmology is given by Longair (2006). The search for dark matter has been a part of this development. The history of the discovery of dark matter and its physical nature have been discussed in numerous conferences and books: Faber & Gallagher (1979), Sanders (2010), Trimble (1987, 1988a,b,c, 1990, 1995, 2010), and van den Bergh (1999, 2001), to mention only the most important sources.

A personal view of the development of modern cosmology was recently published by Peebles (2012), see also a collection of views of other leading cosmologists, collected by Peebles et al. (2009). Such personal views are extremely interesting to read.

In this book I try to present a discussion of the Dark Matter Story as seen from my perspective. The work done elsewhere is described in less detail, since I am not familiar with the 'kitchen' aspects of these studies. I try to give balanced credit to results of other investigators when the topic is related to our studies. Our attempts to understand the dark matter problem brought us to the study of Large Scale Structure in the Universe. I shall describe the history of the development of these two topics. They are related because dark matter is the dominant population in the Universe and properties of dark matter particles determine details of the structure. In preparing of this book I used my previous reviews of the history of studies of dark matter and large scale structure of the Universe (Einasto, 2001a,b, 2005, 2009).

The first steps in the search, which ultimately led us to the understanding of the dark matter problem, were made by the founder of the modern Tartu astrophysics school, Ernst Öpik, and his student Grigori Kuzmin, my first mentor. This search has been continued by the present generation of astronomers. The whole group of galactic and cosmology studies of Tartu Observatory participated in this search.

I thank all of them: Heino Eelsalu, Maret Einasto, Mirt Gramann, Urmas Haud, Jaak Jaaniste, Mihkel Jõeveer, Lev Kofman, Sergei Kutuzov, Grigori Kuzmin, Dmitry Pogosyan, Enn Saar, Erik Tago, and Peeter Tenjes, and the younger generation of astronomers Gert Hütsi, Lauri Juhan Liivamägi, Ivan Suhhonenko, Antti Tamm, and Elmo Tempel.

During the search we had many contacts with astronomers in other centres. The first and longest contacts were with Viktor Ambartsumian, Evgeny Kharadze, Kirill Ogorodnikov, and Pavel Parenago. We collaborated closely with Yakov Zeldovich and his team, including Andrei Doroshkevich, Anatoly Klypin, Igor Novikov, Sergey Shandarin, and Rashid Sunyaev. This played a very important role in our studies, as well as our collaboration with Arthur Chernin, Igor Karachentsev, Andrei Linde, Josif Shklovsky, Alexei Starobinsky, and many others.

I benefitted from contacts with astronomers from other countries. The most fruitful collaboration was with John Huchra. In studying the problems of dark matter and the structure of galaxies and Universe I had discussions with George Abell, Heinz Andernach, Neta and John Bahcall, Ed Bertschinger, Peter Brosche, Margaret and Geoffrey Burbidge, George Contopoulos, Gerard de Vaucouleurs, Margaret Geller, Wilhelm Gliese, Bernard Jones, Rocky Kolb, Dave Latham, Donald Lynden-Bell, Vicent Martinez, Dick Miller, Volker Müller, Jan Henrik Oort, Jerry Ostriker, Changbom Park, Jim Peebles, Luboŝ Perek, Joel Primack, Martin Rees, Mort Roberts, Vera Rubin, Remo Ruffini, Alex Szalay, Gustav Andreas Tammann, Beatrice Tinsley, Alar Toomre, Virginia Trimble, Sidney van den Bergh, Rien van de Weygaert, Simon White, and many others. My sincere thanks to all my friends and colleagues — interactions with them helped to develop the present concept of dark matter and large scale structure of the Universe.

My special thanks to my family for support and practical help. My wife Liia and daughter Maret participated in this study in many ways; my grandchildren Peeter, Triin and Stiina helped in the preparation of the book. I thank Changbom Park, Remo Ruffini, Mikk Sarv, Rein Taagepera, Virginia Trimble, and Rien van de Weygaert for suggestions made after reading preliminary versions of the book. Many thanks to the editors of the book Roh-Suan Tung and Lerh Feng Low.

The book is accompanied by a website which contains additional material: copies of originals of some crucial papers, astronomical movies, and also movies which show our life.[1]

---

[1] http://www.aai.ee/~einasto/DarkMatter

Contents

Preface   vii

1. Prologue   1

2. Classical cosmological paradigm   3
   - 2.1 Astronomy in the first half of the 20th century 3
     - 2.1.1 The nature of spiral nebulae 4
     - 2.1.2 The expansion and age of the Universe 5
     - 2.1.3 The mean density of matter in the Universe 8
     - 2.1.4 The distribution of galaxies 8
     - 2.1.5 Structure of the system of stellar populations 9
     - 2.1.6 The evolution of stars 11
   - 2.2 History of Estonia, my family roots, and Tartu Observatory . . . 15
     - 2.2.1 A short history of Estonia 15
     - 2.2.2 My roots 23
     - 2.2.3 My early life and first steps in astronomy 28
     - 2.2.4 Liia 35
     - 2.2.5 Tartu Observatory after the war, and the building of the new observatory 37

3. Galactic models and dark matter in the solar vicinity   45
   - 3.1 Early Galactic models 45
     - 3.1.1 Early Galactic models and first hints of the presence of dark matter 45
     - 3.1.2 Density of matter in the Solar vicinity 47
     - 3.1.3 Galactic models by Parenago, Kuzmin, and Schmidt . . 50

	3.2	New Galactic models	54
		3.2.1 Search for better models	54
		3.2.2 Generalised exponential model	56
		3.2.3 Our Galaxy, system of galactic constants	57
		3.2.4 Mass-to-luminosity ratios of stellar populations	61
		3.2.5 Evolution of galaxies	63
		3.2.6 Models of galaxies of the local group and M87; mass paradox in galaxies	67
	3.3	Tartu Observatory in the 1960's	70
		3.3.1 New observatory	70
		3.3.2 Philosophical seminars and New Year parties	72
		3.3.3 Space studies	76
4.	Global dark matter		81
	4.1	The discovery of global dark matter	81
		4.1.1 Galactic coronas	81
		4.1.2 Clusters and groups of galaxies	85
		4.1.3 Dynamics and morphology of companion galaxies . . .	88
		4.1.4 Tallinn and Tbilisi dark matter discussions	95
	4.2	The confirmation of the presence of global dark matter	97
		4.2.1 Rotation curves of galaxies	97
		4.2.2 Mass-to-luminosity ratios of galaxies	99
		4.2.3 X-ray data	100
		4.2.4 Gravitational lensing	101
	4.3	Dark matter in galaxies	102
		4.3.1 The density distribution of dark matter	102
		4.3.2 Distribution of luminous and dark matter in galaxies . .	103
		4.3.3 Universal rotation curve of galaxies	105
		4.3.4 The formation of galaxies	106
		4.3.5 Modern models of galaxies	109
	4.4	Tartu Observatory in the 1970's	112
		4.4.1 Computer revolution	112
		4.4.2 Life in the Observatory	116
5.	The cosmic web		119
	5.1	Early studies of spatial distribution of galaxies	119
	5.2	The discovery of the cosmic web	120
		5.2.1 Zeldovich question	120

		5.2.2	The Tallinn symposium on large scale structure of the Universe 127
		5.2.3	Superclusters, filaments and voids 134
	5.3	Tartu Observatory in the early 1980's 140	
		5.3.1	Southern base of Tartu Observatory 140
		5.3.2	Studies of ancient astronomy 143

6. **The nature of dark matter**   147

    6.1 Baryonic dark matter 147
       6.1.1 Early discussions on the nature of dark matter 147
       6.1.2 Stellar or gaseous dark coronae 148
       6.1.3 Nucleosynthesis constraints of baryonic matter 151
    6.2 Non-baryonic dark matter 152
       6.2.1 Cosmic microwave background radiation 152
       6.2.2 Fluctuations of the CMB radiation 153
       6.2.3 Neutrinos as dark matter candidates 154
       6.2.4 Cold dark matter 157
       6.2.5 Dark matter in dwarf galaxies 159
       6.2.6 Missing satellite problem and warm dark matter 159
       6.2.7 Searches for dark matter particles 161
    6.3 Alternatives to dark matter 164
    6.4 Tartu Observatory in the late 1980's 165
       6.4.1 The singing revolution 165
       6.4.2 Academy of Sciences 168
       6.4.3 Towards an independent Estonia 169

7. **The structure of the cosmic web**   175

    7.1 Quantitative characteristics 175
       7.1.1 The search for quantitative characteristics 175
       7.1.2 Topology of the cosmic web 182
       7.1.3 Fractal properties of the cosmic web 188
       7.1.4 Physical biasing 192
       7.1.5 Power spectra of galaxies 199
    7.2 Redshift surveys and catalogues 203
       7.2.1 Redshift surveys 203
       7.2.2 Catalogues of groups and clusters of galaxies 207
       7.2.3 Catalogues of superclusters 211
    7.3 Elements of the cosmic web 217

	7.3.1	Galaxies in different environments 217
	7.3.2	Groups and clusters of galaxies 223
	7.3.3	Chains, strings and filaments 224
	7.3.4	Walls 227
	7.3.5	Superclusters 231
	7.3.6	Voids and supervoids 235
	7.3.7	Cosmic web — cells and the cosmic foam 238
	7.3.8	Regularity of the cosmic web 241
	7.3.9	Baryonic acoustic oscillations 243
7.4	Tartu Observatory in the 1990's 247
	7.4.1	Estonian path to independence 247
	7.4.2	Science reform 250
	7.4.3	Participation in international organisations 251

8. Cosmic inflation, dark energy and the evolution of the Universe   253

8.1	The birth of the Universe and inflation 253
	8.1.1 The classical inflation theory 253
	8.1.2 The new inflation theory and the birth of the Universe . 254
8.2	Structure formation in Hot, Cold and Lambda models 257
	8.2.1 Initial conditions 257
	8.2.2 HDM and CDM simulations 261
	8.2.3 Simulations with cosmological constant 262
	8.2.4 Modern cosmological simulations 264
8.3	The formation and evolution of the cosmic web 265
	8.3.1 The luminosity density field of the SDSS 266
	8.3.2 The role of density waves of various scales 269
	8.3.3 The phase coupling of density perturbations of various scale 273
	8.3.4 The fine structure of the cosmic web 275
8.4	Dark energy 280
	8.4.1 The discovery of dark energy 280
	8.4.2 The role of dark energy in the evolution of the Universe 281
	8.4.3 Cosmological parameters 281
	8.4.4 New cosmology paradigm is ready: What next? 284
8.5	Remembering contacts with colleagues 285
	8.5.1 Encounters with astronomers from other centres 285
	8.5.2 Collaboration with other centres 295
8.6	Tartu Observatory and my life in the 2000's 299
	8.6.1 Transition years 299

| | 8.6.2 | Center of Excellence . . . . 301 |
| | 8.6.3 | Egeri . . . . 303 |

9. Epilogue 305

Bibliography 311

General Index 339

Name Index 345

Chapter 1

Prologue

Once I happened to read Thomas Kuhn's book *The Structure of Scientific Revolutions* (Kuhn, 1970). It was in mid 1970's, and the presence of dark matter in galaxies had just been reported. The discussion between supporters and opponents of the dark matter concept was at its peak. Then I realised that the dark matter story seems to be a good example of a scientific revolution. Ten years later, in concluding remarks of the IAU Symposium on "Dark Matter in the Universe", Scott Tremaine also pointed to the development of the dark matter concept as a classic example of a scientific revolution (Tremaine, 1987), see also Binney & Tremaine (1987)).

In the present book I shall give a personal view of the study of dark matter and large scale structure of the Universe. There are not so many areas in modern astronomy where the development of ideas can be described in terms of paradigm changes, thus I shall discuss the dark matter story from this point of view. Tartu astronomers have participated in the study of dark matter for a long time, starting from Ernst Öpik — the founder of the contemporary astronomy school in Estonia — and followed by Grigori Kuzmin, his most talented student. The present generation of astronomers has continued the investigation of dark matter. Unexpectedly, this work led us to study the distribution of galaxies and clusters on large scales and peaked with the discovery of the cosmic web with voids and filaments. I hope that our story is of interest to the cosmology and perhaps also to the physics community, as some aspects of it have yet to be well documented.

First I give a short review of the classical world view on galaxies and the Universe, as it was when I started as a young scientist my work in Tartu Observatory. The astronomical community and, in particular, my mentors Prof. Taavet Rootsmäe and Grigori Kuzmin in Tartu, and Prof. Pavel Parenago in Moscow had certain views on the structure of galaxies and the Universe. This was the background when I started my studies.

Kuzmin and Parenago developed methods to calculate models of galaxies, and Rootsmäe and Parenago investigated properties of stellar populations. These topics were earlier considered as separate tasks. I saw a possibility to combine both approaches, and to construct models of galaxies, where all possible data on galactic populations are taken into account. This was my first goal. To my surprise I encountered here difficulties, which ultimately led to the discovery of a controversy in data, discussed in detail in Chapter 3. This was our first step towards the development of a new paradigm on the structure of galaxies.

In our attempts to solve the controversy in the structure of galaxies we found that it is not possible to consider galaxies as isolated systems — the environment of the galaxies is also important. We found that galaxies are surrounded by dark massive coronas of unknown origin and nature. This was the second step in our journey towards the solution of the controversy in the understanding of the structure of galaxies.

Next we found that it was not sufficient to consider only the nearby environment of galaxies. To understand the formation and evolution of galaxies their large-scale environment is also important. This led us to the discovery of the presence of a cosmic web with galaxy filaments, filamentary superclusters, and voids between them. This was our third step in the development of a new paradigm.

The nature of galactic coronas was a mystery; no known population fit all data and theoretical considerations. Finally the solution was that coronas must be made of non-baryonic matter, not yet detected experimentally by physicists. In the solution of this problem the whole community of cosmologists played an important role; the understanding came slowly and was made by many scientists.

The next step in the development of the new paradigm was the understanding of the need to accept a rapid expansion of the early Universe, called inflation. Our young collaborator Lev Kofman participated in the development of the inflation scenario together with Alexei Starobinsky and Andrei Linde.

Based on various observational and theoretical arguments we assumed in mid 1980's the presence of the cosmological constant or dark energy, and included this in numerical simulations of the evolution of the structure of the Universe. Direct observational evidence for the existence of dark energy came in late 1990's.

Astronomers are real people who do their work in a certain social environment. Most studies which shall be discussed below were done when our home country Estonia was occupied and annexed by the Soviet Union. This has influenced our work and life. Thus I shall describe shortly our social life and environment. We were surrounded by the "Iron Curtain", and it was not easy to have contacts with the rest of the world.

Chapter 2

Classical cosmological paradigm

In this Chapter I shall describe what we knew of the structure of stars, galaxies and the Universe in the middle of the 20th Century, when I began my astronomical studies. Actually during this period the modern classical cosmological paradigm or world view emerged. It is interesting to note that in the formation of the classical cosmological paradigm one Estonian astronomer, Ernst Öpik, played an important role. This influenced also our view on astronomy.

2.1 Astronomy in the first half of the 20th century

Until the 20th Century cosmology was mostly a philosophical and metaphysical discipline, because very little was known about the actual global structure of the Universe, and the nature of the various astronomical objects within it. In the beginning of the 20th Century most astronomers believed that our Milky Way system is the principal constituent of the Universe. Sir Arthur Eddington (1914) wrote his famous book "Stellar Movements and the Structure of the Universe", where he identified the Milky Way with the whole Universe. The presence of other stellar systems similar to the Milky Way was discussed but there existed no proof for this. Also the birth of the Universe and its age were only objects of speculations.

The modern classical cosmological paradigm was elaborated step by step during the first part of the 20th Century. In the following I shall use the term "Universe" to denote the real physical world around us, and the term "universe" for its mathematical model.

2.1.1 *The nature of spiral nebulae*

In the early years of the 20th Century a hot topic was the nature of spiral nebulae — Are they gaseous objects within the Milky Way system or distant worlds similar in structure to our Galaxy? On 26 April 1920 the Great Debate between astronomers Harlow Shapley and Heber Curtis was held in the Smithsonian Museum of Natural History on the nature of spiral nebulae and the size of the Universe. Arguments in favour of both concepts were serious and it was difficult to decide who was right. The debaters did not know that the correct answer was already available.

This problem interested also Ernst Öpik. In 1918 he delivered a talk at the Meeting of the Moscow Society of Amateur Astronomers, devoted to the study of the structure of the Andromeda Nebula, M31. The paper was published a few years later. Just recently the first relative velocity measurements near the centre of M31 had been published, and Öpik quickly developed a method to estimate distances to spiral nebulae from relative velocities within them. He used Newton's law of gravity, which relates the speed of motion of a test particle around a massive body with the mass of the body and the distance of the test particle from the body. Öpik noticed that it is possible to substitute the mass with the product of luminosity and mass-to-luminosity (M/L) ratio. Apparent luminosity of the central region of M31 can be determined by photometric observations. He accepted for mass-to-luminosity (M/L) ratio a value of 1.54 in solar units, based on measurements of the stellar luminosity function in the solar neighbourhood. From the estimate for M/L, the observed luminosity and the internal rotation speed of M31 he measured the size of M31 and then obtained the distance of 785 kiloparsecs (kpc). This means that M31 is not within the Milky Way and must be an external independent system. A few years later he made a new estimate (Öpik, 1922a) using new determinations of the luminosity function by Kapteyn & van Rhijn (1920), and data by Jeans (1922) on the mass density in the solar neighbourhood. His new value for the mass-to-luminosity ratio is 2.63 in solar units which gives for the distance of the Andromeda nebula 440 kpc.

Hubble (1925, 1926, 1929b) found cepheids in spiral nebulae NGC 6822, M33 and M31, using the 100-inch telescope of the Mount Wilson Observatory. This confirmed the large distance and the extragalactic nature of spiral nebulae. The existence of the world of galaxies was accepted by the astronomical community.

2.1.2 *The expansion and age of the Universe*

The modern era of understanding the global structure of the universe began with the publication of the Einstein (1916) theory of general relativity. In 1917 Einstein considered a static universe with cosmological term Λ. Based on this theory de Sitter (1917) suggested a model with Λ but with zero or negligible matter density. The zero matter density universe is called Milne's universe. A few years later Friedmann (1922, 1924) and Lemaître (1927) discovered solutions to Einstein's equations that contained realistic amount of matter. Einstein & de Sitter (1932) proposed a model with the critical cosmological density. This model contains two parameters — the mean expansion rate of the universe, and its mean density. The expansion rate and density also determine the age of the universe.

Actually Friedmann found three solutions for the cosmic evolution, one with ever-accelerating expansion, one periodic scenario with evolution from and back to zero radius (the oscillating universe), and the third, where initially the universe is decelerating due to gravity, but after some time the expansion accelerates due to the influence of the cosmological constant. Recent observations indicate that just the third scenario corresponds to the real Universe.

In the 1920's radial velocities of some tens of galaxies were measured, and almost all of them showed a shift of spectral lines to the red part of the spectrum — i.e. lines were redshifted. The larger the shift is, the fainter galaxies are, and soon the hypothesis was made that the whole Universe is expanding, the expansion velocity being proportional to the distance to the galaxy. The discovery of the expansion of the Universe is often ascribed to Hubble (1929a). Actually the story of the discovery of the expansion of the Universe is more complex.

The first steps in this discovery were made by Wirtz (1922, 1924) who found that redshifts and distances of galaxies are related. Wirtz (1924) suggests a clear relationship of this phenomenon with the de Sitter (1917) cosmological model. Lundmark (1924) discussed the curvature of the space-time in the de Sitter universe, and the relationship between distances and redshifts. Lundmark (1925) compares various methods to determine distance to spiral nebulae, and found that distances estimated using novae, cepheids and the dynamical method by Öpik are the most reliable and give comparable results.

Lemaître (1927) presented his new idea of an expanding Universe, derived the velocity–distance relation, and provided the first observational estimate of the constant of proportionality in this law. In 1931 he proposed that the Universe expanded from an initial point, which he called the "Primeval Atom". Presently the theory is known as the Big Bang theory. This term was used first by Sir Fred Hoyle in one of his popular radio broadcasts in 1949. Hoyle (1948) and

Bondi & Gold (1948) preferred a different theory of the origin of the Universe, called the Steady State theory, where matter is continuously created and the mean density of matter remains constant. According to this theory the Universe has no beginning and will have no end.

The constant of proportionality of the velocity–distance relation is now called the Hubble constant; it is one of the principal constants not only in astronomy but in physics in general. The reciprocal value of this constant has a dimension of time, and measures the time from the beginning of the expansion if the expansion speed is constant.

In the first decades of the 20th century most astronomers accepted the view that the whole stellar Universe is very old, of the order of 10^{14} years. This age estimate was based on the observation that stellar orbits in our Milky Way system are well mixed and relaxed. The relaxation time of this process by star–star encounters is very long, of the order mentioned above, and this estimate was taken as the possible age of the Universe.

Öpik (1933) realised that the expansion time (called the Hubble time in modern cosmology) is approximately equal to several other completely independent fundamental age estimates. He finds an age $\approx 2 \times 10^9$ years, and writes: *"if we regard the observed motion of the spirals as real, and trace the changes observed at present backwards, we find that a few thousand million years ago the universe was in a peculiar, more concentrated state, from which it started expanding, possibly as a result of some cataclysm"*. The second independent age estimate is the age of the Earth as derived from the decay of heavy radioactive elements, which is up to 5 billion years. Meteorites also have an age of the same order. The third age estimate comes from Öpik's studies of double stars and related questions of stellar structure.

Summarising the results of these completely independent age estimates Öpik (1933) writes: *"we may say that the combined evidence presented by meteorites, by statistical data relating to wide double stars, by the distribution of stellar luminosities in globular clusters, and by the observed recession of spiral nebulae, all this evidence points to an age of the stellar universe of the same order of magnitude as the currently accepted age of the solar system: not much more than 3000 million years"*.

Modern data yield for all three ages larger values, from 5 to 14 billion years. But the method is the same as suggested by Öpik in the early 1930's.

The expanding Universe can be described by two fundamental constants: the mean expansion rate of space, measured by the Hubble constant, and the mean density of the Universe, expressed in units of the critical cosmological density. The critical density is the amount of matter/energy required to make the Universe spatially flat. A flat Universe has no curvature. If the density is less than the critical density, then the Universe will expand forever according to the classical picture. In the opposite case the density is greater than the critical one, and gravity is strong enough to make the Universe collapse back, the so-called "Big Crunch".

Very large efforts have been made to measure the value of the Hubble constant. First measurements by Lundmark and Hubble yielded a value about 500 km/s per megaparsec. The first major correction to this value came in the 1950's when Walter Baade discovered that there are two types of cepheids. Some of them belong to Population II, which dominate in galactic halos and have a different luminosity–period relation. Also it was found that stars in the most distant galaxies, observed by Hubble, were actually star clusters. A detailed description of efforts to determine the Hubble constant is given by Huchra[1]. A special role in these efforts was played by the 200-inch Hale telescope in the Mount-Palomar observatory (Sandage, 1961; Sandage & Tammann, 1976).

By the end of the 1970's there were two schools debating on the correct value of the Hubble constant. Allan Sandage and his longtime collaborator Gustav Andreas Tammann favoured a value about 50 km/s/Mpc, whereas Gerard de Vaucouleurs (1978) and Sidney van den Bergh (1972, 1973) obtained values around 100 km/s/Mpc. This debate was settled when new methods were used, in particular the Hubble Space Telescope Key Program, a number of high-resolution observations of fluctuations of the cosmic microwave background (CMB) radiation, and observations of the spatial distribution of galaxies in the Sloan Digital Sky Survey.

An important property of the expansion of the Universe is its smoothness. Sandage & Tammann (1975) write: *"The local velocity field is as regular, linear, isotropic, and quiet as it can be mapped with the present material. The lack of measurable velocity perturbations, in spite of the observed density inhomogeneities, suggests that the gravitational potential energy is small compared with the kinetic energy of the expansion (provided that there is no high-density, uniform intergalactic medium), and hence that $q_0 < 1/2$".* The expansion parameter $q_0 = 1/2 \times \Omega$, where Ω is the mean matter/energy density of the Universe in units of the critical density.

For an "empty" Milne's model universe with density parameter $\Omega \ll 1$, the age of the universe is $1/H_0$, or 9.7 gigayears (Gyr) for $h = 1$, and 19.4 Gyr for $h = 0.5$. Here and in the following text we use the Hubble constant in dimensionless unit h, defined as follows: $H_0 = 100\, h$ km s^{-1} Mpc^{-1}. For a universe with critical density (Einstein–de Sitter model) the age is 2/3 of that for the empty universe.

---

[1] https://www.cfa.harvard.edu/~dfabricant/huchra/hubble/

For the Hubble constant $h = 0.7$, and an empty universe, as assumed in 1960's, the age of the universe is 13.5 Gyr. This age is considerably less than the age of oldest globular clusters, as estimated from the theory of stellar evolution in 1960's. Thus there was an inconsistency between various cosmological parameters.

2.1.3 *The mean density of matter in the Universe*

Known objects in the Universe which contribute to the matter/energy density are galaxies, intergalactic gas and radiation. This was the common understanding in the middle of the 20th century. The contribution of radiation to the matter/energy density is in the CMB which makes up about 5×10^{-5} of the total density. Thus the basic constituents are galaxies and intergalactic matter.

The mean density due to galaxies can be determined using the mean luminosity density calculated from the luminosity function of galaxies, and the mean mass-to-luminosity ratio of galaxies. Estimates available in the 1950's indicated a low-density Universe, $\Omega \approx 0.05$.

2.1.4 *The distribution of galaxies*

Already in the New General Catalogue (NGC) of nebulae, composed from observations by William and John Herschel, a rich collection of nearby galaxies in the Virgo constellation was known. de Vaucouleurs (1953) called this system the Local Super-galaxy; presently it is known as the Virgo or the Local Supercluster. Detailed investigation of the distribution of galaxies became possible when Harlow Shapley started in the Harvard Observatory a systematic photographic survey of galaxies in selected areas, up to 18th magnitude (Shapley, 1935, 1937, 1940). Shapley discovered several other rich superclusters, one of them is presently named the Shapley Supercluster. These studies showed also that the *mean* spatial density of galaxies is approximately independent of the distance and of the direction on the sky. In other words, the Harvard survey indicated that galaxies are distributed in space more-or-less homogeneously, as expected from the general cosmological principle.

A photographic survey was made using the 48-inch Palomar Schmidt telescope. Abell (1958) used the Palomar survey to compile a catalogue of rich clusters of galaxies for the Northern sky; later the catalogue was continued to the Southern sky (Abell et al., 1989). Using apparent magnitudes of galaxies approximate distances (distance classes) were estimated for clusters in both catalogues. Zwicky et al. (1968) used this survey to compile for the Northern hemisphere a catalogue of galaxies and clusters of galaxies. The galaxy catalogue is complete up to 15.5

photographic magnitude. Both authors noticed that clusters of galaxies also show a tendency of clustering, similar to galaxies which cluster to form groups and clusters. Abell called these objects superclusters; Zwicky called them clouds of galaxies.

A deeper complete photographic survey of galaxies was made in the Lick Observatory with the 20-inch Carnegie astrograph by Shane & Wirtanen (1967). Galaxy counts were made in cells of size $10' \times 10'$, and the distribution of the number density of galaxies was studied.

The Lick counts as well as galaxy and cluster catalogues by Zwicky and Abell were analysed by Jim Peebles and collaborators to exclude count limit irregularities (Peebles, 1973; Hauser & Peebles, 1973; Peebles & Hauser, 1974; Peebles, 1974a). To describe the distribution of galaxies Peebles introduced the two-point correlation (or covariance) function of galaxies (Peebles & Groth, 1975; Groth & Peebles, 1977; Fry & Peebles, 1978). This function describes the probability of finding a neighbour at a given angular separation on the sky from a galaxy. On scales $\geq 25\ h^{-1}$ Mpc the correlation function is very close to zero, i.e. the distribution of galaxies is essentially random.

The conclusion from these studies, based on the apparent (2-dimensional) distribution of galaxies and clusters on the sky, confirmed the picture suggested by Kiang (1967) and de Vaucouleurs (1970), among others, that galaxies are hierarchically clustered. However, this hierarchy does not continue to very large scales as this contradicts observations, which show that on very large scales the distribution is homogeneous. A theoretical explanation of this picture was given by Peebles in his hierarchical clustering scenario of structure formation (Peebles & Yu, 1970; Peebles, 1971a).

2.1.5 *Structure of the system of stellar populations*

The presence of various populations (sub-systems) of stars with different kinematical and spatial properties was clarified gradually. Strömberg (1924) noticed that there exists an asymmetric drift of velocity centroids with respect to the Sun's motion: the centroid motion of a certain class of stars is larger the higher the velocity dispersion of stars of this class. This phenomenon is known as the Strömberg asymmetric drift. Lindblad (1927) and Oort (1927, 1928) interpreted this phenomenon as evidence for the rotation of the Galaxy. The sub-systems rotate around their common axis, and each one has a different speed of rotation. *"The system of globular clusters has the lowest rotation velocity, it is not excluded that this population is not rotating at all. Bright stars in solar vicinity have the smallest velocity dispersion, these stars can be considered as moving very nearly in circular orbits around the centre"* (Oort, 1927).

Lindblad (1927) and Oort (1927) noticed also that the sub-system of globular clusters has an almost spherical shape, whereas subsystems of stars in the solar vicinity form a rather flattened system. According to Lindblad (1927) the gravitational field of the Galaxy can be expressed as a superposition of a spherical mass, and a mass of a flattened spheroidal population. He found that the mass of the spheroidal components exceeds the mass of the spherical population by a factor of about 5.

This concept of the presence in the Galaxy of a number of stellar populations with various kinematical, spatial distribution and morphological properties has been further developed by many astronomers. At the Sternberg Astronomical Institute Boris Kukarkin and Pavel Parenago suggested using variable stars as markers of different populations to investigate the kinematical and spatial structure of populations in Galaxy. They divided galactic populations into three main classes: spherical, flat, and intermediate, which correspond approximately to halos, disks, and bulges of external galaxies.

Already early studies of stellar dynamics (Eddington, 1914) suggested that kinematical and spatial properties of stellar populations change very slowly. Thus galactic populations contain information on their formation and evolution. This property was used in Tartu Observatory by Taavet Rootsmäe, who started in the mid 1930's a study of the kinematics of various stellar populations in the hope of finding the direction of evolution of stars. He assumed that stellar populations of various age were formed during the collapsing proto-Galaxy. Unfortunately the study proceeded rather slowly, and was published only posthumously (Rootsmäe, 1961). A similar task was realized by Eggen et al. (1962), who found evidence from the motion of old stars that the Galaxy collapsed during its formation.

It is generally accepted that star clusters, both open and globular, formed from a single gas cloud. For this reason all stars of a given cluster have similar age and chemical composition. Eggen (1950) performed accurate photoelectric studies of star clusters of various type and found that the colour-magnitude (or HR-) diagram of clusters is very narrow, if binary stars are excluded. This suggests that HR-diagrams can be used to derive the age of the cluster if star evolution tracks are known with sufficient accuracy. This property of clusters is widely used; see Eggen & Sandage (1964) as an example.

The modern concept of stellar populations was generally accepted after the review talk by Oort (1958) at the Vatican Conference.

2.1.6 *The evolution of stars*

The evolution of galaxies and the Universe as a whole depends on the evolution of stars. The understanding of sources of stellar energy and the path of the evolution of stars are important elements of the classical cosmological paradigm.

In the early years of the 20th century astronomers adopted the Russell hypothesis on stellar evolution: stars are born as red giants, they contract to form blue giants, and then cool and move along the dwarf branch (main sequence) towards red dwarfs. The dominating source of energy according to Russell was gravitation or radioactive decay.

It is well known that the mean mass of stars in the main sequence is not constant — O and A type stars have masses 10–30 solar masses, whereas the masses of red dwarfs are only a fraction of the solar mass. Stellar evolution was one of the topics of interest to Ernst Öpik. He concluded that, if the Russell hypothesis is correct, stellar evolution should be accompanied with mass loss. If mass loss occurs in double stars, the distance between components must increase from blue to red double stars of the main sequence — the expected increase is approximately 20 times. To check this result Öpik (1923, 1924) studied double stars, and found that contrary to the expectation, the mean distance between components of double stars decreases about 2 times when moving from blue to red main sequence stars.

Another fact contrary to the Russell hypothesis comes from geological data which indicate that the mean temperature on the Earth surface has been almost constant during its whole geological history. If the Sun evolved according to the Russell hypothesis its luminosity must decrease along the main sequence by a factor of a thousand, and it is impossible to avoid similar changes of the temperature on Earth.

The first conclusion from these calculations was: the Hertzsprung–Russell diagram is not an evolutionary diagram, but a diagram of various initial conditions in mass and chemical composition. The same conclusion was made by Eddington (1924). The energy production per unit mass of blue giants is much higher than that of red dwarfs, thus the energy production must depend on physical conditions in the star. In faint companions of double stars (i.e. on main sequence stars) the luminosity per unit mass is proportional to the 9th power of the mass (Öpik, 1923). The basic physical parameter which changes among the main sequence stars of different mass is the temperature, thus a similar dependence must be valid also for the temperature. Since the temperature rises inwards, the energy source of stars must be located near the centre (Öpik, 1922b). Öpik (1922b) and Eddington (1924, 1926) concluded that stars obtained their energy from nuclear reactions.

In order to find more accurately the energy source and stellar evolution Öpik waited until more data on the structure of atoms became available. He continued to think on these problems, and immediately after basic facts of atomic transmutations were available suggested a new, much more accurate theory of stellar evolution. In the Introduction to his major paper on the stellar structure and evolution Öpik (1938) explains his approach to the problem as follows: *"stellar structure is a physical, not a mathematical problem. What matters are the premises, not the exact mathematical deductions from given premises; we want to know the actual physical conditions determining stellar structure and evolution; a correct mathematical theory may then easily follow. We believe that a mere qualitative picture, taking into account all the complexity of the conditions in stellar interiors, is still a better approximation to the truth than an exact mathematical theory based on simplifications which do not take into account certain most important factors of stellar structure and evolution"*.

The first important factor to be taken into account is convection. Since the energy source is located in the centre of the star, this leads to convection, similar to the formation of convection in boiling water in a kettle heated from below. Due to convection the active matter is continuously replaced. The source of energy was found — nucleosynthesis, mainly the transmutation of the hydrogen into helium. The energy output of this process was known, and Öpik was able to calculate models of stellar interiors taking into account energy production and transport.

Öpik's calculations suggested that the star is convective only in the central parts. In this case its structure is a composite one — the convective core is surrounded by a radiative envelope. There exists no mixing of stellar matter between the core and envelope, and the chemical composition of the core is not constant. This process continues until all the hydrogen is used. Öpik (1938) concludes: *"A core devoid of hydrogen, thus presumably devoid of subatomic sources of energy, is doomed to collapse on a 'Kelvin' time scale, high densities can be attained, and a super-dense core may be formed. The contraction of the core is a gradual one; instead of blowing up, the envelope gradually expands and adjusts itself to such low values of the effective density and temperature that the release of subatomic energy remains more or less normal. ... A typical giant structure results, consisting of a vast extended envelope of low density in radiative equilibrium, an intermediate zone in adiabatic (convective) equilibrium, ... and a contracting superdense core of zero hydrogen content"*, see Fig. 2.1.

The amount of energy which is produced during the burning of hydrogen to helium is well-known, thus Öpik was able to calculate the maximal age of stars of different mass. For highly luminous early type main sequence stars this age is very short — about 10 million years. *"Thus, the presence of massive and luminous main*

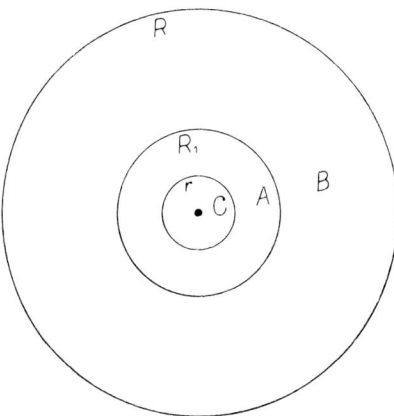

Fig. 2.1 A scheme of a giant star structure. C is the nucleus of radius r, exhausted of nuclear energy; A is the region of release of the subatomic energy, of radius R_1; B is the region of undisturbed radiative equilibrium without energy sources (Öpik, 1938).

sequence stars in the Galactic System we ascribe to stars being continually formed in the place of those which become giants" (Öpik, 1938). The idea of a recent origin for blue main sequence stars was commonly accepted after Ambartsumian's discovery of stellar associations (Ambartsumian, 1958).

The chemical inhomogeneity and composite structure were omitted by previous investigators, but they are the most important factors determining the structure of giants.

Öpik's theory of the evolution of main sequence stars toward giants was not accepted immediately. Alar Toomre recently gave an interview to the Estonian magazine "Horisont", and told a story about Öpik: *"When the famous astrophysicist Chandrasekhar was 65, I was in a conference in Chicago, where one lecture was given by Martin Schwarzschild, a famous astronomer who studied the structure of stars. Remembering Chandrasekhar he told, that the first person who understood the structure of red giants was Ernst Öpik in 1930s, but for personal reasons they decided not to believe him. After the lecture I asked Martin, what kind of personal reasons were. It appears that Öpik was a self-confident person and told to everybody, if you do not believe me, you are fool, you are idiot. Their personal reason not to believe was the unwillingness to believe that Öpik was right. But he was."*

Öpik's theory is now fully accepted. The only basic difference between his theory and modern data concerns the energy source of the giants. According to modern data giants also burn chemical elements to produce energy, first helium to carbon, and thereafter other heavier elements until iron. Only after using all atomic fuel does the core of a giant star collapse under gravity to form white dwarfs or neutron stars depending on the mass.

The modern theory of stellar evolution and the synthesis of the elements in stars was elaborated by Schwarzschild and coauthors (Härm & Schwarzschild, 1955; Hoyle & Schwarzschild, 1955; Oke & Schwarzschild, 1952; Sandage & Schwarzschild, 1952; Schwarzschild et al., 1953), and Burbidge et al. (1957). This theory is well supported by the relative distribution of chemical elements in stars, Earth's core and other cosmic objects. According to this theory all elements heavier than hydrogen, helium and lithium are synthesised in stellar interiors by stellar nucleosynthesis.

In the first stage a main sequence star burns hydrogen to helium. When all hydrogen in the convecting core is exhausted, the core contracts and heats up, and the envelope expands — the star becomes a red giant. As the temperature in the inner region of the star increases, nuclear reactions demanding higher temperature start. Thus a red giant looks like an onion: an outer shell of hydrogen burning, while in inner sheets heavier elements burn to form carbon, neon, oxygen, silicon etc. In the very central zone, where the temperature is high enough, iron is produced. But this is the end of energy releasing nuclear reactions. All previous nuclear reactions produce energy, but when iron fuses into heavier elements, it absorbs energy out of the reaction, slowing it down. The central region of the star no longer produces energy and its gravity pulls outer layers inward. The star collapses quickly and explodes as a supernova.

During supernova explosions large amounts of energy are released and the temperature rises, thus in a short timescale all elements heavier than iron are synthesised. During the explosion outer layers are expelled and enrich the interstellar matter with heavier elements. Thus the next generations of stars form from a medium which already contains elements heavier than hydrogen, helium and lithium. Expanding shock waves from a supernova explosion can trigger the formation of new stars. All atoms in our bodies were formed in stellar interiors, and expelled from a supernova. From remnants of this supernova our Sun and all its planets formed about 4.5 billion years ago.

With the explanation of sources of stellar energy and the evolution of stars the development of the classical cosmological paradigm culminated. This paradigm can be described shortly as follows:

- The Universe formed as a result of the Big Bang about 15–19 billion years ago. Big Bang itself was considered as a mathematical singularity.
- The Universe is presently expanding with a rate which corresponds to the Hubble parameter in the range of 50–100 km/s/Mpc.

- The expansion of the Universe is spatially very uniform.
- The mean density of the Universe is about 0.04 of the critical cosmological density. This density concerns only baryonic matter, but this was recognised later.
- Galaxies are distributed in space almost randomly. About 10% of galaxies form clusters and superclusters of galaxies.
- During the early phase of the evolution of the Universe only light chemical elements formed via the Big Bang nucleosynthesis; all heavier elements formed inside stars by stellar and supernova nucleosynthesis. Nucleosynthesis is the basic source of energy in stars.
- Galaxies consist of populations of various age, composition, kinematics, and spatial structure. HR-diagrams of homogeneous populations (as star clusters) can be used to determine the ages of populations, and to reconstruct the history of the formation and evolution of galaxies.

Almost all observations known in the early 1970's supported this paradigm. There were only a few clouds on the horizon which did not fit into this paradigm. One of these facts was the mass paradox in clusters of galaxies found by Zwicky (1933, 1937). A related theoretical problem was: the Universe has expanded enormously, thus even small deviations from the critical density should increase during this time. How is it possible that the present density is smaller than the critical one, but only by a factor on the order of ten?

From these small clouds and new observational data a new cosmological paradigm formed. But there was a long way to go. In the following chapters I discuss some of these steps which led to the paradigm change in cosmology.

2.2   History of Estonia, my family roots, and Tartu Observatory

2.2.1   *A short history of Estonia*

The Estonian language belongs to the Finno-Ugric family of languages, which covers Northern Europe east of the Baltic Sea to the Ural Mountains and beyond. Finno-Ugric tribes populated this area from the Arctic Ocean to the Southern edge of the forested region. Finno-Ugric languages differ from Indo-European languages in several fundamental aspects. They lack grammatical gender and use one pronoun for both *he* and *she*. Second, Finno-Ugric languages are agglutinative languages, by adding suffixes (parts of the word) instead of prepositions (separate words). There is a smooth transition from languages near the Baltic Sea (Estonian, Finnish) towards the Ural mountains, and further to Samoyedic languages in Siberia.

During the Ice Age most of Northern Europe was covered with ice and only several regions were free of ice, suitable for people to live. These regions had tundra features; there was enough rainfall only near the ice sheets. Regions free from ice are called Last Glacial Maximum refugia. One is today in the Don river region, it reached in the North-East direction almost to the Pechora river near the Ural Mountains. At this time the climate was very dry and tundra type vegetation was available only near the ice sheet. Stone Age people during the Ice Age were nomadic; they travelled to gather and hunt for food, mostly Northern elks, polar bears, possibly also mammoths. When ice melted, animals moved Northwards, and people followed them. Present-day Estonia was populated about 10 thousand years ago. The Finnish Gulf was a hindrance, and people stayed here. Another hindrance was the Peipus lake, one of the largest in Europe. East of the lake there is a continuous transition of Baltic Ugro-Finnish languages from South-East Estonian (Setu) dialect, to Ingrian, towards Karelian.

The Estonian mythology is very similar to the Finnish one, and has many similar features with other Finno-Ugric nations. The mythology is animalistic: animals and trees had souls. Ancient Estonians knew the sky well; constellations had names of everyday life and instruments. The Milky Way is in Estonian *Linnutee*, i.e. "the path of the birds", because birds were believed to use it as a guide during spring and autumn migration.

A meteorite passed over the populated region in North Estonia and landed on the island Saaremaa around 3 thousand years ago, creating a lake, called "Kaali". This cataclysmic event may have influenced Estonian and Finnish mythology, since a "sun" flew over the sky in a wrong direction, and seemed to set in the East. The Kaali lake was considered by ancient Estonians as a sacred place, where the Sun went to rest. According to a theory, proposed by the ethnologist and the first after-war Estonian president Lennart Meri, it is possible that Saaremaa was the legendary "Thule island", mentioned by ancient Greek geographer Pytheas, where the name "Thule" could have been connected to the Finnish/Estonian word tule/tuli (fire).

Genetically the closest relatives of Estonians are Swedes, Latvians, and Russians. The genetic close links to Russians seem surprising, but it stems from the colonisation history of Russia by Slavs. DNA analysis shows that along the maternal line, ancestors of Russians in the whole Northern region took wives from the local Finno-Ugric nations. Along the paternal line only in the North-Western region are the majority of Russians relatives of Finno-Ugric nations. This means that in these regions almost the whole population was originally Finno-Ugric, and a language transfer happened.

At the end of the 1st Millennium there were no states in the present Estonia, Finland, and Latvia. There were two levels of community organisations: smaller units were called "kihelkond" (like present-day parish), larger units were called "maakond" (provincia or counties), ruled by seniors. As all these territories were pagan, thus a christianisation started, both from the East and the West.

The christianisation of the Kievan Rus took place in the 9th century. In 1862 the Millennium of Russia was celebrated, to mark the arrival of Vikings Ruler Rurik to Novgorod. One of the next Grand Princes was Vladimir the Great. He was vice-regent of Novgorod and later Grand Prince of Kiev. In 1030 he founded Yuryev — present day Tartu — and forced the surrounding Estonian province to pay annual tribute. During the next 150 years Estonia was attacked many times by Russian principalities.

The christianisation of the Baltic region from the West was part of the Northern Crusades. Actually the colonisation of Northern Estonia by Danes already took place at the end of the 12th century and was rather peaceful. The starting point for the Northern Crusades was Pope Celestine III's call in 1193. The crusade against the Baltic countries was declared of the same rank as a crusade to the Holy Land. The first crusaders landed in the mouth of the Daugava in 1198. Usually they arrived to fight during the spring and returned in the autumn. To ensure a permanent military presence, the Livonian Brothers of the Sword were founded in 1202. In 1237 the Livonian Order was assimilated by the Teutonic Order, which exercised political control over large territories in the Baltic region. The campaign against Livonians and other Latvian tribes was from 1198 to 1290. The war against Estonians was from 1208 to 1224, against Saaremaa island until 1227.

In 1346 the Danes sold their territory to the Teutonic Order, and Estonia remained under the rule of the Baltic knights until the Order's dissolution in 1561. In the 16th century Estonia was attacked several times by Russia, which devastated the country. From 1561 until 1721 Estonia was a dominion of the Swedish Empire. For a relatively short period (1561–1621) South Estonia was part of the Polish–Lithuanian Commonwealth. The Russian tsar, Peter I (the Great), was finally able to achieve the dream of his predecessors to conquer the Baltic provinces in the Great Northern War (1700–1721).

Trade was controlled by the Hanseatic League, which was an alliance of trading cities and their merchant guilds that dominated trade along the coast of Northern Europe. The League continued the merchant traditions of Vikings. Tallinn, Tartu, Narva and several other Estonian cities became members of the Hanseatic League. Many buildings in Tallinn and Riga have the style of their Hanseatic days. Estonian and Latvian cities reaped huge profits from the Hanseatic trade.

The artisans in the cities were organized by guilds. Guilds helped to raise social values for work, secured education for artisans, and ensured product quality. However, most guilds excluded non-German artisans. This was accomplished by limiting the membership in a guild to citizens of the city, and the Estonian and Russian minorities were excluded from citizenship. This guild system functioned until World War I. Local power in Estonian cities was until the beginning of the 20th century still in hands of Germans. At the turn of the 19th and the 20th centuries the level of education and the economic standing of Estonians had improved. Only in 1904 were elections for the town government of Tallinn won by an Estonian–Russian bloc.

In the 16th century Lutheran reformation reached Estonia. Lutherans promoted the publication of sacred texts in local languages. So the first catechism in Estonian was published in 1535, the whole Bible in 1739. The translation of the Bible into the North-Estonian language strengthened its position to become the common written language in Estonia. During the Swedish period of Estonian history elementary schools were opened all over the country. At the end of the 19th century all Estonians become literate.

Baltic Germans, especially Baltic nobility, played an important role in the Russian Empire. They became high-ranking officials in the Russian bureaucracy, military leaders (Barclay de Tolly — minister of war and field marshal during the Napoleon campaign in 1812), scientists (Karl Ernst von Baer — biologist and a founding father of embryology), and explorers (Fabian Gottlieb von Bellingshausen — admiral and naval explorer, discoverer of Antarctica). For Barclay and Baer very beautiful statues have been erected in Tartu.

During the first century of the Russian period after the Great Northern War the conditions of peasants hit its lowest point. All land belonged to Baltic Barons, and Estonian peasants were transformed to slaves. Barons had the right to sell and buy peasants, and punish them in a rather cruel manner. A hundred years later, under Tsar Alexander I, the peasants of Livonia and Estonia were given the right of private property and inheritance. Other agrarian laws followed: establishing the peasant's right of free movement, abolishing the landowners right to flog, and the corve.

Until the 19th century the ruling elite had remained predominantly German in language and culture. This changed in the middle of the century. The Estonians became more ambitious in their political demands. Also Estonians started to buy their farms from barons. This movement started in Sakala, culturally and economically the most advanced region in Estonia. Significant accomplishments were the publication of the national epic, "Kalevipoeg", in 1862, and the organisation of the first national song festival in 1869. Also Estonian newspapers were founded by Johann Voldemar Jannsen ("Postimees") and Carl Robert Jakobson ("Sakala"), followed at the end of the century by many others.

At the end of the 19th century a period of Russification started. However, in response Estonian nationalism took on more political tones, calling for greater autonomy. This goal was achieved after the February Revolution of 1917, when all Estonian lands were united into one administrative unit — earlier it was divided between Estonia (the former Danish colony), and Livonia, which included South Estonia and North Latvia.

In October 1917 Bolsheviks took power in Petrograd, and also in Estonia. Russian soldiers were tired from the World War and open to Bolshevik propaganda. Russia started peace negotiations with Germany which ended with the Treaty of Brest-Litovsk. Before the treaty was signed, the German army occupied Baltic countries and Ukraine. There was a short time-slot after the retreat of the Red Army and the arrival of German troops. This was used by Estonians to declare independence on February 24, 1918.

When the World War ended in November 1918, the German army was withdrawn, and Soviet Red Army attacked Estonia and Latvia. Soon most of the Latvia and about two thirds of Estonia were occupied. Estonia had initially almost no army. Rapidly volunteers were recruited, the most effective one being a commando type unit, led by Lt. Julius Kuperjanov (a classmate of my father in the Tartu Teachers Seminar). Due to a flow of young but highly motivated volunteers, Kuperjanov's unit became an independent battalion known for its high morale and daring tactics. The battalion played a crucial role in liberating Southern Estonia.

A Royal Navy squadron arrived off Tallinn on the 31st of December, 1918, delivering ammunition, rifles and machine guns. The squadron captured two Russian destroyers and turned those over to Estonia. On 2 January 1919, a Finnish volunteer unit with 2000 men arrived in Estonia. Armored trains were built in Tallinn. These measures helped to start a conter-offensive on January 7, 1919. Within a month the Soviet Army was thrown away from Estonian territory. Estonia had become the first country to repel the Soviet westward offensive. However, heavy fighting continued both in the Eastern front near Narva, and in Northern Latvia.

In summer 1919 a German force was formed in Latvia, consisting of the Baltische Landeswehr of Baltic Germans and the Iron Division of volunteers from Germany. The goal was to form a pro-Germany state in Latvia and Estonia. The Germans captured Riga and continued to move towards Estonia, capturing Cesis, a small city close to Estonian border. On June 23 the Estonians counterattacked recapturing Cesis, and advanced towards Riga. The anniversary of the Battle of Cesis is celebrated in Estonia as Victory Day.

The Soviet Army continued attacks in the Narva front, but without success. The peace talks in Tartu were successful, and on January 3rd, 1920 an armistice took effect. The peace treaty was signed on February 2, the Peace of Tartu.

The most urgent task of the Estonian government was to perform a Land Reform. Large estate holdings belonging to the Baltic nobility were redistributed among the peasants and especially among volunteers in the Estonian War of Independence.

The second important task was to advance cultural development. Tartu University was founded by Swedish king Gustav Adolf in 1632, and continued its activity until the Great Northern War. In 1802 the University was reopened, and it became one of the most important links between German and Russian universities. The teaching was in German, and many world-class scientists worked as professors, among them astronomer Friedrich Georg Wilhelm Struve and biologist Karl Ernst von Baer. During the Russification period teaching was allowed only in Russian. In 1918 almost all previous professors left the University. In 1919 the University was reopened as an Estonian-language institution. One of the actual tasks was the formation of scientific terms in Estonian, and to start lecturing in Estonian.

One of the notable cultural acts of the independence period was a guarantee of cultural autonomy to minority groups comprising at least 3,000 persons, including Jews, Germans and Swedes.

On August 23, 1939, Molotov and Ribbentrop signed a Pact of Non-Aggression between Germany and the Soviet Union. The Pact included a secret protocol dividing East Europe into German and Soviet spheres of influence. This opened the door for German and Soviet forces to invade Poland — World War II had started. Next the Soviet Union forced Estonia, Latvia and Lithuania to host Soviet military bases; next year all three countries were occupied and thereafter annexed as new Soviet Republics. The Soviet Union tried to annex Finland too, and the Winter War between Finland and Soviet Union started. Finnish defences surprisingly held out for over three months. Finally Finland ceded Southern Karelia, about 10% of Finnish territory, but preserved its independence.

One of the main conditions posed by Hitler to Stalin in August 1939 was the transfer of all ethnic Germans living in Estonia and Latvia to Germany. So in 1939–1941 some 20 thousand Baltic Germans were resettled from Estonia, and about 60 thousand from Latvia, mainly to Polish areas annexed by Nazi Germany. At the end of WW II they were resettled from Poland to Germany. The evacuation was delayed until the last moment, so many died on the way. For seven centuries the Baltic Germans formed about 5% of the total population of Estonia and Latvia. But this minority was actually the political, economical and spiritual elite of our countries. Now it is gone, spread over Germany, Canada and many other countries.

World War II had a large impact on Estonian history. We had 3 occupations, the first Soviet occupation 1940–1941, the German occupation 1941–1944, and the second Soviet occupation/annexation 1944–1991. During the first Soviet occupation on June 14, 1941 a mass deportation took place: about 10 thousand people were deported. Additionally nearly 40 thousand Estonian citizens were executed or mobilized into the Red Army, and at least 10 thousand mobilized men died in concentration camps due to starvation before they were sent to front. In September 1944 about 70 thousand people fled abroad for Germany and Sweden to avoid Soviet repressions.

The German occupation was also hard. Over 3 thousand civilians (mostly Estonians) were executed in one concentration camp near Tartu. The Nazi regime established smaller concentration camps on occupied Estonian territory, especially for foreign Jews. The presence of these camps was not known to people except the camp in Tartu. Germany made in 1943–1944 a total mobilisation, and almost 70 thousand young men were forced to join the Waffen SS or other units. Quite often Estonians in the Soviet and German armies had to fight agains each other. This was our tragedy.

Many Estonian boys preferred to fight against the Soviet Union in the ranks of the Finnish army. This appeared a reasonable alternative for those who wished to fight for the freedom of Estonia and against the Red Army, but who for ideological reasons were unwilling to do it in German uniform. The boys crossed the Gulf of Finland in early spring 1943. It was a dangerous journey, as the gulf was ravaged by storms, and the German authorities did not allow Estonians to cross over to Finland. From these boys, called "Soomepoisid" (Finnish Boys) a Finnish Infantry Regiment 200 was formed. The motto of the Regiment was: "For the freedom of Finland and the honour of Estonia". When in September 1944 Finland made peace with Soviet Union, a large fraction of these boys returned to Estonia to fight against the Soviets, who had managed a breakthrough at the Tartu front. Those who survived formed in post-war Estonia our elite — they became scientists, writers etc. I have many friends and colleagues who were in war-time Finnish Boys.

In August 1944 Soviet forces attacked from the South, and were stopped at the river Emajõgi, crossing Tartu. A further advance to take over the whole country was in sight. As Alar Toomre tells in the same interview to 'Horisont', his father had a choice *"between evil and deep see. He chose deep see. It was a dangerous journey, Soviet submarines and planes attacked ships, and several ships ended up in deep see. Our family survived"*. In November 1944 the last piece of Estonia, our largest island Saaremaa, was "liberated".

In March 1949 the second large Soviet deportation followed. About 20 thousand people, mostly peasants, women and children, were sent to Siberia. Almost all military and governmental elite was murdered, similarly to the murder of Polish officers in Katyn. The "Katyn" movie by Andrzej Wajda is one of the best documentaries to show the tragedy of East European nations who happened to be located between the two totalitarian empires (Wajda's father was murdered in the Katyn massacre).

The total losses of the Estonian people during and after the WW II were about 25% of the whole pre-war population. Two other Baltic countries, Poland and Belorussia, suffered similar losses, the countries most influenced by the Molotov–Ribbentrop Pact. To avoid repressions, after the war a large number of people hid themselves in forests: they were called "forest brothers". Some forest brothers fought agains the new rulers as partisans. NKVD chased them ruthlessly, and many were killed or imprisoned. Since for us the war did not end in 1945, Estonia had no babyboom, as happened in most other countries, including Germany and Russia. For Estonians WW II actually ended only in 1994, when the last Soviet troops left Estonia.

The fate of Estonia as well as all East European countries (except Finland) was determined not only by the Molotov–Ribbentrop Pact, but also by treaties between the Soviet Union and Western Allies in Teheran, Yalta and Potsdam. In these treaties the Western Allies actually 'bought' freedom for Western countries by 'selling' East Europe to the Soviet Sphere of interest. Western countries understood this too late, when the "Iron curtain" had already descended to divide East and West. A good description of the fate of nations between the two totalitarian countries is given by Laurence Rees in his book "World War II; Behind Closed Doors; Stalin, The Nazis and The West", BBC Books, 2009.

Similarly to Nazi Germany, the Soviet regime tried to eliminate the cultural and spiritual elite of the nation. A lot of leading people in culture — scientists, writers, teachers — were sent to Gulag, and many died there. The last wave of persecutions was in 1950, the campaign against "bourgeois nationalists". Prominent scientists had to declare openly that in their previous scientific work they had been guided by wrong non-marxist principles, and a number of scientist were arrested. Just one example. My father's colleague from the Tartu Teachers Seminar, the history professor, previous Rector of the Tartu University, and the first post-war President of the Estonian Academy of Sciences, Hans Kruus, was arrested and brutally tortured in the Ljubljanka prison in Moscow. He survived and was released after Stalin's death, but his health was ruined.

Our situation was worse than in so-called "Socialist" countries, since we had not only communist rule and military presence, but also the presence of a "civil garrison". During the Soviet period almost half a million immigrants from other Soviet republics arrived. This made the survival of our own culture and identity difficult. The integration of immigrants was difficult since immigrants felt that they were in the Soviet Union, and so are the masters to dictate the rules of everyday life. There were even plans to change our latin script to cyrillic, but fortunately these plans were abandoned. The official language in the Estonian government was Russian. However, the teaching in schools and University was still in Estonian, except in Russian schools. So it was possible to continue the development of the Estonian science language.

But we had to have our everyday life, and to adapt to the new system. For a small nation the most important aspect is to preserve its identity and to develop the national culture. The Soviet propaganda was so stupid that it had little impact on the people's way of thinking. But to develop the culture, in particular education and science, some cooperation with the new order was needed.

The Soviet system promoted education and science, and this helped to keep education at a rather high level. In science a very rapid development occurred in such sciences as physics, mathematics, chemistry, astronomy, and technical sciences. One reason for this was the fact that in these sciences the political pressure was not so great as in humanities.

Many of our political leaders tried to protect the interests of the Estonian nation and culture as much as possible. Thus, when the Soviet economical system practically collapsed in the mid 1980's, and Gorbachev tried to reform the system by introducing *glasnost* and *perestroika*, Estonian leaders understood how to use this in our interests. Our final goal was to break away from the Soviet Union. But this goal was not declared openly; steps towards independence were made slowly and were carefully planned to avoid open conflict with Moscow authorities. These efforts culminated in the late 1980's, and after the unsuccessful August Coup in 1991 Estonia regained its independence peacefully.

2.2.2 *My roots*

Both my parents' roots are in South Estonia.

My father came from a family in the Southern part of Tartu county near the lake "Pühajärv" (Holy Lake in Estonian). The grandfather Jüri Grossberg of my grandmother Minna participated in 1841 in a revolt against a local baron, known in Estonian history as the "Pühajärve war". It was suppressed by the tsar's army and major participants were punished by gauntlet near an old oak at the lake Pühajärv. Jüri Grossberg survived but was very ill thereafter. My grandmother Minna Grossberg was born in a farm called Türgi. Her sister Marie Koppel was the editor of the daily newspaper 'Olevik' until 1905, when the newspaper was closed by Russian authorities — the first woman in Estonia to work as an editor of a political newspaper. Minna's daughter and my aunt Leida owned the Türgi farm after World War I. She died in 1937 and left the farm to her son Jüri, who was 15 at the time. Jüri with the help of grandmother Minna and his two sisters, Liis and Eva Maaring, was able to keep Türgi farm in good shape. During World War II two of my younger brothers, Rein and Peeter, and I spent all summers in Türgi, helping and learning farming.

My grandfather Karl Eisenschmidt was a son of a nearby farmer too. The family name "Eisenschmidt" is a German translation of the name of the farm "Raudsepa" near the Southern edge of the lake Pühajärv. Family names of Estonian peasants were given by German landowners at the end of the 18th century. Mostly the name comes from the name of the farm they lived in. As names were given by Germans, often the farm name was translated into German. Local people even in the middle of the 20th century called themselves according to the farm they lived. Thus, when I lived in the 1940's in Türgi farm, I was called "Türgi Jaan".

Karl Eisenschmidt married my grandmother Minna in 1891. The first job of the family was lessees of a farm; thereafter Karl served as the secretary of the local parish. Before WW I the family moved to Tartu where their daughter Leida worked in a bank and had a free apartment from the bank. In these years Karl was a journalist, and he wrote articles on the life of farmers as well as reviews on farming. In his articles he was rather critical towards local estate owners (barons of Baltic-German origin). Thus during the German occupation in 1918 he had to hide himself to avoid arrest. After the death of Minna's father he together with Minna went to Türgi to keep the farm.

My father Elmar got his education in Tartu Teachers Seminar. Thereafter he was a teacher in a school, which was moved in 1917 to Siberia's Altai region. After the Estonian War of Independence he was able to return to Estonia in 1921, and soon became the director of a girl's school in Tartu. He was a very good teacher, and worked in this school until his retirement. He was interested in geography and participated in the preparation of a series of monographs on Estonian counties. During one of research trips he visited Egeri farm and become acquainted with the daughter of the owner of the farm, Eva Lammas. They married on June 24 (Jaan-Day) 1927. I was their first son. Thereafter almost every few years new children were born, so I have brothers Vello, Mart, Rein, Peeter and Andres, and sisters Tiina and Kersti.

My great-grandfather from my mother's line, Wiilip Lammas, bought a farm Egeri not far from the Latvian area of Livonia (at this time Southern Estonia and

Fig. 2.2 Pühajärve "war-oak" — under this oak 1841 Estonian farmers were punished by gauntlet. The grandfather Jüri Grossberg of my grandmother Minna also was punished by 500 beats. In this picture our family was celebrating the 100th birthday of our grandmother Minna, and we visited places of interest to our family history (author's photo).

Northern Latvia formed one Livonian governorate in the Russian Empire). He gave to his first son, my grandfather Jaan, a very good education. For several years Jaan was a guest of a local German family who had hired a private teacher. The family was German-speaking, so soon Jaan was fluent in German. Next Jaan studied in a Latvian school to learn Latvian. Later during the Russification period he learned Russian, so he was fluent in all four local languages. First he was a teacher in a local school, but soon took over the main duties of farming Egeri.

In 1879 Jaan Lammas participated in the Second National Singing Festival in Tartu, and acquainted himself with leaders of the Estonian national awakening movement. The goal of the movement was to acknowledge Estonia as a nation deserving of the right to govern itself. The result of the visit to Singing Festival was that Jaan divided his life both to farming and social activity. He applied new methods of farming, experimented with modern tools, and started to write articles in Estonian, Latvian, Russian and German journals and newspapers on farming. Also he translated papers of cultural interest from Latvian newspapers and journals into Estonian and vice versa. For some time he was the head of the local parish, and served as organist in a local church. In social life he was the founder or cofounder of societies on farming, singing, banking etc. Such initiative from the people was very important in preparing the whole society for an independent country, and helped to win the Independence War against Bolshevik Russia.

The social activity brought Jaan close to the main centre of the awakening movement in Viljandi, where he met his future wife Anna Wiegant. Anna was born in a family of local glassmaster of German origin. Her mother died after the birth, and Anna was adopted by Lilli Suburg, one of the leaders of the Estonian awakening movement. Suburg founded a private girl school in Viljandi, as well as the first Estonian journal for families and women, "Linda". She was active as a journalist and writer, and was the initiator of the Estonian women's rights movement. Anna started to help Lilli as a teacher when she was only 13 years old. Later, during the Russification period, Anna became director of the school as Lilli was not fluent in Russian. The main goal of the school was to cultivate educated Estonian women as creators and holders of families and self-governing citizens.

At the end of the 19th Century the school was closed due to economic difficulties. In 1899 Anna and Jaan married, and moved together with Lilli Suburg to Egeri. Here Lilli and Anna continued to teach children until the birth of Eva, my mother, in 1907. In 1928 Jaan Lammas died, and Anna moved to Tartu to my parent's family, where she lived until her death in 1937. The farming of Egeri was given to lessees. When the Soviet occupation started, my mother was clever enough to give Egeri away to the state. In this way our family was not deported to Siberia as most other owners of big farms in the region.

My parents and grandparents liked music. My mother writes in her memoirs that on Sunday mornings she was awakened by quiet piano music played by her mother Anna, or by organ music played by her father Jaan. In our house in Tartu we also had a piano, and both my mother and father often used it. We had a very large collection of notes, mostly organ music, collected by grandfather Jaan.

Our family lived in Tartu where my father was a teacher and school director. In the 1930's Estonian government started a campaign to change German names to Estonian ones. So our family name was also changed. My father was a real patriot of Estonia, so he invented the name "Einasto", which is a permutation of "Estonia". The name was patented, so nobody else can have this name. In this respect our family name is unique.

In 1937 our parents built a private house near the border of Tartu. The architecture of the house is rather modern, and the house is registered as of cultural importance. In this house I lived for 25 years, including the first ten years when I was married to Liia. Due to the work by Jaan Lammas and Lilli Suburg the Egeri farm is also in the list of objects of cultural importance.

After retirement my father and his sister Erika became interested in our family history and started to search in parochial registers. They were able to reconstruct our family history up to the Great Northern War — older archives and registers were destroyed during the war. Now almost all these registers are scanned and can be accessed via internet. One of my relatives, the husband of my niece, is an active member of the Estonian Genealogical Society, and helped to reconstruct our family history using data presently available. He confirmed most findings of my father and aunt, and generated family trees for my 80th birthday, see the website accompanying this book.

Most recently the genealogy can be reconstructed using the website at Geni.com. Using data available at Geni.com I found that the grandmother of my grandmother Anna was of noble origin. For noble families the genealogical tree can be restored for a much longer period than for peasants, thus through this line my ancestors are much better known. If all ancestors were not related to each other, then the number of ancestors of the nth generation would be 2^n. In this case for large enough n the number of ancestors would be larger than the population of all mankind. Actually many ancestors are relatives, and the number of ancestors grows much more slowly. This effect is well-known in the genealogy; German scientists call it "Ahnenverlust". This is especially valid for small communities where almost everyone are relatives of people from the same community.

People from noble families had a tradition of taking wives also from noble families, thus, step-by-step I found that many Baltic German noble families are my relatives, and some are my direct ancestors. These lines go back to the conquest of Estonia and Latvia by the Livonian Brothers in the 13th century, and through them to noble families of the whole of Northern Europe. In this way I discovered that Rurik of Svealand, the Prince of Novgorod and the founder of the Russian State, is my 28th great grandfather. Further lines go to the antique world: Augustus Caesar, the first Roman Emperor, is my 56th great grandfather; Philip II, King of Macedonia and father of Alexander the Great, is my 64th great grandfather. Geni.com data indicate that my most distant known ancestor lines go back over kings of Armenia, Babylon and Persia to Egypt: my 124th great grandfather was Ro, a Predynastic pharaoh of ancient Egypt (circa 3250 BC).

Actually this long known family tree is not a surprise: once you happen to have among your ancestors some nobleman, you can go very far, since for these people data on births, deaths and relatives are fixed in various documents. My unwritten family tree goes at least 300 generations back — Estonians are one of the most stable people and have lived on the same territory more than nine thousand years. This is the reason why almost all pieces of the landscape have names —

people thought that not only animals but also trees, lakes, and other elements of Nature have souls, and gave them names.

And after all, DNA analyses show that all mankind has one common ancestor, the Mitochondrial Eve, the most recent woman from whom all living humans today descend on their mother's side. Our First Eve lived around 200,000 years ago, most likely in East Africa. My colleague and friend Richard Villems, director of the Estonian Biocenter, leads a team to study the topology of the human mitochondrial tree and the origin and diversity of modern human populations. In collaboration with teams in different countries Villems has studied the DNA history of Estonians and other nearby nations. He is particularly interested in the early phase of the spread of modern humans from Africa to Eurasia, Australia, and to the New World.

2.2.3 *My early life and first steps in astronomy*

I grew up in a teacher's family. We had a large library, and I used this in my school years. First I read Estonian classical authors. In the 1930's many good series of books were translated into Estonian, so I had the chance to read novels of Nobel prize winners, as well as books on modern history (this series was called "From the Marseillaise to the International"). Also I had the opportunity to read popular books on many natural sciences. During the war I graduated from primary school and entered the Tartu Real Gymnasium.

My interest in astronomy is due to a present from my father for Christmas 1942 — a popular book by Roopi Hallimäe, "Astronomical Observation" (in Estonian). The book was so well written that I started to observe various astronomical phenomena. Quite soon I started to visit Tartu Observatory. One of the first persons I met was Roopi Hallimäe, an enthusiastic observer and poet in his spirit. We became good friends and I learned a lot from him.

My physics teacher Osvald Sulla was a friend of the director of the Observatory, Prof. Taavet Rootsmäe, and introduced me to him. So, in the autumn 1943 Rootsmäe invited me to his lectures on general astronomy. His lectures were very clear, and with some help from textbooks on elements of higher mathematics and spherical trigonometry I was able to follow lectures. There were only two students and me in these lectures. Soon the students disappeared — Germany had made a total mobilisation of young men in Estonia. So I was the only listener, but Rootsmäe continued lecturing as before. After the lecture he invited me to his office, and we had long discussions on the topic, as well as on more general problems. In one of these discussions Ernst Öpik entered the office — this was my first meeting with him.

In the Observatory there were copies of older issues of the Calendars of the Observatory, where popular articles on various astronomical topics were published, as well as summaries of papers published in Tartu Observatory Publications. The most interesting of these papers was the fundamental study by Öpik (1938), where the modern theory of stellar structure and evolution was formulated. Another interesting area was the study of the connection between the kinematics of stellar populations and the mean ages of stars by Rootsmäe. Results of this study were published only posthumously (Rootsmäe, 1961).

In January 1944 the Soviet Army broke the German defence around Leningrad. The frontline stopped at the Narva river, and Estonia became arena of military actions. Schools were closed and it was suggested that pupils go to countryside if possible. My grandmother Minna lived with her grandson Jüri in Türgi farm about 60 km south of Tartu. So I and two of my younger brothers went to Türgi. I brought with me a self-made telescope which had an objective made of eyeglass. With this telescope I continued observations of the Moon, planets and other interesting objects. The winters in the war years were cold and the sky mostly clear, so I had the chance to observe almost every evening.

In August I made preparations to observe the Perseid meteor shower, but then an artillery cannonade from the East signalled the start of a new offensive in Pskov towards Southern Estonia. About a week I observed from the roof of our house the air battles about 15 km South of our farm. The gunfire was heard South of us, and it was very rapidly moving westwards. Then in one day it was very quiet and my grandmother suggested it's time to begin to harvest rye. We had just started when from North-West several grenades were shot towards us. This was the signal of a German counteroffensive. It was high time to escape. With two horse-carts we left our farm in the early morning, taking with us 4 cows. With cows our movement was slow, and soon it was time to milk cows, so we had a stop not far from our farm.

We had not finished milking when we saw Russian soldiers approaching from South to set up a gun on a nearby hill. The Germans immediately shot at and destroyed the gun. And then a bitter gun- and mine-fire started, and we were just in the middle between two attacking armies. Several grenades came very close to us, and one cow was wounded, but we were unharmed. In the evening it became quieter and we thought where to hide. But then a Russian officer came and gave an order — to come with our horse-cart and evacuate a wounded soldier. So one of our carts was unloaded and I accompanied the evacuation. The ambulance was about 10 km to South. There our horse was exchanged for a slightly wounded horse and I could go back. During the night it was not easy to find the place where I left my relatives. Finally I found them, but then the battle started again and a group of

Russian soldiers surrounded me. Finally I found the place where our family was hiding — in a deep valley not far away, where numerous other farmers with their families, horses and cows hid.

I had just fallen asleep when a Russian officer came and ordered me to evacuate a wounded soldier. I had to go again. This time not so far. But there I met a farmer who said that he had been there already several days without any food and hope of going back to the others. Then I thought that it is perhaps better to leave the horse and cart there, and to try to go back to the valley where all of us were hiding. However, this was a wrong decision, and soon I was arrested and brought to a command post. I was thoroughly searched, and they found in my pocket a self-made map of the routes in South Estonia. I had forgotten to put it away, and I was suspected to be a German spy. This unit was probably Smersh — the Soviet counter-intelligence agency during the war. After the first questioning they showed me an open grave and threatened to shoot me if I did not provide evidence of my spying tasks. Fortunately I was not shot immediately. In the following nights I was questioned again and again, I had to tell of my evacuations of wounded soldiers, what my father thinks of the Soviet regime etc. During the questioning I always said what actually happened, because I had nothing to hide.

The Smersh unit moved with the attacking army, and a lot of suspect people were arrested and questioned. Among them were Estonians who had served in the German army, Soviet prisoners of war who had worked in Estonian farms, and randomly picked up people, young and old. After about two weeks they apparently trusted my answers and set me free. But they ordered a young boy to accompany me. The boy told me that he was tasked with informing Smersh what I said when I returned home.

So, at the end of August I was back in our farm. All my relatives were there, and the joy of reunion was unforgettable. I told my story, the boy left next morning, and Smersh did not take any actions against me. Later I learned that usually if you are already in the hands of Smersh (or KGB after the war), they have methods to press out whatever accusations they want. I know many such cases, both among my Estonian friends as well as my colleagues in Soviet observatories. Thus I had been very fortunate.

The area around our farm was a place of fierce battle; several grenades had hit our houses but the damage was not very great. Also several corpses of Soviet soldiers were found, but on my arrival they were buried. The farm about one kilometre West from ours was burned down, there bodies of German soldiers were found. My relatives had been allowed to return home only a few days before my arrival. The farm was occupied by a Russian command-unit. The commander of the unit was a very educated man, he had graduated from the Leningrad conservatory and often played our piano. Under his command almost nothing was stolen by soldiers.

Now we had a problem — we had only one horse, and with one horse it is impossible to use our harvester (it had been left exactly at the place before our escape). Thus we had to harvest everything with scythes. Autumn 1944 was nice, almost no rain, so we managed to collect everything before October rains began. We started early in the morning and worked until the first stars appeared in the dark sky. The view of autumn evening stars with Lyra, Cygnus and Pegasus are forever in my memory. By the end of October it was possible to return to Tartu to continue studies in high-school.

My first task was to visit the Observatory. For several weeks Tartu was a front-city. The German army held the Northern side of the river Emajõgi (Mother River), and the city was under permanent gunfire. The centre of the city was almost completely destroyed, but the main buildings of the University had suffered only little damage. The Observatory was hit and its dome started to burn, but the keeper of the Observatory was able to quench the fire. When he ran down from the dome another shell hit the dome, and through the window splinters wounded the keeper's leg. Fortunately, his bones were not hit, and he could come down.

Prof. Rootsmäe told me that small equipment such as chronometers had been stolen by soldiers, but major instruments were undamaged. Also the library, which was hidden in the basement, was more-or-less preserved. Only some books and pictures were befouled. The windows were already mended, and routine work had started. In summer 1945 the Solar eclipse was observed there by Prof. Kipper and his assistants.

My gymnasium in the old-town had not been damaged. One of the oldest churches just next to our school was burned out. When I arrived at school, routine lessons had already begun. But I noticed changes. Earlier in the city center there were two gymnasiums for boys, the Real Gymnasium and the Treffner Gymnasium (a basically humanitarian one). In both schools at every level there were at least two parallel classes. Now both gymnasiums were united to form the 1st Tartu High-School (now it has the old name — Treffner Gymnasium). Instead of four classes from the two joined schools there were two classes, one level higher there was one class, and two levels higher only one half-full class. So great were the losses of Estonian young men during WW II. Boys in my class were too young to be mobilized into the German Army; the losses were mostly due to the escape of a large fraction of educated families to the West to avoid Soviet occupation. In higher classes additional losses were due to the war — boys were either killed, imprisoned by the Soviets or escaped to the West after the end of the war.

Fig. 2.3 With my classmates around the Kalevipoeg statue in 1947. Kalevipoeg was considered as the symbol of freedom for Estonia. The statue was destroyed by Soviets in 1950 (author's photo).

In my class some boys formed a coterie who started to learn English in their free time (in our class foreign languages were Russian and German). I also joined this group. We organised during vacations excursions to interesting places, celebrations of birthdays and holidays. Among ourselves we used English names, so I was called John. We organised also a physics club and had reports on various topical problems. In 1947 I graduated from the school and started my studies at the University.

In the University my first mentor was Grigori Kuzmin, a student of Ernst Öpik. He advised me on how to solve a new problem. First the problem must be simplified, so that only the main factors are taken into account. On the basis of this, one can find a preliminary answer. Thereafter other factors influencing the process

Fig. 2.4 My first mentors Prof. Taavet Rootsmäe (left) and Dr. Grigori Kuzmin (right) (author's archive).

under study can be taken into account, step by step. This allows one to select all important factors, and eliminate less important ones. He told me how he solved problems when he was a student. When the professor formulated the main problem during the lecture, and started to develop further steps, Grigori did not look at what professor was doing, but tried to solve the problem independently. Usually the professor finished first, as he already knew the answer, but sometimes Grigori was quicker. This habit of solving problems independently was characteristic of Ernst Öpik too. When I met Ernst Öpik during the General Assembly of the International Astronomical Union in 1970 in Brighton, Öpik was just looking for the place where a meeting of interest was to take place. I suggested asking somebody for help. Öpik's reply was: *"it is more interesting to find the location myself"*.

There was an active social life in the University. Students interested in sciences joined the Students Science Society (SSS). When I became a student I organised a Students Astronomical Society, which was later transformed to the Estonian Branch of the All-Union Society of Astronomy and Geodesy. I was also rather active in the SSS, and was elected the secretary of the Society. I continued my activity in this Society for several years. But then I understood that this takes too much time, and when starting to write my diploma thesis, I stopped all my activity in the SSS.

The other activity was among my own course-mates in the mathematics group. We continued close friendships after the University, celebrated family birthdays, and organised walking and bike excursions.

In Tartu University I was the only student in my group interested in astronomy, so I had to make a choice to join either the physics or the mathematics group. I prepared an individual plan for study, taking as basis the program in the Moscow University at this time, where astronomy was in the mathematics faculty. Most astronomical disciplines I learned from textbooks, such as spherical astronomy, geodesy, stellar statistics and dynamics etc. Also I often visited the Sternberg Astronomical Institute in Moscow, the Institute where Ernst Öpik studied as a student of the Moscow University. There I met Professor Pavel Parenago, the leader of the Soviet School of stellar statistics and dynamics. When discussing possible topics for my diploma thesis he suggested a detailed study of the kinematics of stars of the main sequence. He had just discovered that the main sequence is kinematically inhomogeneous, and wanted to have more detailed information on this effect. This problem was very close to my own interests, as well as to the topic of the research of Prof. Rootsmäe, so I agreed. This resulted in my diploma thesis (Einasto, 1952), as well as my PhD thesis (Einasto, 1955).

Together with students from the Sternberg Institute in 1951 I visited the Abastumani Observatory in Georgia. Here we had lot of time to discuss various topics of interest. The supervisor from the Sternberg Institute was Alla Massewich. Once, during observations, we discussed the stellar evolution problem, which was the main subject of her studies. Alla was a supporter of the old Russell concept of stellar evolution and criticised the work by Öpik as too complicated. Probably the modern stellar evolution concept was generally accepted only after Martin Schwarzschild (1958) confirmed earlier results by Öpik, Gamow and other authors. In his early papers (Hen & Schwarzschild, 1949; Oke & Schwarzschild, 1952) the first reference is to the classic Öpik (1938) paper.

In 1952 I graduated from Tartu University and started my work in Tartu Observatory as junior scientist. Here I stayed for my whole astronomical life. My first work was to help Grigori Kuzmin in his calculations of a new model of the Galaxy (Kuzmin, 1952a). During this work I found a way for a better extrapolation of the mass distribution to larger distances, to avoid infinite or zero total mass and negative spatial densities. This simple trick was the starting point for my future work on galactic modelling which led finally to the understanding of the dark matter phenomenon, a work which is not finished even today.

2.2.4 Liia

There were not many physics and mathematics students after the war, so we had some lectures common to students of various grades, and had common new-year celebrations. Among the students one grade higher from my own was a very nice and charming maiden, Liia Tiit. First I did not catch sight of her. Then, one evening we met in the hall of the University in a classical music concert. The hall has very good acoustics and is used as a concert hall for chamber music. We started to discuss the concert, and I accompanied her home. We discovered many common interests.

In summer 1950 Liia rented a room in Elva, a small town near Tõravere within a forest. Once I visited her — and stayed there for the whole summer vacation. We had long walks in the pine forest of Elva and bathed in the Elva river. We discovered that we have a lot of common interests and dispositions. We found that we wanted to live together the whole coming life.

On March 10, 1951 it was the day to register the marriage. It was a usual working day, I was in the Observatory, and had a discussion with Grigori Kuzmin. He talked at length on a topic. I looked at my watch, it was already time to go to meet Liia, but Grigori did not stop talking. Then I said quietly that I have to go to register my marriage, and I am already late. Grigori looked at me with a surprised face, apologised, and let me go. I hurried to Liia's home, she was already a bit excited and thought that something had happened with me. Together with her parents we walked through the Toome hill to the Office of Marital Status. It was a sunny day, and the first spring water rills trickled down the hill. My parents and brother Mart were already there, together with the son of Liia's uncle Valdur Tiit. After the official procedure we had a joint lunch in Liia's parents' home. Thereafter we came together to our house.

My parents gave for us two rooms in our family house, and our joint life started. At this time we were both students and lived from our stipend, which was rather low. Thus we lived very modestly. After we both graduated our financial situation improved a bit. However, in order to buy something needed for the household, we had to collect money for at least a full year. In the first year we bought a queen-size bed, in the next year a radio, in the third year bikes for the both of us. On summer weekends we made long bike-trips together.

Here our first children, daughters Riina and Maret, were born. Our two rooms were incommodious, and we decided to build our own house in the garden of my parents. I spent two years to build the house. However, my financial possibilities were very limited, and the building was slow. Then it was decided that for astronomers apartments will be built in the new observatory in Tõravere. So I decided to make use of this possibility, and sold the half-built house. On May 1961 we moved to our apartment in Tõravere. Here our family has spent the rest of our life.

In Tõravere our third child was born, son Indrek. A nursery school was organised, first in the apartments of one of the houses, and a few years later a special house for 50 children was built. Conditions for children in Tõravere are very good — after nursery school they can stay in fresh air and play until dusk, on the Observatory grounds and in the nearby forest.

Liia's first job was teaching mathematics at Tartu University. In the early 1960's in the Observatory a computer center was organised, and programmers were needed. We already had three children, and it was not so easy for Liia to go daily to Tartu, so she applied for a job in the computer center. She got the job. In the 1960's and early 1970's she participated in our study of the structure and evolution of galaxies.

We continued in summer time bike-trips, first with Liia, and later together with Maret and Indrek. Also we had a garden in the Observatory. Here we planted apple and pear trees, raspberry, blackcurrant and redcurrant, so we could eat these fruits fresh and make jam for winter. Liia loved gardening very much and spent much time thinking about how to do it all better.

Our joint hobby was music. Liia played the piano, and we both liked to listen to classical music. In 1955 I bought our first vinyl record player, and started to buy records. Good classical recordings were not available in Estonia, so I used my trips to Moscow to find good records. Soon the manager of one Moscow record shop asked where I was from, as it was clear that I was not Russian. When the manager hear that I am from Estonia, she started to reserve for me good records, which usually disappear from the shelf rapidly. When I was next in Moscow and visited the shop, I was able to buy the best examples of classical music, played by the best musicians. So I have the full set of Mozart's piano concerts, played by Daniel Barenboim, the full set of Beethoven's piano sonatas, played by Maria Grinberg, and many other very good records. Maria Grinberg, Svjatoslav Richter and other top Soviet musicians often gave concerts in Tartu University hall, and Liia and I always listened.

In 1972 I had my first visit to West Germany, and was able to buy a stereo player. Since then I collected only stereo records. In 1985, also in a visit to Germany, I was able to buy my first CD-player. At that time they were very expensive, I think that this player was the first in Estonia, so that even Estonian Radio people once came to me to make a copy of a rare CD (Mozart's 'Requiem').

In the late 1990's I started to convert my collection of vinyl records into CDs. This is a time consuming work — all records must be played one at a time to bring the music to the computer, then cleaned from noise, clicks and other defects using special programs. Initially I converted vinyl records into CDs, but then I discovered that it is possible to compress the recordings to the mp3 file format, which takes much less room in the computer. Now I have almost my whole music collection on my iPhone, and can listen to it everywhere.

After retirement Liia's health worsened. She had problems with memory, and needed my help in the household. So I cancelled all my planned visits and stayed home. We had a lot of time to be together, and enjoyed this very much. One winter day she fell down so badly that a strong brain hemorrhage occurred. She was hospitalised but lost consciousness. On May 1, 2003 she died. She is buried at Elva cemetery, located in a pine forest. In the same forest we walked and got to know each other. Here I shall be buried when it is my time.

2.2.5 Tartu Observatory after the war, and the building of the new observatory

All classical universities have an astronomical observatory. Tartu University was founded in 1632 by the Swedish king Gustav Adolf, but was closed during the Great Northern War between Sweden and Russia. Estonia and Livonia became provinces of the Russian Empire in 1710. In 1802 Tartu University was reopened, and in 1810 the building of the Observatory was finished. Soon it became one of the most advanced university observatories of the Russian Empire. Its director, Friedrich Georg Wilhelm Struve, was an outstanding astronomer, the pioneer of double star studies, one of the first to determine the distance to a star (Vega). But its location near the centre of the city made astronomical observations difficult, and several times during the first half of the 20th century plans were made to build a new observatory outside the city limits.

In 1946 the Estonian Academy of Sciences was reopened after the war, and many scientific institutes were founded on the basis of departments of Tartu University and Tallinn Technical University. So in 1950 Tartu Observatory was moved to the Academy, to the institute which was soon named the Institute of Physics and Astronomy. The chair of astronomy in Tartu University was joined with the chair of geophysics, and Taavet Rootsmäe continued as before as the head of the chair and professor of astronomy. But most of his former collaborators joined the new Institute, among them also Grigori Kuzmin and Vladimir Riives, an expert on comets. Prof. Harald Keres, an expert on theoretical physics, was appointed as head of the astronomy division.

The director of the Institute, Prof. Aksel Kipper, initiated plans to build a new observatory outside the town. To discuss the problem he invited Viktor

Fig. 2.5 Viktor Ambartsumian visiting Tartu to discuss the building of a new observatory (1948). From left Prof. T. Rootsmäe, Viktor Simm (management director of Institute of Physics and Astronomy), Viktor Ambartsumian, Vladimir Riives, Harald Keres, Aksel Kipper (author's photo).

Ambartsumian to Tartu. Ambartsumian was the director of the recently built Byurakan Observatory in Armenia, and a leading astrophysicist of the USSR. Soon after this visit two Tartu astronomers, Harald Keres and Grigori Kuzmin, made a trip to the Krim, Abastumani and Byurakan Observatories to see how a modern observatory functions.

To find the finances, Kipper considered it necessary to secure the support of the astronomical community. For that, a meeting of the Astronomical Council of the USSR Academy of Sciences was held in Tartu in spring 1953 (Fig. 2.6). During the preparations I was in the Sternberg Astronomical Institute in Moscow to collect data for my research. On Moscow's side one of the organisers was professor Pavel Parenago. We discussed the program of the meeting, in which reports by the astronomers of Tartu and Moscow were assigned about equally. The main speaker on Moscow's side was Parenago himself, who had just finished a cycle of studies that was very close to Kuzmin's. On our side the main speakers were professor Aksel Kipper and Grigori Kuzmin.

The presentations by Tartu astronomers left a deep impression. Kipper spoke about the theory of the radiation of gaseous nebulae that he developed, which

Fig. 2.6 Participants of the Meeting of the Astronomical Council 1953. This Meeting was crucial for the future development of the Tartu Observatory (author's archive).

enabled him to explain the structure of planetary nebulae with much more accuracy than before. But the focus of interest were the presentations by Parenago and Kuzmin which were dedicated to very similar problems. Parenago was a brilliant lecturer and presented his results with characteristic clarity and simplicity. Kuzmin, on the other hand, was at this time a young researcher. It seemed that for him, competing with the leading astronomers of Moscow was hopeless. Indeed, Kuzmin's presentation (see Fig. 2.7) lacked the shine that was so characteristic of Parenago. In a quiet voice, he elaborated his model step by step, explaining his innovations and comparing the model to the previous ones, especially Parenago's. As the presentation progressed, it became apparent that his approach was much wider in reach and more insightful, and that his model was at a completely new level in studying the structure of the Galaxy. Ultimately, everyone had the feeling that astronomy at Tartu Observatory is top class. To the credit of the men of Moscow it should be said that they did not become envious. They acknowledged the high level of the astronomy in Tartu and gave strong support for our plans to develop the new observatory.

After the plenum Kipper invited all the Tartu astronomers to a dinner at his home. After the first glass Kipper glanced at the astronomers gathered around the table (there were not many of us at the time so we fit around it well) and

Fig. 2.7 Grigori Kuzmin delivering his talk at the Meeting of the Astronomical Council 1953 (author's photo).

said, "Boys, now it's time to work". A special board was formed, which started preparations for the construction of the new observatory. One of the first ventures was to find the location. For that I made bicycle trips North and South of Tartu with my spouse Liia and a student, Ene Humal, looking for uplands the observatory could be built on. An interesting coincidence: our first rest stop was on Tõravere hill, where the observatory was later actually built. Ene soon married Valdur Tiit, the son of my wife's uncle, and became Professor of Mathematical Statistics at Tartu University.

In the following years meteorological observations were done in chosen places both North and South of Tartu. We did not expect to find differences in the quality of the images of the stars, but we tried to ascertain possible differences in the number of clear nights, the frequency of autumn mists, et cetera. There were no significant differences between the North and South, so the decision was made in favour of South, where the connection with Tartu was better. To make the final decision for the location, the entire board of the observatory made additional bus rides to compare places. The chosen construction site was the upland of Tõravere, about a kilometre away from the Tartu–Riga road and the Tõravere railway stop. Kipper thought that important considerations were a suitable distance from major highways, so

Fig. 2.8 With my wife Liia searching for the location of the new observatory, summer 1953 (author's photo).

that traffic would not affect our work, especially the astronomical observations, but that on the other hand, bus stops and train stations should still be within a reasonable walking distance. As life in the new observatory has indeed shown, these standpoints by Kipper were very foresightful.

For the new observatory young astronomers were needed, so I started to think about how to prepare students in Tartu University as future astronomers. There were no textbooks on astronomy at University level, so it was evident that there is a need for such a monograph. At this time I had no experience in teaching, so I thought that to write a completely new textbook would be too difficult. Then I looked for existing textbooks in Russian, and found one which was more-or-less suited for our University. With several other young astronomers we translated the book into Estonian, I acting as the Editor. During the editing I discovered that in many places the Russian book is outdated. Thus about a half of the original text of the book was replaced with a new text written from scratch.

In 1957 a Satellite (Sputnik) Observing Station in Tartu University was organized. The first head of the station was Valdur Tiit; a year later I was appointed to this post. At this time orbits of satellites were known only approximately, thus a large team of students was needed for observations. Students were invited basically

Fig. 2.9 With Ene Humal (Tiit) climbing to a geodetic tower during the search for the location of the new observatory, summer 1953 (author's photo).

from the physics department of the University. Then I realized that it is possible to make use of the presence of the Station to give physics students a bit more education in astronomy. I gave regular courses on general astronomy and astrophysics to physics students. Both the practical observing experience and lectures gave good results — more than half of our present staff in Tartu Observatory came to us through the Satellite Station.

The actual construction building work began in spring 1957, and in May 1961 the first astronomers moved to Tõravere, including our family. The main building was not yet finished and for several years our temporary work space was in two apartments.

When I think back to my young years then two aspects in my education were especially important. First, my summers spent in Türgi farm. My grandmother was an exceptionally good organiser of housework. She never gave commands,

Fig. 2.10   A. Kipper, H. Keres and V. Riives overlooking possible locations for the new observatory, autumn 1953 (author's photo).

just some hints on what needed to be done. Also I learned that work must be done whatever your mood is: in farming if some work is not done in a proper way and time, you do not get any harvest. The habit to work systematically came from this experience.

The second important lesson came from my first supervisors at the Observatory, Prof. Rootsmäe and Grigori Kuzmin. They always gave you freedom to think yourself and independently. Their role was to help you to find errors in your work and to give hints on how to do things better. This freedom of thinking is extremely important in science. Tartu Observatory has always been a place where people had the freedom to think without any pressure.

Chapter 3

Galactic models and dark matter in the solar vicinity

In the 1960's I was rather heavily involved in administrative work related to the preparations to build our big 1.5 m telescope and its dome. But parallel to these duties I started to think about how to improve models of galaxies, so that all available observational data on galaxies and their populations could be used in model construction. The general background of these efforts was a dream to learn more about the formation and evolution of galaxies. The first object for a more detailed analysis was our own Galaxy. Soon I understood that the overall structure of galaxies is better seen in external galaxies, so my next object for detailed study was the Andromeda galaxy M31.

3.1 Early Galactic models

3.1.1 *Early Galactic models and first hints of the presence of dark matter*

In the beginning of the 20th century little was known about the overall structure of the Milky Way system. Star counts suggested that it has a flattened shape, and that it consists of stellar populations of different type. Many astronomers tried to describe the structure of the Galaxy in a more quantitative way.

One unsolved problem was the possible existence of absorbing material near the plane of the Galaxy, which distorts distance estimates of stars. This problem interested Öpik (1915), who tried to estimate the possible density of the absorbing material by calculating the dynamical density of matter near the Galactic plane. For this purpose he calculated a simple dynamical model of the Galaxy. The Galaxy is a flattened system, thus the density in the Solar vicinity can be determined using vertical oscillations of stars around the Galactic plane. Öpik determined the vertical distribution and velocity dispersion of flat stellar populations, assuming a normal distribution of velocities and spatial coordinates. From these data he estimated the total dynamical density. He found that the density is approximately equal to the density calculated from star counts. In other words: the amount of invisible matter near the Galactic plane is so small that it does not influence the dynamics. This paper seems to be the first determination of the possible presence of dark matter in the Solar vicinity. However, it was published in Russian in a not well-known journal, and remained unnoticed by the astronomical community.

Soon the same problem was studied by Kapteyn (1922). He used the latest observational data to compute a dynamical model of the Galaxy. To calculate the gravitational potential, the Galaxy was represented by 10 concentric ellipsoids of constant density and axial ratio $1/5.1$. These ellipsoids were not related to any stellar population, and used only to express the changes of the mean density of the Galaxy. The Sun was placed near the centre of the Galaxy. Using kinematical data and star count Kapteyn was able to estimate the total spatial density of visible stars, as well as the total dynamical density. He noticed that these two quantities can differ due to the possible presence of some dark matter or faint stars. He writes *"We therefore have the means of estimating the mass of dark matter in the universe. As matters stand at present it appears at once that this mass cannot be excessive"*. The total density of matter near the Sun is 0.099 solar mass per cubic parsec. This is probably the first use of the term "Dark Matter" in its present meaning.

The analysis by Kapteyn was repeated by Jeans (1922) using the same data, but a different method of analysis. He comes to a conclusion that indicates the presence of two dark stars to each bright star. Both authors use the overall mean velocity dispersion of stars, not the dispersion in the vertical direction.

In the 1920's new data on the structure of our Galaxy as well as on the Universe as a whole accumulated rapidly. The rotation of the Galaxy was discovered, as well as the position of the Sun far from the Galactic centre. Also it was found that the Galaxy consists of many stellar populations with different kinematical, spatial and physical properties (spectral class).

All these data were used by Oort (1932) in his study of the force exerted by the Galaxy in the vertical direction. Oort discussed in detail Kapteyn's study and found that the dynamical density can be determined essentially by the vertical gravitational acceleration. Oort accepted the Sun's distance from the Galactic centre as 10 kpc, and the circular velocity as 300 km/s. He calculated several dynamical models. In all models a centrally located massive population was assumed with a mass value, which corresponds to the observed rotational velocity near the Sun. The second population was a flat disk-like population. All models provided fairly consistent results for the dynamical density. Oort accepted as the most probable value 0.092 Solar masses per cubic parsec, very close to the value found by Kapteyn. He also calculated the density due to visible stars, and found a value 0.038 Solar masses per cubic parsec. This difference is often considered as an indication for the presence of dark matter. However, Oort estimated the total expected mass of faint stars, extrapolating the luminosity function. The extrapolated total mass gets very near to the value found from vertical motions of stars.

3.1.2 *Density of matter in the Solar vicinity*

Problems of the structure and evolution of stars and stellar systems were a central issue at Tartu Observatory. Öpik (1938) developed the modern theory of stellar evolution based on the burning of hydrogen in stellar cores, which leads to the formation of red giant stars after the main sequence stage. Rootsmäe (1961) applied these ideas to kinematics of stars to find the sequence of formation of different stellar populations; similar ideas were developed independently by Eggen et al. (1962).

Grigori Kuzmin was a student of Ernst Öpik. He graduated from Tartu University just before World War II, and got a position as an assistant in Tartu Observatory. Just before the war the first rotation curve of the Andromeda galaxy M31 was published by Babcock (1939), and Kuzmin started to think about how to use this information to calculate a dynamical model of M31. Andromeda is rather similar to our Galaxy, and Kuzmin hoped to use data from M31 to find a better model of our Galaxy too. The local structure is better known for our Galaxy, but the general structure of the system can be easier studied in the Andromeda galaxy. A similar approach was used by Öpik in his determination to the distance of the Andromeda galaxy.

Next Kuzmin turned his attention to our Galaxy. He studied in detail the Oort (1932) paper, and thought about how to get a better solution. Here the central problem was the determination of the dynamical density near the Sun. Kuzmin soon realised that it is not needed to calculate the whole gravitational potential for the cylinder, perpendicular to the plane of the Galaxy at the Sun's location, as Oort did. The dynamical density is determined by the Poisson equation, which connects the *local* density and the *local values* of the second derivatives of the gravitational potential (accelerations). Assuming rotational symmetry of the density distribution, and applying cylindrical coordinates, Kuzmin found that the radial and tangential accelerations can be expressed through the Oort constants of Galactic rotation A and B. The constant A describes the shearing motion in the Galactic disk near the Sun, while B describes the angular momentum gradient in the solar neighborhood, i.e. the vorticity.

The vertical acceleration can be expressed through a similar constant C, which has the same dimension as Oort's constants. Subsequently this constant was called

Fig. 3.1 Grigori Kuzmin explaining his formula to calculate the dynamical density of matter in the Galaxy, late 1970's (author's archive).

the Kuzmin constant. Simple estimates showed that $C \gg A, B$, thus the local density is determined essentially by the constant C (a similar conclusion was obtained already by Öpik (1915)).

Kuzmin (1952b) found that the constant C is equal to the ratio of the vertical component of the velocity dispersion of stars to the vertical dispersion of star positions. This equality is more accurate the flatter the respective population. There were further problems related to random and possible systematic errors in the determination of both dispersions. Oort calculated the vertical velocity dispersion using radial velocities of stars near Galactic poles, and the thickness of the population using stars near the Galactic plane, i.e. different stars. Instead, Kuzmin used *identical* stars to calculate the ratio of both dispersions. For velocities he used vertical components of proper motions, and for spatial positions galactic latitudes; both quantities were calibrated using identical parallaxes of stars. In this way errors in distance influence both dispersions in the same way, and cancel each other out. The card catalogue of galactic-equatorial A and gK stars perpendicular to the Galactic plane was collected during the war time. This means that the method to calculate and find the dynamical density was elaborated already in the early 1940's. However, there was a long way ahead to finish the study (Kuzmin, 1952b). His result was $C = 56 \pm 5$ km/s per kpc, which corresponds to a density 0.05 ± 0.01 Solar masses per cubic parsec.

A few years later Kuzmin (1955) turned to the problem again, and made a reanalysis of his own data, as well as data by Oort (1932) and Parenago (1952). In the new analysis he found a value for the parameter $C = 68 \pm 3$ km/s per kpc, which corresponds to the dynamical density 0.077 ± 0.008 Solar masses per cubic parsec. His student Eelsalu (1959) applied a slightly modified method, and found a value $C = 67 \pm 3$ km/s per kpc for the dynamical parameter.

Approximately at the same time Leiden astronomers worked hard to calculate a new model of the Galaxy, and to find the dynamical density of matter near the Sun. The model by Schmidt (1956) shall be described in more detail below. He accepted the mass density 0.093 Solar masses per cubic parsec.

A very detailed analysis of the vertical acceleration was performed by Hill (1960). He used the same method as Oort (1932), i.e calculated the vertical acceleration for a wide range of distances from the Galactic plane, using radial velocities and density dispersion of K-stars. His result was 0.13 Solar masses per cubic parsec. In the same issue of the Bulletin of the Astronomical Institutes of the Netherlands Oort (1960) analysed recent determinations of the vertical acceleration, comparing Kuzmin (1955), Hill (1960), and Eelsalu (1959) analyses. Again the whole range of vertical accelerations was used as in previous Leiden determinations. His conclusion was that the Hill analysis is the most accurate one, and arrived at an even higher mean density of matter in the Solar vicinity — 0.15 Solar masses per cubic parsec.

So there was a large discrepancy between Tartu and Leiden results. The Leiden studies hint at the presence of a large amount of dark matter near the Galactic plane. The Tartu results showed that, if there is dark matter in the Solar vicinity, the amount should be small. Our collaborator Mihkel Jõeveer studied the problem again. First he reanalysed the motions of K and B stars (Jõeveer, 1968b,a), and confirmed the earlier results by Kuzmin (1955) and Eelsalu (1959). Thereafter Jõeveer (1972, 1974) applied a completely new method to find the vertical acceleration. I gave a detailed overview of this work in my talk at the IAU symposium on dark matter (Einasto et al., 1987).

Jõeveer (1972, 1974) noticed that very young populations are not in a stationary state, but oscillate in the z-direction; the oscillation period is inversely proportional to the dynamical parameter C. Such oscillations are expected if stars form in gas clouds slightly away from the Galactic plane. After their birth stars are not bound to the gas cloud and start to fall toward the minimum of the gravitational potential in the plane of the Galaxy. The period of oscillations can be determined using ages, velocities and z-positions of young stars. He used data on early B8–B9 stars and cepheids, and found for the dynamical parameter a value $C = 70$ km/s per kpc, and for the density 0.09 Solar masses per cubic parsec.

Bahcall & Soneira (1980) calculated a very detailed model of the Galaxy. A central problem in the mass model was the total amount of matter near the Sun, which was investigated separately (Bahcall et al., 1983; Bahcall, 1984b,c,a; Bahcall & Soneira, 1984; Bahcall et al., 1985; Bahcall & Casertano, 1986). Bahcall (1987) give an overview of earlier determinations of the density determinations, and finds that the total dynamical density exceeds the density due to known stars by a factor of 0.5 ... 1.5. At the same IAU symposium I gave my review of our work on dark matter. In the discussion after my talk Bahcall raised the question of why the Soviet (i.e. Tartu) studies of the local "missing mass" have given a different answer to those made in the West. My first thought was to explain that based on my experience at Tartu Observatory, Kuzmin had the same ability as Öpik to find in a complicated situation the proper path to a correct answer. However, I did not want to discuss the matter in the context of the East–West controversy, thus I simply expressed my view that in such a complicated situation further detailed work is needed to find a better value of the local density.

Modern studies by Kuijken & Gilmore (1989c,a,b) have confirmed the early results by Oort (1932), and the results by Kuzmin and his collaborators. Gilmore et al. (1989) gave a very detailed analysis of the problem and concluded that the mass density near the Sun is $\simeq$0.10 Solar masses per cubic parsec. Using Hipparcos satellite data Creze et al. (1998) and Holmberg & Flynn (2000) found the local dynamical mass density 0.102 ± 0.010 $M_\odot$ pc^3, very close to the estimate of the density of visible disc matter. Thus we can say that *there is no evidence for the presence of large amounts of dark matter in the disk of the Galaxy*. If there is some invisible matter near the galactic plane, then its amount is small, of the order of 15 percent of the total mass density. The Gilmore et al. (1989) review as well as the papers by Creze et al. (1998) and Holmberg & Flynn (2000) have no references to Kuzmin or other Tartu astronomers papers, thus their results are completely independent.

3.1.3 *Galactic models by Parenago, Kuzmin, and Schmidt*

In the 1950's several authors elaborated models of the Galaxy using new observational data on the kinematics and structure of stellar populations of various types. For our understanding of the structure of the Galaxy most important were models by

Parenago (1948, 1950, 1952), Kuzmin (1952a, 1953, 1954, 1956a,b), and Schmidt (1956).

Galactic models are given by functions which describe the gravitational potential, the spatial density, the shape of the Galaxy, the circular velocity, and the projected density. In dynamical models some other functions are used, such as the phase density, which describes the velocity distribution of stellar populations. Functions are connected by equations, which follow from their definitions, laws of gravitation and dynamics. The density is connected to the gravitational potential by the Poisson equation, the circular velocity and the projected density are integrals of the spatial density, etc. If one of the principal functions and equations connecting principal functions are given, then all other functions can be calculated. So the modelling can be reduced to the choice of the principal function and the connecting equations.

Several authors have chosen the spatial density as the principal description function. It is well-known that the density decreases when we move from the centre of a galaxy to its outskirts. Oort (1932) in his model of the Galaxy used a number of ellipsoids of constant density to represent the change of density with changing distance from the Galactic centre. Instead, Öpik (1915) applied a Gaussian law to express the density distribution in the vertical direction. Thinking about the relation between the circular velocity and the spatial density, Kuzmin (1952a) found a simple solution: the square of the circular velocity can be calculated by integrating an ellipsoidal density distribution of an *arbitrary* density law. It is also possible to find the inverse solution — to calculate from the circular velocity the density distribution by solving the respective integral equation. Kuzmin used this approach to calculate from Babcock (1939) rotation data the density distribution for the Andromeda galaxy. It was war time and difficult to publish scientific papers, thus Kuzmin published his results only in the Almanac (Calendar) of the Tartu Observatory in Estonian in 1943. For our Galaxy he used the new method almost ten years later (Kuzmin, 1952a). Models with non-homogeneous ellipsoids were used independently by Wyse & Mayall (1942) for the Andromeda galaxy, and Perek (1948) for our Galaxy.

Parenago (1948, 1950, 1952) in his model of the Galaxy accepted a circular velocity law used by Lindblad (1933). Kuzmin (1954) analysed Parenago's model and came to the conclusion that this circular velocity law can be applied only in a restricted range of the distance from the Galactic centre. If applied to the whole range of distances then this leads to slightly negative densities at the outskirts, and to a zero total mass of the Galaxy. Negative masses can be avoided if the circular velocity law is used only up to the distance which corresponds to the zero spatial density, and to consider the corresponding distance as the outer radius of the Galaxy, taking the density outside this radius equal to zero. Such a model was suggested by Idlis (1956, 1957).

In his model of the Galaxy Kuzmin (1952a) used data on the rotation of young populations. These populations have small velocity dispersions and rotate with almost circular velocity. Accepting for the Galaxy rotational symmetry and flat shape Kuzmin solved the integral equation to calculate the spatial density.

My first experience of galaxy modelling was in 1952, when I made calculations for the Kuzmin model of the Galaxy. I suggested an improvement to the model: to apply the observed rotation velocities only until the galactocentric distance where rotation data were available. From these rotation data the spatial density was calculated. At larger distances the density was smoothly extrapolated to zero value at the probable outer radius of the Galaxy. The circular velocity in this region was calculated from the density. In this way negative spatial densities can be easily avoided.

Kuzmin was not satisfied with his first model of the Galaxy. He wrote (Kuzmin, 1956b): *"It is known that the velocity distribution of stars in all subsystems of the Galaxy is triaxial, i.e. can be represented by triaxial velocity ellipsoid. Theoretical explanation of this phenomenon meets still certain difficulties. The theory of stationary stellar systems, being successful in explaining the basic properties of stellar motions, gave biaxial velocity distribution. According to Jeans it was usually assumed that the phase density of a stationary axisymmetric stellar system is a function of two integrals of motion — the energy integral and the angular momentum integral. In this case the velocity distribution is biaxial. In order to have a triaxial velocity distribution we must suppose that the phase density is a function of three integrals of motion. The existence of such kind of integrals is related with the restrictions on the gravitational potential of the stellar system. For the existence of the energy and the angular motion integrals it is needed the stationarity and the axial symmetry of the potential. For the existence of the third integral, being different from previous integrals, some other additional restrictions are needed."*

I suggested further improvements when I made calculations for the Kuzmin (1956b) model. Kuzmin describes these improvements as follows: *"The extrapolation of the mass function must be done smoothly without jumps. In addition there must be taken into account three conditions. First, the mass function must vanish when the radius a approaches to a certain upper limit R_m which can be considered as the Galactic outer radius. Second, the mean radial gradient of the mass function must be in agreement with the observed radial gradient of stellar density in a wide neighbourhood of the Sun. And third, the potential corresponding to the mass function must be chosen in a way, that the Galactic outer radius R_m*

will be in agreement with the real upper limit of stellar velocities in the vicinity of the Sun. In other words, stars moving in the vicinity of the Sun with maximum galactocentric velocities (Oort's limiting velocity) must reach their apogalaxy just at the boundary of the Galaxy. Usually it is wrongly supposed that the upper limit of velocities corresponds to the escape velocity, but this does not take into account the finite dimensions of the Galaxy." I used this improvement also in my models of the Galaxy, see Fig. 3.3.

In his new model Kuzmin (1953, 1954, 1956a) added one more restriction, posed by the existence of the third integral of motion. He found differential equations for the gravitational potential. Solving Poisson equations, he derived a formula to find the matter density, following conditions that the model has finite mass and the density is nowhere negative. He found a simple density law where the isodensity surfaces differ somewhat from ellipsoids of revolution. Their meridional sections are more curved near their extremes than the ellipses with the same eccentricity. The mass distribution can be approximately expressed by a sum of two components, one almost spherical, and the other as a flat disk. In this limit the density distribution can be considered as a flat disk with the property that the gravitational potential above the plane is equal to that of a point mass at a distance $-z_0$ on the z-axis below the plane, while the potential below the plane is equal to that of a point mass at a distance z_0 on the z-axis above the plane. This mass distribution is called the Kuzmin disk, and it has found wide use in the dynamics of galaxies.

Models by Parenago, Kuzmin and Idlis were published in Russian and were known only to a small circle of astronomers outside the Russian speaking community. A more popular model of the Galaxy was elaborated by Schmidt (1956). This was the first model where new rotation data based on observations of neutral hydrogen at 21-cm wavelength were used. The density of the Galaxy was approximated by 4 non-homogeneous and 9 homogeneous spheroids. Homogeneous spheroids represent 4 main populations: population I, F-M stars, high-velocity F-M stars and unknown objects. The main Galactic constants used in the model were as follows: the distance of the Sun from the Galactic centre, $R_0 = 8.2$ kpc, and the circular velocity near the Sun, $V_c = 216.5$ km/sec. Schmidt calculated the gravitational potential for the whole Galaxy, and the escape velocity near the Sun, $V_e = 286.5$ km/sec. The difference between these velocities is approximately equal to the limiting velocity 65 km/sec, obtained by Oort in the direction of the rotation of the Galaxy. The matter density near the Sun was found to be 0.093 Solar masses per cubic parsec.

For many years the Schmidt model was used as a standard in Galactic studies.

3.2 New Galactic models

3.2.1 *Search for better models*

I wrote my PhD thesis on kinematics of stars and returned to galactic modelling again in the late 1950's. Soon a good review on models of galaxies by Perek (1962) was available and I read it carefully. What surprised me was the approach of most authors to the modelling — to model the structure of galaxies of different morphological type, completely different methods were used. For spiral galaxies rotation curves were used to find the mass distribution, but photometric data were mostly ignored. In models of elliptical galaxies the projected surface density profile was the basic source of information, and the dynamical calibration was made either from the central velocity dispersion or from relative motions of companion galaxies. In modelling of both external galaxies as well as our Galaxy, data on stellar populations were mostly ignored.

Actually all galaxies consist of similar basic populations: disk, bulge, halo, in spiral galaxies also a population of gas and young stars forming the spiral structure. The differences between galaxies of different morphological type lie basically in the proportion of various populations. When populations are explicitly used in models, then galaxies of all morphological types can be modelled in a similar way. The comparative study of galaxies of different morphological type can give information of the evolution of galaxies.

Thus I had the feeling that in order to make optimal use of available data one has to proceed as follows:

- To use all data of interest to construct models of galaxies of various morphological type;
- To apply identical methods to model galaxies of all types.

To model the structure of galaxies in more detail several problems must be solved. First: Which descriptive functions are to be used? In mass distribution models the spatial density, the projected density, and the circular velocity are the basic functions. The gravitational potential (and escape velocity) can be easily calculated on the basis of these functions. If we include data on velocities of individual populations then we can calculate hydrodynamical models.

The second basic problem is: What connection formulae are to be used which relate different description functions? The forms of these functions depend on the shape of galaxies. When we look for the structure of regular galaxies, the natural assumption is that galaxies are flattened systems having rotational symmetry and

Fig. 3.2 Discussing galactic models with Sergei Kutuzov and Grigori Kuzmin, late 1960's (author's photo).

symmetry with respect of the plane of the galaxy. Just these assumptions were made by most authors, including Oort, Kuzmin, Schmidt and others.

The third problem is connected with the choice of the principal descriptive function. If we consider only regular galaxies of axial and planar symmetry, all basic description functions are connected by simple integrals. If one function is given, all the other functions can be calculated. The question is: Which function is to be used as the principal one? If the choice is not optimal, one can have problems like Parenago had getting a model with negative densities on the outer regions. Our experience told us that the optimal choice is the spatial density, since all other functions are integrals of the spatial density. If some other function is chosen as the basic one, then other functions can be found by solving integral equations, which makes the problem more complicated.

And the final question: Can we formulate some simple physical criteria that the principal description function must satisfy? In the first calculations for the Kuzmin model in 1952 we used direct observational values of the rotation velocity to calculate the spatial density, and extrapolated it graphically, demanding that the function itself and its radial gradient be smooth functions of the distance. It is clearly better to use some mathematically defined expression to represent the density law. But how to make a choice between various expressions? After some thinking I came to the conclusion that some physically motivated criteria for the density law are needed. Mathematically these criteria can be expressed as criteria for the density and for integrals over the density multiplied by the radial distance at some power (moment of the density). It seems natural that at least the following criteria must be fulfilled:

- The spatial density is non-negative and finite;
- Some moments of the spatial density are finite, at least moments which define the central gravitation potential, the mass and the effective radius of the system;
- At large distances from the centre the density smoothly approaches zero value;
- The description functions have no breaks.

Together with my collaborator Sergei Kutuzov we analysed various aspects of galactic modelling (Einasto, 1968a,b,c,d; Kutuzov & Einasto, 1968; Kutuzov, 1968). A summary in English of methods of galactic modelling was published separately (Einasto, 1969a).

3.2.2 Generalised exponential model

Our analysis of various expressions for galactic description functions led us to the conclusion that the best approximation of the spatial density of all stellar populations can be obtained by a generalised exponential model (Einasto, 1965):

$$\varrho(a) = \varrho_0 \exp\left(-(a/a_c)^{1/N}\right), \tag{3.1}$$

where ϱ_0 is the central density, a is the semi-major axis of the equidensity ellipsoid, a_c is the core radius, and N is the structural parameter, which allows one to vary the shape of the density profile. The cases $N = 1$ and $N = 4$ correspond to the conventional exponential and the de Vaucouleurs models, respectively. This model satisfies all conditions mentioned above, allows a natural extrapolation of the density for large distances from the centre of the galaxy, and fits observed density profiles of known galactic populations very well.

In comparing various density profiles used to express the same observational density distribution I noticed that we get very different values of scaling parameters to measure the mass and the radial extent of the model, if we use parameters of the density law, such as the central density, ϱ_0, the core radius, a_c, or similar other parameters as the virial radius. Then I started to think whether it would be possible to define such scaling parameters, which characterise the mass and the radius of the model in a way that is less dependent on the particular form of the profile.

I came to the conclusion that the most stable results give the parameters defined by moments of the mass density profile — the mass M, and the harmonic mean radius, a_0. These parameters are related to parameters shown in 3.1 as follows: $\varrho_0 = hM/(4\pi\epsilon a_0^3)$, and $a_c = ka_0$; here $\epsilon = b/a$ is the axial ratio of the equidensity ellipsoid, and h and k are dimensionless normalising constants depending on the shape parameter N. Tamm et al. (2012) calculated the relations between the harmonic mean radius, and various other characteristics (the core radius a_c, and the radius a_{-2}, where the logarithmic slope of the profile equals the isothermal value -2). For large N values the ratios of these radii to the harmonic mean radius vary between 0.1 and 10.

In all our models of galaxies, starting from the models of M31 by Einasto (1969b) and Einasto & Rümmel (1970c), we have always used the harmonic mean radius to characterise the radial extent of the model, and the density profile 3.1.

Trial calculations with this profile were made already in the late 1950's, and used in my first model of the Galaxy (Einasto, 1965). A similar expression was used by Sérsic (1963) to represent the projected (surface) densities of galactic populations. In all subsequent models of galaxies we used this profile to represent the spatial density of stellar populations. In the 1960's and early 1970's computers were not very powerful, and it was customary to make preliminary estimates of density profile parameters by graphical comparison of observed profiles with a series of standard profiles. For this purpose tables of the generalised exponential profile were published for a wide range of the structural parameter N by Einasto & Einasto (1972b).

This profile is now called the "Einasto profile", and the shape parameter N the "Einasto index". For discussions of properties of this profile see, among others, Dhar & Williams (2010); Chemin et al. (2011), Retana-Montenegro et al. (2012).

3.2.3 *Our Galaxy, system of galactic constants*

Detailed local structure is known only for our own Galaxy, thus I computed a model for the Milky Way first. The Galaxy was represented by a sum of three main populations: the flat disk consisting of young stars and interstellar gas, the disk consisting of stars of medium age, and the halo, consisting of metal poor stars, like stars in globular clusters. The structural parameter N of these populations was estimated on the basis of analogy with similar external galaxies. Also I used radial gradients of the density in the solar vicinity of main populations. These gradients are related to velocity dispersions of populations, and can be taken into account in the calculation of the model, using the modified Strömberg equation (Einasto, 1961). In the calculation of the model I considered the main populations as representatives of a number of similar populations which have close spatial structure, kinematics, and physical properties (mean age, chemical composition and colour).

The basic problem in the modelling of the Galaxy is the choice of galactic parameters. The most important parameters are the distance of the Sun from the centre R_0, the Oort constants A and B, the Kuzmin constant C, the circular velocity near the Sun V_0, and the matter density in the solar vicinity ϱ_0. Additional parameters are the ratios of velocity dispersions in the radial, tangential and vertical directions, and some other observed quantities. The circular velocity as a function of the distance from the galactic centre, $V(R)$, can be found from radio observations of the neutral hydrogen through the function of differential rotation $U(x)$ as follows:

$$V(x) = U(x) + xV_0, \qquad (3.2)$$

where $x = R/R_0$, and R is the distance from the centre. Radio observations allow us to derive directly the gradient of the differential rotation near the Sun, $W = -1/2(dU/dx)$ at $x = 1$. This gradient can also be considered as a galactic parameter.

In earlier models of the Galaxy (Kuzmin, 1952a; Schmidt, 1956) some fixed values for principal galactic parameters were used, such as R_0, V_0, A, B, and ϱ_0. As noted above, there are actually more observable quantities available. All these quantities are connected by certain formulae which follow either from the definition of parameters or from the theory. To make use of these additional observed quantities, the system of galactic parameters can be found using all quantities and connection formula, and a balanced system of parameters can be found by the method of least squares. This method was developed in detail by Kutuzov (1965). The theory of the determination of the balanced system of galactic parameters was the subject of his PhD thesis. I calculated a preliminary system of parameters in the model of the Galaxy, using data published until the end of 1961 (Einasto, 1965).

Observational data accumulated rapidly, and I prepared together with Sergei Kutuzov a new system of galactic parameters for the Commission 33 Meeting of the XII General Assembly of the International Astronomical Union (IAU) in Hamburg 1964 (Einasto & Kutuzov, 1964). In the new system we used 10 independent parameters and 6 relations between them, which were used to find by the method of least squares the optimal set of parameters. I applied to attend the Assembly, but my application was not approved by Soviet authorities. Our text was printed, and I made a trip to Moscow to hand over our preprint to Professor Ogorodnikov, who was allowed to participate in the General Assembly. He presented my results on the new system of galactic parameters.

A few weeks later one of the leading experts on the structure of our Galaxy, Bart Bok, made a visit to the Sternberg Astronomical Institute of Moscow University. I regularly visited Sternberg Institute and was informed on his planned seminar talk. So I made a second trip to Moscow to listen the talk by Bart Bok. He reported scientific news from the IAU General Assembly, and to my surprise gave a rather detailed overview also of my new system of galactic parameters. After the talk I had a long discussion with Bart — this was our first personal meeting. He was interested in the work of Tartu astronomers, in particular that by Kuzmin.

In the late 1960's I prepared a second model of the Galaxy using recent observational data and an improved method to calculate the system of galactic parameters. This new model was presented at the Commission 33 Meeting of IAU General Assembly in Brighton (Einasto, 1970a). The model consists of three populations, flat disk, disk, and halo, using the modified exponential profile. Tables for the gravitation potential, circular velocity and other main parameters as functions of z and R (in cylindrical galactocentric coordinates) were published separately by Einasto & Einasto (1972a).

In this model I used a novel method to find the main galactic parameters, R_0 and V_0, based on ideas applied first by Eggen et al. (1962) in their study of the formation of the Galaxy, as well as ideas used in the calculation of the Kuzmin (1956b) model. As written above, radio observations of the 21-cm line allow us to find the differential rotation function $U(x)$. The circular velocity is then calculated from equation 3.2. But radio observations allow us to find the rotation and the mass distribution function only for inner galactic regions, $R \leq R_0$. For larger distances, $R > R_0$, the mass distribution can be found by smooth extrapolation. Figure 3.3 shows two variants of extrapolation for different values of V_0. In both cases the function $U(x)$ is identical. If we use smooth extrapolation with different V_0 we get also two different values of the limiting radius of the model, $R_{\lim}$; this radius is defined as the distance where the spatial density is a hundred times lower than near the Sun.

It is well-known that there are no stars with velocities in the direction of galactic rotation which exceed the solar velocity by about 65 km/s or more (Oort, 1928). This velocity, Δv, is often called the Oort's limiting velocity. Oort (1928) assumed that this velocity corresponds to the escape velocity near the Sun. However, as shown by Kuzmin (1956a,b), this assumption ignores the finite dimensions of the Galaxy, and should be interpreted as the velocity needed to reach the boundary of the Galaxy, $R_{\lim}$. Using the mass distribution model we can calculate the gravitational potential and find apogalactic distances, R_{apogal}, of stars moving with the velocity $V_a = V_0 + \Delta v$ in the direction of the rotation of the Galaxy.

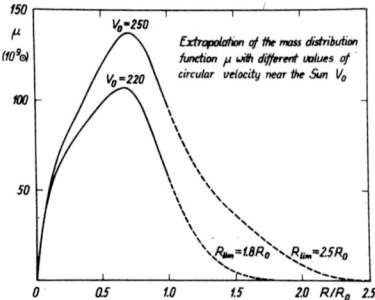 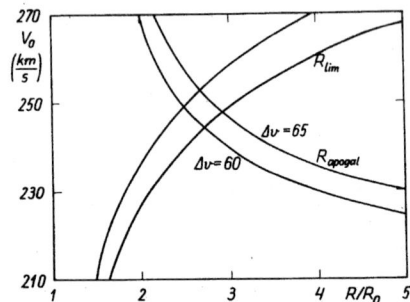

Fig. 3.3 Left panel: the extrapolation of the mass distribution function beyond the Sun's distance $R > R_0$ (dashed lines) with different values of V_0. The limiting radii of models, $R_{\lim}$, are indicated. Right panel: the dependence of the limiting radii, $R_{\lim}$, and the apogalactic distances, R_{apogal}, on the circular velocity near the Sun, V_0. Two cases of smooth extrapolation of the mass function with different $R_{\lim}$ are shown. Apogalactic distances are given for two values of the Oort's limiting velocity, Δv (Einasto, 1970a).

The left panel of Fig. 3.3 shows the mass distribution function (the mass of an equidensity ellipsoidal layer of unit thickness), $\mu(a) = 4\pi\epsilon a^2 \varrho(a)$, where a is the semi-major axis of the equidensity ellipsoid with the axial ratio $\epsilon = b/a$, and $\varrho(a)$ is the spatial density. The function is calculated for two values of V_0. The right panel of Fig. 3.3 shows the dependence of the limiting radii, $R_{\lim}$, and the apogalactic distances, R_{apogal}, on the circular velocity near the Sun, V_0. The circular velocity V_0 is calculated for two values of the limiting radii, $R_{\lim}$, and of the Oort's limiting velocity, Δv.

If the adopted circular velocity is too small, then the Galaxy has a rather sharp edge, and $R_{\lim}$ is small, but the gravitational potential is also small and apogalactic distances R_{apogal} of stars moving with the velocity V_a are large, much larger than $R_{\lim}$. If the adopted circular velocity is too high, then we have the opposite situation, as seen in Fig. 3.3. Both distance coincide when we use the true value of the circular velocity for this mass distribution. This method I applied in the calculation for the Kuzmin (1956a,b) models and my own models (Einasto, 1965, 1970a). In these models no dark halo was included, which led to a rather high value of the circular velocity near the Sun, $V_0 = 250$ km/s.

About ten years later I applied the same method using a much more accurate mass distribution model including a dark halo (Einasto et al., 1979a). The result was $V_0 = 220$ km/s, in very good agreement with values found by other independent methods. This value is presently generally accepted.

A new system of galactic parameters was discussed by Commission 33 of IAU in the late 1980's using the same approach. The accepted system (Kerr &

Lynden-Bell, 1986) was rather close to the system we presented in 1964. It was very difficult to get permission for visits outside the USSR, so I was not able to take part in the preparation of the new system in 1986.

The local structure of a galaxy is best known for our own Galaxy. But our position inside the Galaxy makes it difficult to see its overall global structure. Global information on stellar populations is better known for external galaxies. Thus it was natural to continue the study of physical and dynamical properties of galaxies using as test object our neighbour, the Andromeda galaxy M31. This task was realised in a series of papers (Einasto, 1969b, 1970c; Einasto & Rümmel, 1970b). In the construction of models of external galaxies a number of other problems need to be solved.

3.2.4 Mass-to-luminosity ratios of stellar populations

In the late 1950's and 1960's astronomers having access to large optical telescopes continued to collect dynamical data on galaxies. The most extensive series of optical rotation curves of galaxies was made by Margaret and Geoffrey Burbidge, starting from Burbidge & Burbidge (1959); Burbidge et al. (1959), and including normal and barred spirals as well as some ellipticals. Rubin & Ford (1970) derived the rotation curve of M31. For all galaxies the authors calculated mass distribution models. The problem with these models was the difficulty of extrapolating mass distribution to large distances from the galactic centre.

Also radio observations of the neutral hydrogen 21-cm line accumulated. van de Hulst et al. (1957) found that the neutral hydrogen emitting the 21-cm line extends much farther than the optical image. They were able to measure the rotation curve of M31 up to about 30 kpc from the centre. Roberts (1966) made a new 21-cm hydrogen line survey of M31 using the National Radio Astronomy Observatory's 300-foot telescope. He calculated a mass distribution model for M31, using analytical expressions for the rotation curve, similar to the Bottlinger model and suggested by Brandt (1960), Brandt & Scheer (1965). As in the case of the Parenago model of the Galaxy, results for the mass distribution at large distances from the centre depend critically on the choice of parameters of the velocity law.

In these models data on the distribution of light in stellar populations was not used. My goal in modelling M31 was to make use of these data. In my preliminary model of M31 (Einasto, 1969b) I used the following populations: nucleus, core, disk, and flat disk. Photometric data were available for all of them. The problem was how to get mass-to-luminosity (M/L) ratios to find the mass distribution of populations. In principle, the rotation data can be used for this purpose, since in various regions of the galaxy different populations dominate. But there exists also a second possibility to use independent data: velocity dispersions of the nucleus and the core are measured, and detailed spectroscopic observations of the stellar content are available to find the stellar luminosity function, which allows one to estimate the M/L. For the nucleus of M31 such observations were made by Spinrad (1966), who found $M/L = 16.7$.

Einasto (1969b) calculated mass distribution models for both M/L variants. If rotation data by Roberts (1966) are identified as circular velocities, then for the nucleus and core very small values for the mass-to-luminosity ratio were obtained, $M/L \ll 1$. For the disk rotation the data suggest $M/L \simeq 9$, and for the flat population $M/L \simeq 80$. These values are in conflict with other data, found on the basis of the luminosity function of respective stellar populations. For the nucleus we have every reason to accept Spinrad data, and for the disk and flat disk we can trust the estimate by Öpik (1922a), who found for the solar vicinity $M/L \simeq 3$. On the other hand, if we apply the Spinrad value $M/L = 16.7$ also for the core and the disk, then this leads to too high a value for the circular velocity at small distance from the centre, $V \simeq 380$ km/s at $R \simeq 1$ kpc.

To understand the very low value of M/L found from radio data at small distances from the centre, Einasto & Rümmel (1970a) investigated the radial velocity field of M31. This analysis showed that small rotational velocities, obtained from radio data, are due to the low angular resolution of radio data, which smear also rotation velocity profiles at small distances from the centre.

To avoid these difficulties, I calculated a new model of M31 (Einasto, 1970b), which included the nucleus, bulge, disk, flat disk, and halo. I used new spectral determination ($M/L \simeq 42$) of the nucleus by Spinrad (personal communication), and new rotation data by Rubin & Ford (1970). To find the spatial distribution of the halo I used data on the distribution of globular clusters in M31. The spatial structure of the flat disk was found from the distribution of neutral hydrogen. Mass-to-luminosity ratios for populations were found from the rotation curve in regions where the particular population dominates. The M/L values, obtained for the bulge, the disk, the flat disk and the halo, were still very uncertain.

Thus my experience with these preliminary models of M31 emphasised that the problem of finding proper values of M/L for stellar populations is not so easy. Additional information is needed to find better values.

One possibility is to use additional spectrophotometric data for regions outside the nucleus and in other compact systems — star clusters and nuclei of galaxies. Spinrad et al. (1970) and Spinrad & Taylor (1971) found that star clusters M67, NGC 188 and NGC 6791 are super-metal-rich, which suggests a high M/L value. Spinrad et al. (1969) studied M7 giants in the nuclear bulge of the Galaxy and found that the metal abundance is higher than in stars near the Sun. Spinrad et al. (1971)

investigated colour changes and absorption-line variations over the inner disk of M31, and in central regions of M32 and NGC 4472. They made a preliminary synthesis of the disk population of M31 and found that the mass-to-luminosity ratio of the disk is equal or ever slightly greater than the nuclear value, $M/L \simeq 45$. On the other hand, the metal abundance of the disk of M31 and NGC 4472 is close to the solar level, in contrast to a high level at their nuclei. Thus it is difficult to accept so high values of M/L for the disk.

The second possibility is to use dynamical data on compact stellar systems like star clusters. It is well-known that stars form in various star-formation clouds, which later evolve to stellar associations and clusters. Later associations and loose clusters are dissolved by various perturbations. Probably all field stars of galaxies formed in just such way. So I started to collect dynamical data on star clusters of various type and age. I expected that for a certain set of colour, metallicity and age, star clusters have mass-to-luminosity ratios which are close to galactic populations with similar physical properties.

Data on velocity dispersions in star clusters were just starting to be available. For metal-poor systems like globular clusters very low mass-to-luminosity ratios were obtained, for instance, $M/L_v = 1.7 \pm 0.4$ in solar units for NGC 6388 by Illingworth & Freeman (1974). Freeman & Munsuk (1972) found for relatively young LMC clusters NGC 1835 and NGC 2210 even lower values, $M/L_v = 0.2$. For Omega Centauri Poveda & Allen (1975) found a significantly larger value, $M/L_B = 3.2$, than for other globular clusters.

Using a compilation of velocity dispersion measurements in nuclei of galaxies of various luminosity Einasto & Kaasik (1973) and Einasto (1973) found a clear dependence between the total luminosity and the velocity dispersion: more luminous galaxies have higher values of central mass-to-luminosity ratios.

3.2.5 *Evolution of galaxies*

To summarise, my experience suggested that different data and methods lead to rather different values of M/L, and it is not clear which data can be trusted. Thus there exists a need to bring all estimated M/L values to a coherent system. It is clear that the luminosity of a population depends on its chemical composition and the age. A similar problem exists in compact stellar systems, such as star clusters. To understand how M/L depends on the composition and on the age of the population, evolution models are needed. For this reason I started in the late 1960's to develop my own evolution model. The description of the evolution model is given in my Doctor of Sciences thesis (Einasto, 1972b). The model is similar to the model developed by Tinsley (1968), but my model was developed completely independently from the Tinsley model and some details were different. I started to develop my model before the model of Tinsley was published, thus I had to invent all details of the model calculation myself.

Models of the evolution of stellar populations and galaxies are based on stellar evolution tracks, star formation rates (as a function of time), and the initial mass function (IMF). From literature I found stellar evolution tracks for 16 star mass values between 0.05 and 60 Solar masses. To model the evolution of star systems, tracks were tabulated for 19 evolution stages, either directly from published tracks, or by interpolation. Each stage in all tracks corresponds to a certain phase of the evolution. For each stage the age, the bolometric luminosity and the effective temperature were given. Tracks were found for three values of metal content: direct data were available for stars of normal metal content, $Z = 0.02$. By extrapolation tracks were found for metal-rich stars, $Z = 0.10$, and for extremely metal-poor stars, $Z = 10^{-5}$. This dataset allows through interpolation to find isochrones — the luminosity–temperature diagrams for a stellar population of an arbitrary age and composition. Luminosity functions, colours and mass-to-luminosity ratios were calculated in the UBVRIJKL colour system. Model populations were found for 14 epochs: 0.01, 0.03, 0.1, 0.3, 1, ..., 15, 20 billion years.

For IMF I used the Salpeter (1955) law $F(m) = a\, m^{-n}$, where m is the mass of the forming star, and a and n are parameters. This law cannot be used for stars of arbitrary mass, because in this case the total mass of forming stars may be infinite. Thus I assumed that this law is valid only in the mass interval from m_0 to m_u, the lower and upper limits of the forming stars, respectively. My calculations showed that the mass-to-luminosity ratio M_i/L_i of the population i depends critically on the lower mass limit of the IMF, m_0.

An independent check of the correctness of the lower limit is provided by homogeneous stellar populations, such as star clusters. Here we can assume that all stars were formed simultaneously, the age of the cluster can be estimated from the HR diagram, and the mass derived from the kinematics (velocity dispersion) of stars in the cluster. Such data were already available for some old metal-poor globular clusters, for some relatively young medium-metal-rich open clusters, as well as for metal-rich cores of galaxies. I compared the results of stellar population modelling with direct dynamical data for central regions of galaxies (velocity dispersions) (Einasto & Kaasik, 1973). As a further test I studied the rate of star formation as function of the density of stellar populations in M31 (Einasto, 1972c). The results confirmed the Schmidt (1959) law: the star formation rate is proportional to the density squared.

Using for calibration various data, available at this time, I accepted for stellar populations with normal metal content $m_0 = 0.03\, M_\odot$, for metal-rich populations

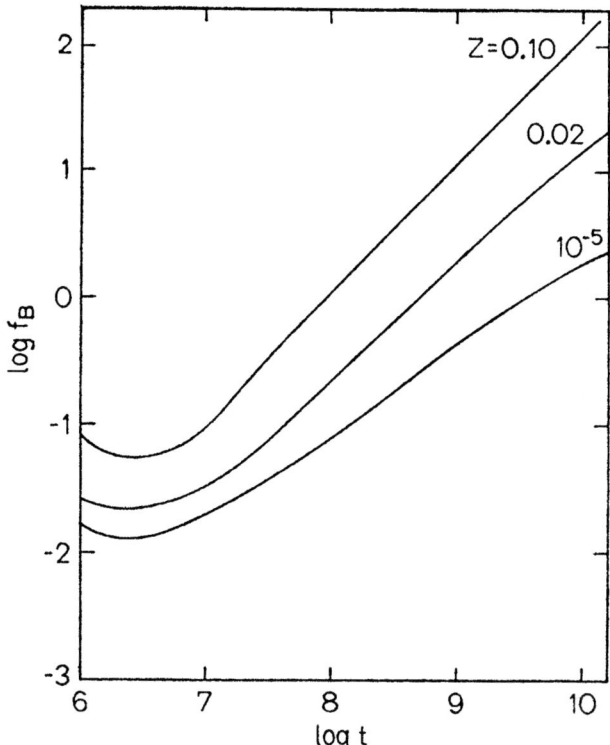

Fig. 3.4 The evolution of the mass-to-luminosity ratio f_B for stellar populations of different metallicity Z and instant star formation. The age t is given in years (Einasto, 1974a).

$m_0 = 0.001~M_\odot$, and for metal-poor populations $m_0 = 0.1~M_\odot$. Using these lower mass limits we get for old metal-poor halo populations $M_i/L_i \approx 3$, for extremely metal-rich populations in central regions of galaxies $M_i/L_i \approx 100$, and for intermediate populations (bulges and disks) $M_i/L_i \approx 10$, see Fig. 3.4. Using these data I obtained for galaxies M31, M32, and M87 mean total values $M/L_B = 9, 4, 22$, respectively. M/L_B of old galactic populations depends also on the total mass of the galaxy, see Fig. 3.5 (Einasto, 1974a).

As I found later, so wide a range of m_0 values for various metal content is not needed. Modern dynamical data yield for all populations lower values of M/L, due to more accurate measurements of velocity dispersions in star clusters, and in central regions of galaxies. As suggested in pioneering studies by Faber & Jackson (1976); Faber et al. (1977), the bulge of the Sombrero galaxy has a mass-to-luminosity ratio $M/L = 3$, and the mean mass-to-luminosity ratios for elliptical

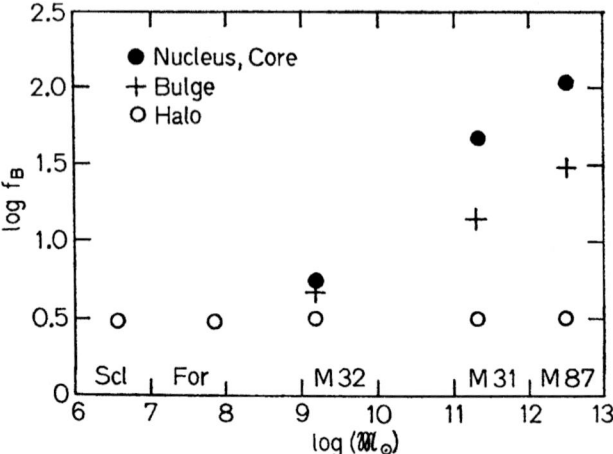

Fig. 3.5   The dependence of the mass-to-luminosity ratio f_B of old galactic populations on the total mass of the galaxy (Einasto, 1974a).

galaxies is about 7, close to the ratio for early type spiral galaxies. New data suggested that central velocity dispersions of giant galaxies are by a factor of 2 lower than accepted earlier. This reduces M/L values for metal-rich cores by a factor of up to 4. For extremely metal-poor globular clusters new data suggest a value $M/L \approx 1$. Thus in retrospect I can say that my first set of population evolution models is my version of 'maximum disk' models.

New data solved one discrepancy in my earlier models — if mass-to-luminosity ratios of central regions are taken from early velocity dispersion data, then circular velocities, calculated from mass distribution models, are too high, much higher than suggested from rotation data in central regions of galaxies, see Fig. 4.3 and the discussion below. Earlier I thought that such large deviations are due to the dominance of random motions in central regions, and that rotation velocities cannot be used to find the mass distribution in these regions. According to new velocity dispersion data, M/L values found for central regions are considerably lower, and observed (optical) and model rotation data are in good mutual agreement, as found in our later model of M31 by Tenjes et al. (1994), shown in the right panel of Fig. 4.3. But this suggests that corrections are needed to my previous galaxy evolution models. The lower limit m_0 of masses in the Salpeter law must be higher than accepted earlier, for all galactic populations it should be in the range of $0.05 - 0.1$ $M_\odot$. This lower limit is close to the limit found from independent data for stars where nuclear burning of hydrogen is possible.

New data increased the discrepancy between earlier classical galaxy models and new ones. In other words, high mass-to-luminosity ratios of groups and clusters of galaxies cannot be explained by known galactic populations.

My results were similar to those of Tinsley (1968) with one important difference. Tinsley used much lower values of $m_0 \sim 10^{-6}\,M_\odot$ to get high values of $M/L_B \simeq 250$–500 for elliptical galaxies, as suggested by the dynamics of companions of luminous elliptical galaxies. In other words, in Tinsley models of elliptical galaxies most stars were Jupiter-like objects without internal sources of nuclear energy.

Modern calculations suggest that the first generation of metal-free population III stars have large masses ($\sim 100\,M_\odot$) due to the large Jeans mass during the initial baryonic collapse (for a discussion see Reed et al. (2005) and references therein). Thus population III stars are not suitable to represent a high M/L halo population at the present epoch.

3.2.6 Models of galaxies of the local group and M87; mass paradox in galaxies

In the early 1970's I collected from all possible sources data on our Galaxy, M31, M32, Fornax and Sculptor spheroidal dwarf galaxies, and the giant elliptical galaxy M87 in the Virgo cluster. These data were used to calculate population models of these galaxies. The most detailed model was found for M31, where parameters for the nucleus (a massive single object at the center, now we know that it is a massive black hole), core, bulge, disk, halo, and young disk were found. Dwarf spheroidal galaxies were represented only by the halo, for other galaxies at least three populations were used. In these models I used results of my calculations of the evolution of populations. These population models as well evolution models were presented as part of my Doctor of Sciences thesis (Einasto, 1972b).

For M31 and our Galaxy I calculated hydrodynamical models for all main galactic populations. These models included densities, density gradients, velocity dispersions, and a number of other parameters (Einasto & Rümmel, 1970b; Einasto, 1972b, 1974a). As an example, for one flat population of our Galaxy these functions are shown in Fig. 3.6.

In the modelling of M31 I encountered a serious problem. If rotation data were taken at face value, then it was impossible to represent the rotational velocity with the sum of known stellar populations. The local value of M/L increases towards the periphery of M31 very rapidly, if the mass distribution is calculated directly from the rotation velocity, see Fig. 4.2 in the next Chapter. All known old metal-poor halo-type stellar populations have a low $M/L \approx 1-3$; in contrast on the basis of

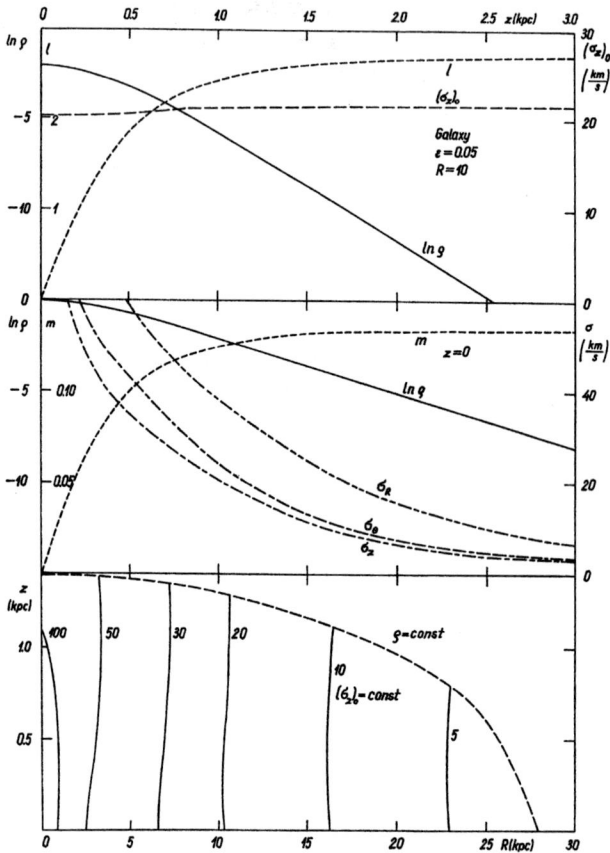

Fig. 3.6 Various description functions of a flat population of the Galaxy with the effective radius $a_o = 7.4$ kpc, axial ratio $\epsilon = 0.05$, and structural parameter $N = 1$. In the upper panel the z-dependence of the density ϱ is shown in units of the central density of the population. Additionally the vertical density gradient $l = -\delta \log \varrho/\delta z$, and the velocity dispersion $(\sigma_z)_0$ are given for the distance from the center of the Galaxy $R = 10$ kpc. In the middle panel the R-dependence of the density ϱ, of the radial density gradient $m = -\delta \log \varrho/\delta R$, and of the three velocity dispersions σ_R, σ_θ, σ_z are shown, at the Galactic plane $z = 0$. In the bottom panel one line of the constant density ϱ is shown, as well as lines of constant velocity dispersion, σ_z, for the meridional plane of the Galaxy; velocities are expressed in km/s (Einasto, 1974a).

rotation data I got $M/L > 1000$ on the periphery of the galaxy near the last point with measured rotational velocity.

There were two possibilities to solve this discrepancy: to accept the presence of a new population with very uncommon properties, or to assume that on the periphery of galaxies there exist large non-circular motions. I considered both possibilities. My conclusion was that if there exist only stellar populations with known properties, then the first alternative has several serious difficulties.

If the hypothetical population is of stellar origin, it must be formed much earlier than all known populations, because known stellar populations form a continuous sequence of kinematical and physical properties (Rootsmäe, 1961; Einasto, 1974b), and there is no place to put this new population into this sequence: the estimated velocity dispersion of stars of the new population is too high for a conventional stellar population. In other words, the new population must be well separated from all known populations, both in velocity space and spatially.

Secondly, the star formation rate is proportional to the square of the local density (Schmidt, 1959; Einasto, 1972c), thus stars of this new population should have been formed during the contraction phase of the formation of the population near its central denser regions, and later expanded to the present distance. The only source of energy for expansion is the contraction of other stellar populations. The estimated total mass of the new population exceeded the summed mass of all previously known populations. Estimates of the energy needed for the expansion demonstrated that the mass of the new population is so large that even the contraction of all other stellar populations to zero radius would not be sufficient to expand the new population to its present size. This means that the new population must be formed prior to the formation of known stellar populations, before the collapse of the gas from which all known stellar populations formed.

And, finally, it is known that star formation is not an efficient process: usually in a contracting gas cloud only about 1% of the mass is converted to stars. Thus we have a problem of how to convert, in an early stage of the evolution of the Universe, a high fraction of primordial gas into this population of first generation stars.

I also found it psychologically difficult to accept the first alternative. From rotation data it follows that the mass-to-luminosity ratio of this hypothetical population must be very high, thus it should be essentially a dark population. Our studies of the local density of matter near the Sun by Kuzmin and his students have shown that there is no room for the presence of large amounts of dark matter in the plane of the Galaxy. Thus I was not ready to accept the presence of a dark population elsewhere in galaxies.

Taking into account all these difficulties I accepted the second alternative — the presence of non-circular motions (Einasto, 1969b; Einasto & Rümmel, 1970c). A similar decision was made by many other astronomers, as discussed during the Third European Astronomy Meeting by Materne & Tammann (1976).

As I soon realised, this was a wrong decision.

One more episode from this period. In the late 1960's a young astronomer from the Leningrad University published two papers in the Armenian Journal Astrofizika, where he developed a model of M31 (Sizikov, 1968, 1969). He used both rotation and photometric data, while the density was expressed by one spheroidal component of variable flatness as in the Kuzmin (1956a) model. These papers were the basis of his PhD (candidate) thesis, and I was appointed as the reviewer (opponent). Our main discussion during the defence was devoted to the mass distribution on the outskirts of M31. At this time I already had preliminary results of my galaxy evolution model calculations, and knew that it is very difficult to assume the presence of a large amont of very faint dwarf stars in outer regions of galaxies, as explicitly assumed by Sizikov. However, my results were only preliminary and not yet published, thus it was difficult for me to show any error in the Sizikov model. I gave the thesis high opinion, and we agreed that further study was needed to understand the difference between our models.

3.3 Tartu Observatory in the 1960's

3.3.1 *New observatory*

The first scientific conference in the new observatory was a cosmology summer school in July 1962, where we met for the first time Yakov Zeldovich. He had just returned from military to civil science, and started to study the physics of the most powerful explosion — the Big Bang. At this time we had no idea that in years to come our group would have a very close collaboration with him.

The new observatory was opened with a conference in 1964 under the motto *"Science is carried by search for truth that is as sincere and honest as the Nature itself" (T. Rootsmäe)*. A number of leading astronomers attended, among them the director of the Pulkovo Observatory, academician A. Mikhailov, the director of the Byurakan Observatory, academician V. Ambartsumian, the chairman of the Astronomical Council of the USSR Academy of Sciences, academician E. Mustel, and also a great-granddaughter of F.G.W. Struve. After the conference trees were planted by many participants. The trees grew well and now there is a beautiful park around the main building.

After defending my PhD thesis I spent several years teaching at the University and working in the station to observe artificial Earth satellites. The observation of satellites (or sputniks) gave the students a certain experience in astronomical observations and thus popularised studying astronomy at university. Among other things I found a way to construct a four-axis mount for satellite tracing telescopes (Liigant & Einasto, 1960), together with my student M. Liigant we got an author's certificate for this telescope and a medal at the All-Union Exhibition. A similar mount was invented by the Carl Zeiss factory in Jena, and a series of satellite-tracking

Fig. 3.7 The cosmology school in the new observatory, July 1962. In the center of the second row with beret is Yakov Zeldovich, in the first row in front of Zeldovich are Alla Massevich and Bruno Pontecorvo (author's archive).

telescopes were built. Zeiss representatives invited me to Jena to inspect their telescope. This visit was in summer 1962. We did not patent our version, so Zeiss was not obliged to pay us any royalties. This was my short excursion away from galactic modelling.

After moving to Tõravere, together with our students and young collaborators I spent nearly half a year exploring the astronomical literature and studying new trends of development in astronomy. We had a series of astronomy seminars where we discussed the new directions in astronomy. According to earlier plans, it was intended to buy a Schmidt camera as the main telescope in order to continue stellar-statistical observations. As a result of discussion we came to the conclusion that in our climatic conditions it would be rational to pay more attention to stellar physics, particularly to spectral observations of stars, where much more useful information can be obtained in relatively short periods of clear weather.

After these discussions we prepared a new program of the observatory's development, in which we planned to construct a reflecting telescope with a mirror of diameter 1.5 meters. This plan received the support of the Astronomical Council and became the basis of our future development. In the mid-sixties I worked with

Fig. 3.8 The opening ceremony of the new Tartu Observatory in 1964. In the first row sitting from left are: Mrs. Naan, G. Naan, A. Mikhailov, E. Mustel, V. Ambartsumian, A. Kipper, K. Ogorodnikov, Mrs. Kuzmin, G. Kuzmin (author's archive).

the design of the telescope and its dome for nearly five years. Among other things I studied the thermal regime of the telescope dome to avoid micro-turbulence of air during observations. Results of this study were published (Einasto & Laigo, 1973) and used in the construction of the dome of our 1.5-m telescope, as well as the dome of the large Soviet 6-m telescope in North Caucasus.

The 1.5-m telescope was completed in 1975. At first we tried to use it for spectral observations of galaxies too, but the climatic conditions were unstable, so the new telescope remained mainly for spectral observations of stars, as originally planned.

The sixties were the years of rapid growth for Tartu Observatory. Many talented young people had received their initial training at the satellite station of the Tartu University. The satellite station provided the students with good observation practice which was highly useful in the new observatory. In accordance with our development plans, most of the students moved to stellar physics.

3.3.2 *Philosophical seminars and New Year parties*

In the new observatory we continued philosophical seminars. The foundation for this kind of seminar has already been laid in the fifties by professor Aksel Kipper. The invited speakers were distinguished scientists from other fields, as well as cultural and political figures. The atmosphere at the seminars was quite free. For

Galactic models and dark matter in the solar vicinity 73

Fig. 3.9 Grigori Kuzmin and Viktor Ambartsumian planting a tree at the opening ceremony of the new Tartu Observatory in 1964 (author's archive).

example, one of the most interesting presentations was professor Uku Masing's "Religion in the History of Mankind". In the presentation Uku Masing developed the thesis that all major religions have taken the form of a political movement, and vice versa, all political movements acquire the characteristics of a religion. As an example, Uku Masing brought up Hitler's Germany. Everyone of course realised which country he really meant. According to Masing's thesis, the main task of a religion is to establish a system of conventions and beliefs, without which no society can exist. If such a system of beliefs and customs is destroyed, the society would become very unstable. Once again it was clear which society the talk was of.

Fig. 3.10 The 1.5-m Telescope dome of the new Tartu Observatory in 2011. In the foreground are Laurits Leedjärv and Tim de Zeeuw (author's photo).

I once appeared at the seminar myself, on the topic "The Effectivity of Scientific Research". Studies had recently begun on this topic so I could rely on some earlier work. But most of the analysis had to be done myself without using examples. Among other things, I looked at how many articles appeared in different publications in the field of astronomy, and how often they were cited. Thus it was possible to determine the impact factor of a magazine or a publication series. I did not know that the impact factor has already been used. It turned out that at that time (1965) nearly half of all astronomical papers were published in the observatories' own publications, but there were almost no references to those papers. The journals "Nature" and "Annual Review of Astronomy and Astrophysics" had the highest impact factors, followed by "Astrophysical Journal". The effectivity of scientific research also varied; sometimes some astronomers had many publications but they were not cited. Citations of the publications of Tartu Observatory were very rare, as were that of most journals in Russian.

I drew my own conclusions and thereafter refrained from publishing my papers in our own Observatory publications. I published a few papers in Russian-language journals, but soon I started sending my papers to English-language journals. Another reason to avoid Russian was because, when translated from Russian to English, my name became distorted (it had the shape Ya. E. Ejnasto). For a while I pondered over the best way to write my address, because I did not want to use "Estonian SSR". Finally I decided to use as my address "Tartu Observatory, Estonia, USSR". There was a hidden message in this address — Estonia is still alive, but occupied by the USSR; curiously enough, all our papers were censored, but the censors did not notice the message. Later, when the time was right, the "USSR" could be omitted; the rest of the address was already correct.

I repeated the statistical analysis of astronomical publishing and referencing in 1975 and 1985. The analyses showed that people had understood the inefficiency of the publications of observatories, which had practically disappeared, whereas during these years the volume of journals had increased more than tenfold. I have also advised my students to publish their work in English-language journals, a suggestion that has been of benefit.

In the 1960's some physics theorists of the Institute of Physics worked in our new Observatory at Tõravere. Among them was Madis Kõiv, who wrote plays and tractates in his spare time. Now his dramas are played in several theatres in Estonia, but at the time we could not imagine that one of the innovators of Estonian drama was working beside us. At that time we used to prepare plays and gag stories for the observatory's New Year's parties, the scenarios authored by Madis Kõiv and Arved Sapar, one of our best in the field of theoretical astrophysics. One of our most successful parodies was based on the movie "Supernova". The outdoor shots for this movie were taken in Tõravere, and for a couple of months the filmmakers lived with us, dined in the same canteen and, if they chose to, could take notice of the way we lived and worked. What surprised us the most was their complete lack of interest in what we were actually doing. The filmmakers had their own idea of how science is done and they did not let the reality faze them. The movie was finished and the premiere was held in Tõravere. We watched it and marvelled that while everything seemed to be right, it felt completely phony. Madis Kõiv was among the first to speak up, pointing out that the movie was very artificial. So the idea came about to prepare a parody at the New Year's party on this subject. It was made as a pantomime and titled "Superprima". Arved Sapar played a genius young astronomer, who discovers a new star — the Superprima. One young astronomer

Fig. 3.11 New observatory in autumn 1995 (author's photo).

depicted the telescope; four other young men danced to the music of the little swans from Tchaikovsky's "Swan Lake". It was the best parody I have encountered to this day.

The directing of New Year's party presentations was later continued by climatologist Ain Kallis, who produced a piece on defending a dissertation in the eighties. Back then we had a council where our own astronomers as well as guests from other centres could come to defend their PhD theses. Now and then there were also mishaps due to our defective knowledge of Russian — these found their way in the play. It panned out well and word of it spread in the Estonian community. A recording of the play was made by a TV team, and aired on Scientists' Day. The humour was so subtle that many viewers did not realise it was a parody until towards the end. The matter ended with Rein Ristlaan (secretary of the Estonian communist party responsible for ideology) getting angry and ordering the recording to be destroyed. So it came to be that for us only memories of those nights remain.

3.3.3 *Space studies*

When the new observatory was planned, one of the new research directions was space studies. Similar plans were made in the Sternberg Astronomical Institute in

Moscow, in the Crimean Astrophysical Observatory and in the Byurakan Astrophysical Observatory. Our astronomers had discussions with leaders of these observatories, and an agreement was reached that the role of Tartu Observatory is to develop UV-sensitive detectors and UV-calibration systems, needed to observe astronomical objects in ultraviolet light. In our observatory Valdur Tiit was the initiator of this research, and soon a special laboratory was formed to develop and build the equipment needed.

This work was rather successful. Very sensitive UV-detectors were developed and tested first in rocket flights, and then installed on the first astronomical satellite, Kosmos 215, launched in April 18, 1968 at the Soviet site at Kapustin Yar. The goal was to examine ultraviolet and X-ray radiation from stars. This was a common enterprise of the Sternberg Institute, Crimean and Tartu Observatories. Valdur Tiit was the head of the team at the launch site. Preliminary results of the mission were described by Dimov (1970). Over a period of 40 days spectral regions from 1250 to 2700 Å were recorded by identical telescopes of aperture 70 mm, see Fig. 3.12. The telescopes scanned the sky and recorded all stars which crossed the fields of view. UV-photometry was obtained for 36 A and B stars, in addition an X-ray telescope was used to measure radiation between 0.05 and 0.5 nanometers. Several X-ray sources were detected.

After this very good experience Valdur Tiit had an agreement with space authorities that a launcher shall be reserved to put a larger telescope into space, of aperture 350 mm. The mirror was made in Leningrad and there was an agreement with the

Fig. 3.12   One of the eight telescopes used in the astronomical satellite Kosmos 15 (author's archive).

Crimean and Sternberg Observatories on how to put all the equipment together. However, quite unexpectedly, Prof. Kipper announced that Tartu Observatory shall not continue with this project. Because Tartu Observatory was the only one with capabilities to build and test UV equipment, the whole project was cancelled. Years later, Kipper explained that he feared that if we are very seriously involved in space studies, then the whole Observatory could be surrounded by barbed wire, and civil scientific work would be difficult. The laboratory led by Valdur Tiit was moved to the Institue of Physics, where he continued to develop UV equipment and other modern devices.

However, this was not the end of space studies in Tartu Observatory. During the International Geophysical Year 1957 our astronomers participated in one of the projects — the study of noctilucent (or night) clouds. These clouds are located at altitudes of around 80 km, and are visible only when illuminated by sunlight while the lower atmosphere is in the Earth's shadow during summer nights. Their formation is not clear; they are observed only at Earth latitudes between $50°$ and $70°$ North and South of the equator.

The leader of the study of noctilucent clouds in Estonia was Charles Villmann, a former amateur astronomer. He was a very good organiser and was invited in the mid 1960's to the Observatory as vice-director to help organise the building of the new observatory in Tõravere. Soon he understood that it is much easier to observe noctilucent clouds from space. He contacted the manned flight center in Moscow and got permission to install on space stations Salyut 6 and Salyut 7 equipment to scan the upper atmosphere near the Earth limb. He managed to form a team who constructed and built photometers Mikron and Faza (see Fig. 3.13), which were used on these space stations by Soviet kosmonauts Georgy Grechko and Vitaly Sevastyanov. Both kosmonauts often visited Tartu Observatory to discuss results of observations; one picture of such a visit is shown in Fig. 3.14.

Georgy Grechko wrote his PhD thesis on the basis of data collected in collaboration with the Tartu team by Villmann, Enn Saar acting as consultant in theoretical physics. Initially visits of kosmonauts were made secretly, but in the last years of 1980's they visited us quite openly. Once kosmonauts even participated in our New Year party where the play on the thesis defence was performed. For our guests the play was performed in Russian.

Actually the fear of Prof. Kipper concerning the freedom of scientific thinking in the Observatory was not completely unjustified. One office in the Observatory was reserved for a 'special department', its door was covered with an ironplate and window with an iron grid, and a KGB officer had her office there.

Galactic models and dark matter in the solar vicinity

Fig. 3.13 The photometer Faza used in Space Station Mir to observe Earth's atmosphere from space (author's archive).

Fig. 3.14 Soviet kosmonauts visiting Tartu Observatory in 1981. From left the wife of Georgy Grechko, Charles Villmann, Georgy Grechko, Aksel Kipper, Vitaly Sevastyanov, Väino Unt (author's archive).

All correspondence on the space program went through this office, and only people with special permissions could read and write the correspondence. For these 'trusted' people this possibility was a hindrance for foreign travel, because these people knew 'state secrets'. This office had one more function in the observatory — Soviet officials made all possible efforts to prevent any kind of demonstration on May Day and October Revolution anniversaries. For this reason all typewriters and xerox-machines were locked, in addition all offices of the observatory were locked and sealed up. All these measures were rather stupid, but we accepted this with humor. We were aware what was possible and what not, we did not have any open hostility against the rulers, and continued our scientific studies and cultural activities as before.

Chapter 4

Global dark matter

During the 1960's I elaborated, step by step, the main principles on how to calculate models of galaxies which make use of as much observational data as possible. To bring physical data on various populations to a coherent system, models of physical evolution of galaxies and their populations were calculated. And then I ran to difficulties — no combination of stellar populations was able to explain rotation data of galaxies. The solution came in the early 1970's when my collaborator Enn Saar suggested abandoning the idea that only known populations exist in galaxies. This brought us to the dark matter problem. But then we had another difficulty — there were no suitable candidates for the nature of the dark matter. The whole decade of the 1970's was needed to finally find a possible candidate for dark matter particles. The candidate was found elsewhere, as we had no good experts on particle physics in our team, but we were able to test its role in the formation of the cosmic web. In this Chapter I shall concentrate on our efforts to find the amount of dark matter, its distribution, and connection with ordinary matter.

4.1 The discovery of global dark matter

4.1.1 *Galactic coronas*

In 1970 I had the chance to attend the IAU General Assembly in Brighton. I reported my models of galaxies at the meeting of the Commission 33 of IAU on the Structure and Dynamics of the Galaxy. These were essentially models I calculated for my Doctor of Sciences thesis. Galactic evolution was already included, but dark coronas not. I had a chance to meet Ernst Öpik and his wife. I had a lot of discussions also with other astronomers.

In spring 1972 George Contopoulos invited me to give a review on Galactic models at the First European Astronomy Meeting in Athens. At this time population

Fig. 4.1 Ernst Öpik with his wife and author in Brighton during the IAU General Assembly 1970. Presentation of new galactic models, but without dark coronas (author's photo).

models of galaxies had been calculated already for 5 galaxies of the Local Group and for the giant elliptical galaxy M87 in the Virgo cluster. More and more data accumulated on rotation velocities of galaxies. New data suggested the presence of almost flat rotation curves on the periphery of galaxies, thus it was increasingly difficult to accept the previous concept of large non-circular motions. On the other hand, recently finished calculations of the physical evolution of stellar populations confirmed our previous view that it is extremely difficult to accept stellar origins for the hypothetical population, responsible for flat rotation curves.

In summer 1972 I discussed the problem with my collaborator Enn Saar. He suggested abandoning the idea that only known stellar populations exist in galaxies, to assume that there is a population of unknown nature and origin, and to look at which properties it should have using available data on known stellar populations and galaxy rotation data.

This discussion was one of the decisive moments of the whole dark matter story. It was immediately clear that the assumption of the presence of a new population demands a radical change in our understanding of the structure of galaxies, and that we are dealing with a completely new phenomenon of unknown origin.

First of all — the extended dark population cannot be of the same nature and origin as the dark population in the Solar vicinity near the Galactic plane. In other words, there are two dark matter problems: one of the dark matter in the Solar vicinity, and the other of the dark matter surrounding galaxies and clusters of galaxies. The dark matter in the Solar vicinity is strongly concentrated in the plane of the Galaxy, thus dissipation is needed to form this population. This population probably consists of very faint stars or Jupiter-like objects with no hydrogen-burning in their interiors. This assumption is not new; Jeans (1922) already argued that there are several invisible stars per visible one.

What concerns the nature of the new extended population then are arguments against its stellar origin. These arguments were mentioned in the previous Chapter and shall be discussed in more detail in the next Chapter. Taking all these considerations into account I realised that we are dealing with a new very extended population of unknown nature, well segregated from known populations. To avoid confusion with the known halo population consisting of old metal-poor stars, I called the new population "corona". Data also indicated that the presence of dark coronas is a general property of galaxies, at least of giant ones. It may be of the same origin as the dark matter in clusters of galaxies.

When I understood all this, I had the feeling that we have reached a peak of a mountain, and that behind the mountain there is a new terrain, completely unknown before, with other mountains and valleys, and a new horizon far away. The peak itself had been visible already for some time, but everything on the other side of the peak was unknown.

Quickly I calculated a second set of models for galaxies, assuming, as a first approximation, that the total mass of the new population is equal to the mass of the sum of known stellar populations. The central density of the new population can be easily found using observed rotation curves and data on known stellar populations. Already first calculations showed that this assumption is too modest, and it improves the rotation velocity law only a little. Indirect arguments, applied to the giant elliptical galaxy M87, suggested that the mass and the radius of the dark population can exceed the total mass and the mean radius of known populations even tenfold. The distribution of mass-to-luminosity ratios of models is shown in Fig. 4.2 for all galaxies studied; the rotation curve for the Andromeda galaxy is given in Fig. 4.3. In both Figures two variants of models are given, the variant A without, and the variant B with the dark corona. For comparison, the right panel of Fig. 4.3 shows the rotation of M31 according to our later model by Tenjes et al. (1994).

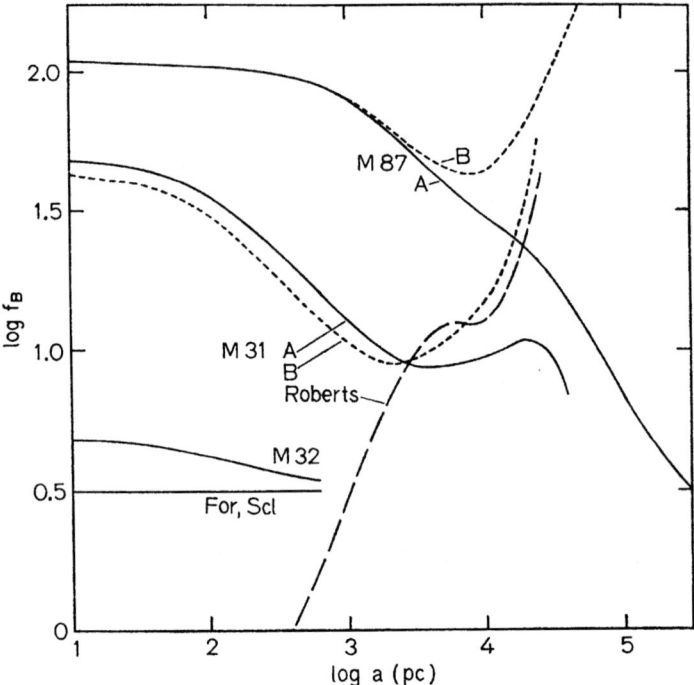

Fig. 4.2 The distribution of mass-to-luminosity ratio, $f_B = M/L_B$, in galaxies of the Local Group and M87: models without (A) and with (B) dark corona (Einasto, 1974a).

My talk in Athens was on September 8, 1972. The main results were (Einasto, 1972a, 1974a):

(1) There are two dark matter problems: the local and the global one;
(2) Local dark matter, if it exists, must be of stellar origin, since it is strongly concentrated in the Galactic plane, and dissipation is needed to form such a flat population (in the gaseous phase of the evolution of the Galaxy);
(3) Global dark matter is of non-stellar origin; it has very low concentration in the plane and centre of the galaxy; its dynamical and physical properties are different from properties of all previously known stellar populations; to avoid confusion with known stellar populations I called the new population 'corona';
(4) Available data are insufficient to determine radii and masses of coronae.

The nature of the corona was unclear. In the Athens talk I wrote: *"The matter in question cannot be in the form of neutral gas, since this gas would be observable. The matter cannot be in the form of stars too. Luminosity decreases in outer galactic*

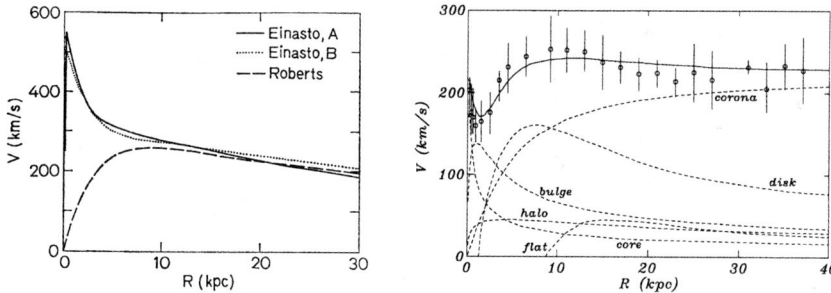

Fig. 4.3 Left: the rotation curve of M31 according to the model by Einasto (1974a). Variants A and B correspond to models without and with corona. For comparison the rotation curve by Roberts (1966) is shown. Right: the rotation curve of M31 according to the model by Tenjes et al. (1994). Open circles mark observations, thick line — the best-fit model, dashed lines — contribution of populations to the rotation curve.

regions rapidly, therefore, if the matter is in the form of stars, the latter must be of very low luminosity to be invisible. The presence of low-luminosity stars in outer galactic regions without bright ones would require a powerful process of a large-scale segregation of stars according to mass (low-luminosity stars have smallest masses), but this is highly improbable. There remains the possibility that the unknown matter exists in the form of rarefied ionised gas."

Kahn & Woltjer (1959) in their study of the dynamics of the system M31-Galaxy also suggested hot gas as the possible form of the unseen matter in the Local group of galaxies.

My Athens report did not give rise to special excitement. The main reason for this lukewarm reception was probably the absence of a solid proof for the existence of the corona, of its main parameters (mass and radius), and of its nature. Thus I continued the search for further evidence.

4.1.2 *Clusters and groups of galaxies*

In the search for further evidence of dark matter I spent a lot of time searching in the literature and looking at what other people had already done. I noticed that the problem of galactic coronae is the same as discussed already long time ago in clusters and groups of galaxies, starting from the pioneering work by Fritz Zwicky (1933, 1937). He measured redshifts of galaxies in the Coma cluster and found that the velocities of individual galaxies with respect to the cluster mean velocity are much larger than those expected from the estimated total mass of the cluster, calculated from masses of individual galaxies. The only way to hold the cluster

Fig. 4.4 Author with Vladimir Nikonov and Evgeny Kharadze in Sounion near Athens during the First European Astronomy Meeting in 1972. Galactic models with dark coronas announced (author's photo).

from rapid expansion is to assume that the cluster contains huge quantities of some invisible dark matter. According to his estimate the amount of dark matter in this cluster exceeds the total mass of cluster galaxies at least tenfold, probably even more.

Smith (1936) measured radial velocities of 30 galaxies in the Virgo cluster and confirmed the Zwicky result that the total dynamical mass of this cluster considerably exceeds the estimated total mass of galaxies. This conclusion was again confirmed by Zwicky (1937), who discussed masses of galaxies and clusters in detail. As characteristic for papers which change our world view, early indications of problems in the current world view are ignored by the community; this happened also with the Zwicky's discovery. Astronomers were interested in the structure and evolution of stars, and Zwicky's work seemed to be remote and uninteresting.

However, slowly more dynamical data on clusters of galaxies were collected, and the discrepancy between the cluster galaxy measured velocities and expected velocities for a stable cluster could not be ignored. To explain high velocities of cluster galaxies Ambartsumian (1961) suggested the idea that clusters are recently formed and are now expanding. In 1961 during the International Astronomical Union (IAU) General Assembly a special meeting to discuss the stability of clusters of galaxies was organised (Neyman et al., 1961).

I applied to attend the 1961 IAU General Assembly. Recently in sorting old documents I discovered a letter from the Astronomical Council where it was confirmed that I am included in the team of Soviet delegates to the Assembly. However, as characteristic for this time, somewhere my application was stopped, and from Tartu Observatory only Professors Kipper and Keres attended. Thus I missed the opportunity to participate in the first wide discussion of the mass discrepancy in clusters of galaxies.

When reading the Proceedings of the instability of clusters of galaxies I was impressed by the talk of Sidney van den Bergh (1961), who drew attention to the fact that the dominating population in elliptical galaxies is the bulge consisting of old stars, indicating that cluster galaxies are old. It is very difficult to imagine how old cluster galaxies could form an unstable and expanding system. The background of this meeting and views of astronomers supporting these and some alternative solutions were described by Trimble (1995), van den Bergh (2001), Sanders (2010) and Trimble (2010). I fully agree with arguments by van den Bergh (1961, 1962) that clusters of galaxies are old and stable systems.

A similar problem exists in double elliptical galaxies. The mean mass-to-luminosity ratio of double elliptical galaxies is $M/L \approx 66$ (Page, 1952, 1959). A certain discrepancy was detected also between masses of individual galaxies and masses of groups of galaxies (Holmberg, 1937; Page, 1960). The conventional approach for the mass determination of pairs and groups of galaxies is statistical. The method is based on the virial theorem and is almost identical to the procedure used to calculate masses of clusters of galaxies. Instead of a single pair or group often a synthetic group is used, consisting of a number of individual pairs or groups. These determinations yield for the mass-to-luminosity ratio (in blue light) values $M/L_B = 1 \ldots 20$ for spiral galaxy dominated pairs, and $M/L_B = 5 \ldots 90$ for elliptical galaxy dominated pairs (for a review see Faber & Gallagher (1979)). These ratios are larger than found from local mass indicators of galaxies — velocity dispersions at the centres of elliptical galaxies and rotation curves of spiral galaxies.

A completely new and innovative approach in the study of masses of systems of galaxies was applied by Kahn & Woltjer (1959). The authors paid attention to the fact that most galaxies have positive redshifts as a result of the expansion of the Universe; only the Andromeda galaxy M31 has a negative redshift of about 120 km/s, directed toward our Galaxy. This fact can be explained if both galaxies, M31 and our Galaxy, form a physical system. The negative radial velocity indicates that these galaxies have already passed the apogalacticon of their relative orbit and are presently approaching each other. From the approaching velocity, the mutual distance, and the time since passing the perigalacticon (taken equal to the present age of the Universe), the authors calculated the total mass of the double system. They found that $M_{tot} \geq 1.8 \times 10^{12}$ $M_\odot$. The conventional masses of the Galaxy and M31 were estimated to be of the order of 2×10^{11} $M_\odot$. In other words, the authors found evidence for the presence of additional mass in the Local Group of galaxies. The authors suggested that the extra mass is probably in the form of hot gas of temperature about 5×10^5 K. Using modern data Einasto & Lynden-Bell (1982) made a new estimate of the total mass of the Local Group, using the same method, and found for the total mass an even higher value, $4.5 \pm 0.5 \times 10^{12} M_\odot$.

Materne & Tammann (1974b,a) tested a number of nearby groups of galaxies for stability. They found that of 15 groups tested about half had either small or very uncertain values of M/L_{pg}. Of the remaining groups most had $10 < M/L_{pg} \leq 30$, and only a few had $M/L_{pg} > 30$. The galaxy membership in groups with high M/L_{pg} is not very certain. Thus the authors conclude that there is little evidence for the presence of a large discrepancy between mass-to-luminosity ratios of individual galaxies and groups of galaxies.

These relatively high M/L_{pg} ratios are in conflict with well-known values calculated from the luminosity function of stars in the Solar vicinity (Kapteyn & van Rhijn, 1920; Öpik, 1922a). The high M/L_{pg} ratios are also in conflict with values calculated from rotation and photometric data (the Burbidge's series of papers on rotation curves), which suggest $M/L_{pg} \approx 3$ for main bodies of galaxies. My own recent population models suggested that most stellar populations have $1 \leq M/L \leq 30$. In particular, for bulges of galaxies our composite galactic model gave $M/L \approx 3-10$ in good agreement with models of physical evolution of stellar populations.

On the other hand, already early rotation curves of galaxies suggested very high values of M/L on the periphery of galaxies (Babcock, 1939; Oort, 1940; Roberts, 1966; Rubin & Ford, 1970). However, these rotation curves were not long enough to calculate correctly mass distribution models for larger distances from centres of galaxies. Thus it was not clear how serious the discrepancy between the total masses and mass-to-luminosity ratios, found using different methods, is.

4.1.3 *Dynamics and morphology of companion galaxies*

Reading these papers on the mass discrepancy in clusters, groups and galaxies I realised how it is possible to check the presence of dark coronae around galaxies. If galactic coronae are large enough, then in pairs of galaxies the companion galaxy can be considered as a test particle to measure the gravitational attraction of the

Fig. 4.5 Left: Enn Saar. Right: Mihkel Jõeveer in the 2000's (author's archive).

main galaxy. Mean relative velocities, calculated for different distances from the main galaxy, can be used instead of rotation velocities to find the mass distribution of giant galaxies for a much larger range of distances from the center of the main galaxy.

Quickly I collected data for pairs of galaxies. To avoid inclusion of optical pairs only double galaxies with some sign of mutual interaction were chosen. The analysis was ready on January 11, 1974. It showed that radii and masses of galactic coronae exceed radii and masses of visible parts of parent galaxies by an order of magnitude! Together with Ants Kaasik and Enn Saar we calculated new models of galaxies including dark coronae. This time we had enough data to find *total* masses and radii separately for visible parts of galaxies and for their dark coronae.

In those years Soviet astrophysicists had the tradition of gathering in Caucasus Winter Schools. My results from galactic mass modelling were reported in the Arkhõz Winter School in 1972. In 1974 the School was held near the Elbrus mountain in the Terskol winter resort. I had my report on the masses of galaxies on January 29, 1974. I had a chance to discuss the results before the talk with my friend from the Ioffe Institute in Leningrad, Arthur Chernin. He suggested excluding all other stuff from the talk (initially the talk was on new models of galaxies), and to concentrate to the main result and its consequences to our world view on the structure of the Universe. I followed his suggestions.

My message was:

(1) Data suggest that all giant galaxies have massive coronae, exceeding the mass and the radius of known populations about tenfold;

(2) The presence of massive coronae around galaxies may solve the problem of high masses of clusters of galaxies. X-ray data suggest that the mass of hot gas is not sufficient to stabilise clusters — clusters must be stabilised by dark matter;
(3) According to new estimates the total mass density of matter is 20% of the critical cosmological density, thus dark matter is the dominant population in the whole Universe.

Also I stressed my arguments suggesting that the corona is probably not a stellar population. What impressed me most was the beauty of the dark matter concept. Once you accept the presence of dark matter, a number of problems are solved — the Zwicky paradox of masses in clusters of galaxies, a similar paradox in groups, and flat rotation curves of galaxies.

In the Winter School prominent Soviet astrophysicists like Zeldovich, Shklovsky, Novikov and others participated. After the talk the atmosphere was as if a bomb had exploded. Everybody realised that, if true, this is a discovery of principal importance. Two questions dominated: What is the physical nature of the dark matter? and What is its role in the evolution of the Universe? Zeldovich and his group had been working over 15 years to find the basic physical processes of the formation and evolution of the structure of the Universe. For them the possible presence of a completely new, massive non-stellar population was a great surprise.

Zeldovich had a habit of waking up very early in the morning — these early hours were the most productive to think about new problems. He had a very good knowledge in physics but did not have all the important astronomical facts in his head. So he often called his students or collaborators almost in the middle of the night asking for some astronomical number or other question. Thus his team of young collaborators was trained to give quick answers to any question. In particular, they had the ability to do very quickly order of magnitude calculations. In such calculations there were only numbers 1 and 3, multiplied by a factor of 10 to some power.

Already during the School Zeldovich's boys started to make estimates on the possible nature of dark matter. My own opinion was that the coronae are not of stellar origin for reasons discussed above. I also had arguments against neutral gas, thus my preliminary guess was hot gas (Einasto, 1974a).

Komberg & Novikov (1975) studied in more detail the hypothesis that massive coronae around spiral galaxies are composed of hot ionised gas. They found that this is in conflict with X-ray data on the amount of hot gas, already available in early 1970s (see the introduction of my paper on dark coronas (Einasto et al., 1974b)). Moreover, hot gas in such quantities would ionise neutral gas in the galactic disk, but

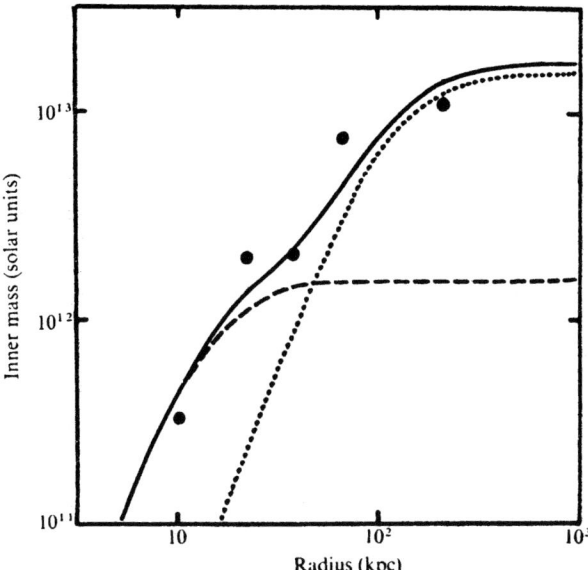

Fig. 4.6 The mean internal mass $M(R)$ as a function of the radius R from the main galaxy in 105 pairs of galaxies (dots). Dashed line shows the contribution of visible populations, dotted line the contribution of the dark corona, solid line the total distribution (Einasto et al., 1974b).

neutral gas exists, thus this process is not dominant. Neutrinos were also considered, but rejected, since quick estimates showed that they can form only supercluster-scale coronas, about 1000 times more massive than coronae around galaxies are.

The School caused an avalanche of new studies to find the properties and physical nature of dark coronae. Ozernoi (1974) pointed out that, within rich clusters, the dark matter must mostly belong to the cluster as a whole and not to the individual galaxies, because the galaxy separations were smaller than the dark matter corona sizes. Bobrova & Ozernoi (1975) found that the mass of the hot gas is much smaller than the mass of hidden (dark) matter. If the hidden matter were confined to coronae of galaxies, then one would observe a more spotty structure of the X-ray emitting gas than observed. Chernin (1976) studied possible consequences of gaseous coronae, and Jaaniste & Saar (1975) investigated the possible stellar nature of the corona in more detail (see below). Antonov et al. (1975b,a); Antonov & Chernin (1975) investigated dynamical properties of galactic coronae. Zeldovich (1975) suggested a mechanism to explain the nucleosynthesis constraints on the amount of baryonic matter with the density due to dark matter (see below).

We had to hurry with the publication of our results, since large masses of halos were already discussed by Ostriker & Peebles (1973). Preliminary results of our analysis were published in February 1974 as Einasto et al. (1974d). But Zeldovich insisted that this is not enough: *"Major results must be published in major journals"*. Thus a more detailed report was sent to "Nature" (Einasto et al., 1974b) and, for the first time, a preprint was made and sent to all observatories. Soon we realised that it was just in time: Ostriker et al. (1974) got similar results using similar arguments; their paper was published several months after our "Nature" paper and has a reference to our preprint.

Since our paper and the paper by Princeton astronomers played a crucial role in the development of the dark matter concept, I shall discuss in a bit more detail the similarities and differences between the two papers.

The basic similarity is that the presence of massive halos/coronae was suggested using rather similar arguments. Both papers suggest that the total cosmological density of the matter in galaxies is about 0.2 of the critical cosmological density. But more interesting are the differences.

First, in our paper it was clearly stated that we have here a *new population — corona*. The principal goal of our paper was *to find the main parameters of the corona*. Estimates of radii, masses, and central densities of coronae were given in two Tables, which by mistake were missed in the "Nature" version. Because these Tables contain very important data, they are reproduced here as Table 4.1 and 4.2. Here L_s is the total luminosity of stellar populations, M_s — the total mass of stellar populations, M_c — the mass of the corona, R_{cc} — the effective radius of the corona, and ϱ_c — the central density of the corona. Table 4.2 shows that mean radii and mean masses of coronae exceed mean total masses and mean radii of known populations about tenfold. The segregation of the corona from known stellar populations was not stressed in the text, but is clearly evident from data shown in Tables.

Table 4.1 Parameters of Galactic Populations; Individual Galaxies.

Galaxy name	L_s $10^{10} L_\odot$	M_s $10^{10} M_\odot$	M_c $10^{10} M_\odot$	R_{cc} kpc	$-\log \varrho_c$ g cm^{-3}
NGC 224	2.0	17	> 35	> 14	24.58
NGC 300	0.45	2.0	> 2.9	> 6	24.60
NGC 598	0.32	1.5	> 2.8	> 6	24.54
NGC 3031	1.8	12	> 21	> 11	24.52
IC 342	4.9	12	> 34	> 13	24.55

Table 4.2   Parameters of Galactic Populations; Pairs of Galaxies.

Type of Primary Galaxy	$<L_s>$ $10^{10} L_\odot$	$<M_s>$ $10^{10} M_\odot$	$<M_c>$ $10^{10} M_\odot$	$<R_{cc}>$ kpc	Number of pairs
Spiral (intermediate)	3.8	38	350	25	33
Spiral (bright)	15	150	1600	47	32
Elliptical	12	250	≥ 1500	≥ 46	40

Second, opinions about the nature of dark matter were different. In our paper in the introduction it was noted that dark matter in clusters cannot be explained by hot gas, since its mass is insufficient to stabilise clusters. Thus dark matter cannot be identified with hot X-ray emitting gas. The paper ends with a statement that a further discussion of the nature of galactic coronae shall be published elsewhere. Our additional studies were the previous review paper by Einasto (1972a, 1974a), the ongoing analysis by Einasto et al. (1974c) on morphological properties of systems of satellite galaxies around giant ones, and the study by Jaaniste & Saar (1975) on the possible stellar nature of galactic coronae. Ostriker et al. (1974) did not notice that dark matter forms a new population of unknown nature; authors write in the discussion that *"the very great extent of spiral galaxies can perhaps most plausibly be understood as due a giant halo of faint stars"*.

Soon the first reaction to the results of both papers appeared: Burbidge (1975) formulated difficulties of the dark corona concept. The main problem is in the statistical character of the dynamical determination of masses of double galaxies. If companion galaxies, used in mass determination, are not real physical companions but random interlopers, then the mean velocity dispersion reflects random velocities of field galaxies, and no conclusions on the mass distribution around giant galaxies can be made.

These three publications initiated the dark matter boom. Our paper was written by a previously almost unknown group of astronomers, thus the publication of papers on the subject by leading astronomers was important. From now on the possible presence of dark matter in and around galaxies was taken more seriously, which initiated further studies and discussions of the problem by the astronomical community. As noted by Kuhn, a scientific revolution begins when leading scientists in the field start to discuss the problem and arguments in favour of the new over the old paradigm.

Difficulties connected with the statistical character of our arguments were discussed already in the Winter School, thus we started immediately a study of properties of companion galaxies to find evidence for some other regularity in the satellite system, which surrounds giant galaxies. Soon we discovered that companion

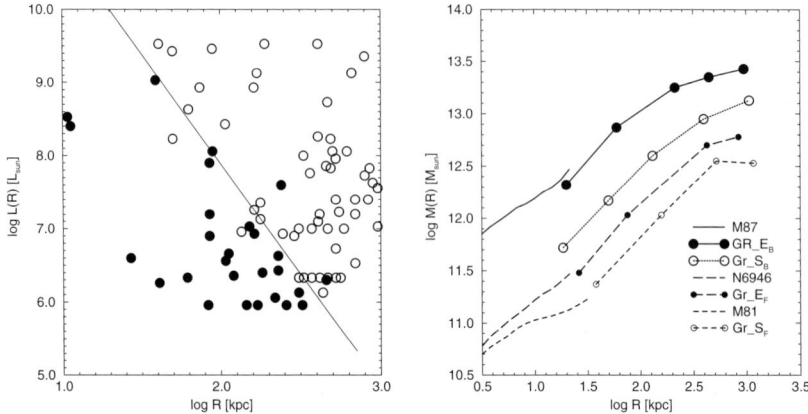

Fig. 4.7 Left: Distribution of luminosity of companion galaxies of different morphology vs. distance from the central galaxy; spiral and irregular companions are marked with open circles, elliptical companions with filled circles (Einasto et al., 1974b). Right: Distribution of internal mass in the giant elliptical galaxy M87, giant spiral galaxy NGC6946, and medium luminous spiral galaxy M81, compared with mass distribution in groups of galaxies derived from relative motions of companions of giant and medium bright elliptical and spiral galaxies (Einasto et al., 1976d).

galaxies are segregated morphologically. Elliptical (non-gaseous) companions lie close to the primary galaxy whereas spiral and irregular (gaseous) companions of the same luminosity have larger distances from the primary galaxy. The distance of the segregation line from the primary galaxy depends on the luminosity of the satellite galaxy, see Fig. 4.7.

This result shows, first of all, that companions are real members of these systems — random by-flyers cannot have such properties. Second, this result demonstrated that diffuse matter can have a certain role in the evolution of galaxy systems. The role of diffuse matter in galactic coronae was discussed in detail by Chernin et al. (1976). Morphological properties of companion galaxies can be explained if we assume that at least part of the corona is gaseous.

Also we found that dynamical and morphological properties of primary galaxies are well correlated with properties of their companions (Einasto et al., 1976d), see right panel of Fig 4.7. This suggests the presence of a physical link between primary and satellite galaxies.

A further evidence of the large mass of the corona of our Galaxy came from the study of the dynamics of the Magellanic Stream (Einasto et al., 1976a). On the other hand, as already discussed at the Winter School, coronae cannot be fully gaseous (Komberg & Novikov, 1975). Thus the nature of coronae remained unclear.

4.1.4 *Tallinn and Tbilisi dark matter discussions*

To discuss the existence and the physical nature of dark matter, we organised in January 28–30, 1975 a conference in Tallinn, Estonia, devoted solely to dark matter (Doroshkevich et al., 1975). One of the goals was to prepare for the Third European Astronomical Meeting in Tbilisi. The rumour on dark matter had spread around the astronomical and physics community and, in contrast to conventional regional astronomy conferences, leading Soviet astronomers and physicists attended. This conference is not well known, so I give here the list of major talks:

- Zeldovich: "Deuterium nucleosynthesis in the hot Universe and the density of matter";
- Einasto: "Dynamical and morphological properties of galaxy systems";
- Ozernoi: "Theory of galaxy formation";
- Zasov: "Masses of spiral galaxies";
- Novikov: "Physical nature of galactic coronas";
- Saar: "Properties of stellar halos";
- Doroshkevich: "Problems of the origin of galaxies and galaxy systems";
- Komberg: "Properties of central regions of clusters of galaxies";
- Vorontsov-Velyaminov: "New data on fragmenting galaxies".

The list shows that central problems discussed in Tallinn were: What is the physical nature of the dark matter? and: What is its role in the evolution of the Universe? Two basic models were suggested for coronae: faint stars or hot gas. It was found that both models have serious difficulties (Jaaniste & Saar, 1975; Komberg & Novikov, 1975).

I was a member of the Scientific Organising Committee of the Third European Astronomical Meeting, and was invited to Abastumani Observatory by Evgeny Kharadze, the Chairman of SOC, to discuss the program. The SOC meeting was held in October 1974. We had our first results of the study of dark matter, but the topic was still rather new and controversial. Initially Evgeny Kharadze and Richard West, another member of SOC present in Abastumani, had doubts of including this topic to the program. The Meeting was aimed at *observational* aspects of the structure of stars, galaxies and Universe, but dark matter was initially considered as a theoretical problem. It took some time to convince other members of SOC that the problem is presently an observational one. Finally we agreed that a full session shall be devoted to the dark matter problem. However, my suggestion to include a talk by Zeldovich on the theory of galaxy and structure formation was not accepted. As a compromise an agreement was achieved that Zeldovich can give a special lecture, but after the official program is finished.

In the dark matter session the principal discussion was between the supporters of the classical paradigm with conventional mass estimates of galaxies, and of the new one with dark matter. Statistical arguments against the dark matter concept were presented by Fesenko (1976). Arguments for the presence of dark matter were presented in my lecture (Einasto et al., 1976e), arguments favouring the classical paradigm by Materne & Tammann (1976).

Historically, the Tbilisi Meeting was the first well documented international discussion between supporters and opponents of the dark matter concept. Using their own statistical data Materne and Tammann concluded that systems of galaxies are stable with conventional masses. However, their most serious argument was: *Big Bang nucleosynthesis suggests a low-density Universe with the density parameter* $\Omega \approx 0.05$; *the smoothness of the Hubble flow also favours a low-density Universe.* Allan Sandage in collaboration with Gustav Andreas Tammann has made great efforts to develop the classical cosmological paradigm. This cosmological paradigm is sometimes called "the science of two constants" — the Hubble constant and the density parameter. If one excludes inconvenient data by Zwicky, Kahn and Woltjer, and recent data on flat rotation curves of galaxies and dynamics of double galaxies, then everything fits well into this classical cosmological paradigm.

Tammann is a world class astronomer, and his arguments were serious. In the framework of the classical cosmological paradigm it is really impossible to accept the presence of dark matter in quantities as suggested by Tartu and Princeton analyses (assuming the baryonic nature of DM). In the report to the Astronomical Council of the results of the Tbilisi Meeting the Chairman of the Organising Committee Kharadze noticed that the dark matter concept did not find support. Even Zeldovich started to doubt in the existence of dark matter. He asked *"show me at least one galaxy where the existence of dark matter is proven reliably"*. If dark matter exists in quantities as suggested by new data, then the arguments by Materne and Tammann must be explained in some other way.

In Tbilisi I defended a view opposite to the view by Gustav Andreas Tammann. This did not violate our good relationship. He has Estonian roots: his grandfather Gustav Tammann was a professor of physical chemistry at Tartu University before World War I. Their roots are in Sangaste, South Estonia, their relatives still live in Estonia. In 1903 Gustav Tammann emigrated to Germany and became director of the Inorganic Chemistry Institute of the Göttingen University. Gustav Andreas Tammann was the director of the Astronomical Institute of the Basel University. In 1990 the first Meeting of the European Astronomical Society was held in Davos, Switzerland. At this time it was already easier to travel, and I attended the Meeting together with my younger collaborators, my daughter Maret, and Mirt Gramann. After the Society Meeting we drove with my first western car (an old Mercedes) to

Basel, following the invitation by Gustav Andreas Tammann. We gave talks at the Astronomical Institute and had very interesting discussions. After Estonia declared independence Gustav Andreas visited Estonia several times to meet his relatives and to see Sangaste.

The dark matter problem was also discussed during the IAU General Assembly in Grenoble, 1976, at the Commission 33 Meeting. Here arguments for the non-stellar nature of dark coronae were again presented (Einasto et al., 1976c). I remember that after my talk Ivan King quietly said from the audience *"Perhaps really there are two halos of galaxies"* — the conventional halo of old metal-poor stars and the extended and non-luminous corona. After the lecture Ivan came to me and asked me to repeat main arguments against the stellar origin of dark matter. A year later at Yale a conference was held on the topic "Evolution of galaxies and stellar populations", and Ivan in his introductory talk listed the dark matter or "missing mass" problem as one of the most disquieting ones (King, 1977). However, no new independent data were presented in Grenoble nor at Yale. It was clear that by dispute only it is not possible to solve the problem — new data were needed.

4.2 The confirmation of the presence of global dark matter

4.2.1 *Rotation curves of galaxies*

As discussed above, a problem with the distribution of mass and mass-to-luminosity ratio was detected in spiral galaxies. Already Babcock (1939) obtained spectra of the Andromeda galaxy M31, and found that in the outer regions the galaxy is rotating with an unexpectedly high velocity, far above the expected Keplerian velocity. He interpreted this result either as a high mass-to-luminosity ratio on the periphery or as strong dust absorption. Oort (1940) studied the rotation and surface brightness of the edge-on S0 galaxy NGC 3115, and found in the outer regions a mass-to-luminosity ratio ≈ 250.

Further evidence came from radio observations of the rotation of galaxies using neutral hydrogen. Strom (2012) describes how the atomic hydrogen 21-cm line in space was discovered. The information on radio emission from space became known to Dutch astronomers in late 1940, and Jan Oort started to think about how to use this emission to investigate astronomical objects. He organized a colloquium on radio astronomy in 1944 in Leiden, insisting that *radio astronomy can really become very important if there were at least one line in the radio spectrum*. His student van de Hulst calculated that hydrogen emits radio waves at 21-cm, and that this emission can be used to detect interstellar hydrogen, as well as to measure its velocity.

After World War II there were numerous abandoned German Würzburg radar dishes in the Dutch territory. The head of the Post Office Radio Division, de Voogt, appropriated a number of these radar dishes to set up the first radio astronomy observatory near Kootwijk. De Voogt suggested that Oort might use these antennas for the 21-cm line search. The main problem was building a low-noise receiver able to detect the 21-cm line. The Dutch team detected the line 6 weeks after Harvard astronomers Ewen and Purcell; both results were published simultaneously. Ewen and Purcell did not follow up their detection, but for Oort and his team, it was the beginning of systematic studies.

The first goal was to measure radio emission from our own Galaxy (van de Hulst et al., 1954). The next goal was the Andromeda galaxy M31. van de Hulst et al. (1957) found that the neutral hydrogen emitting the 21-cm line extends much farther than the optical image. They were able to measure the rotation curve of M31 up to about 30 kpc from the centre, confirming the global value of $M/L \approx 20$ versus $M/L \approx 2$ in the central region.

About ten years later Morton Roberts (1966) made a new 21-cm hydrogen line survey of M31 using the National Radio Astronomy Observatory's 300-foot telescope. The flat rotation curve at large radii was confirmed with much higher accuracy. He constructed also a mass distribution model of M31.

Astronomers with access to large optical telescopes continued to collect dynamical data on galaxies. The most extensive series of optical rotation curves of galaxies was made by Margaret and Geoffrey Burbidge, starting from Burbidge & Burbidge (1959); Burbidge et al. (1959), and including normal and barred spirals as well as some ellipticals. For all galaxies the authors calculated mass distribution models; for spiral galaxies rotation velocities were approximated by a polynomial. They found that in most galaxies within visible images the mean $M/L \approx 3$.

Subsequently, Rubin & Ford (1970) and Roberts & Rots (1973) derived the rotation curve of M31 up to a distance ~ 30 kpc, using optical and radio data, respectively. The rotation speed rises slowly with increasing distance from the centre of the galaxy and remains almost constant over radial distances of 16–30 kpc.

The rotation data allowed us to determine the distribution of mass, while the photometric data determined the distribution of light. Comparing both distributions one can calculate the local value of the mass-to-luminosity ratio. On the periphery of M31 and other galaxies studied the local value of M/L, calculated from the rotation and photometric data, increases very rapidly outwards, if the mass distribution is calculated directly from the rotation velocity. On the periphery old metal-poor halo-type stellar populations dominate. These metal-poor populations have a low $M/L \approx 1$ (this value can be checked directly in globular clusters which contain similar old metal-poor stars as the halo). On the peripheral region the luminosity of a galaxy drops rather rapidly, thus the expected circular velocity should decrease according to the Keplerian law. In contrast, on the periphery the rotation speed of galaxies is almost constant, which leads to very high local values of $M/L > 200$ near the last points with a measured rotational velocity.

All available rotation data were summarised by Roberts (1975) in the IAU Symposium on Dynamics of Stellar Systems held in Besancon (France) in September 1974. Extended rotation curves were available for 14 galaxies; for some galaxies data were available up the galactocentric distance $\approx 40\ h^{-1}$ kpc. About half of galaxies had flat rotation curves; the rest had rotation velocities that decreased slightly with distance. In all galaxies the local mass-to-luminosity ratio on the periphery reached values over 100 in Solar units. To explain such high M/L values Roberts assumed that late-type dwarf stars dominate the peripheral regions.

In the mid-1970's Vera Rubin and her collaborators used new sensitive detectors to measure optically the rotation curves of galaxies at very large galactocentric distances. Their results suggested that practically all spiral galaxies have extended flat rotation curves (Rubin et al., 1978, 1980). The internal mass of galaxies rises with distance almost linearly, up to the last measured point.

At the same time measurements of spiral galaxies with the Westerbork Synthesis Radio Telescope were completed, and mass distribution models were built for 25 galaxies (Bosma, 1978). Observations confirmed the general trend that the mean rotation curves remain flat over the whole observed range of distances from the centre, up to ≈ 40 kpc for several galaxies. The internal mass within the radius R increases over the whole distance interval.

4.2.2 *Mass-to-luminosity ratios of galaxies*

Another very important measurement was made by Faber & Jackson (1976); Faber et al. (1977). Sandra Faber and her collaborators measured the central velocity dispersions for 25 elliptical galaxies and the rotation velocity of the Sombrero galaxy, a S0 galaxy with a massive bulge and a very weak population of young stars and gas clouds just outside the main body of the bulge. Their data yielded for the bulge of the Sombrero galaxy a mass-to-luminosity ratio $M/L = 3$, and a mean mass-to-luminosity ratio of about 7 for elliptical galaxies, close to the ratio for early type spiral galaxies.

Sandra Faber presented her results at the IAU General Assembly in 1976. I had the fortune to be present at her presentation. It was clear how important these results were. After her presentation I congratulated her for this work. Her results showed that our earlier models, based on velocity dispersions in central regions of galaxies, need correction, since our accepted dispersions were too high. Faber used better equipment; new measurements of velocity dispersions indicated that actual velocity dispersions in central regions of galaxies are about a factor of two lower than values accepted earlier. A review of masses and mass-to-luminosity ratios in galaxies is given by Faber & Gallagher (1979).

New observational results by Vera Rubin and the Westerbork group confirmed the presence of dark halos of galaxies with high confidence. Observations by Sandra Faber suggested that mass-to-luminosity rations of optically visible populations of galaxies are rather low, thus the discrepancy between rotation and photometric data is serious. Now all new results were taken seriously.

4.2.3 X-ray data

Hot intra-cluster gas emitting X-rays was detected in almost all nearby clusters and in many groups of galaxies by the UHURU and Einstein X-ray orbiting observatories. Observations confirmed that the hot gas is in hydrodynamical equilibrium, i.e. gas particles move in the general gravitation field of the cluster with velocities which correspond to the mass of the cluster. The distribution of the mass in clusters can be determined if the density and temperature of the intra-cluster gas are known. This method of determining the mass has a number of advantages over the use of the virial theorem. First, the gas is a collisional fluid, and particle velocities are isotropically distributed, which is not true for galaxies used as test particles to find the cluster mass (uncertainties in the velocity anisotropy of galaxies affect mass determinations). Second, the hydrostatic method gives the mass as a function of radius, rather than the total mass alone as given by the virial method.

Using UHURU satellite data the method was applied to estimate the masses of several clusters of galaxies (Forman et al., 1972; Gursky et al., 1972; Kellogg et al., 1973). The results confirmed previous estimates of masses made with the virial method using galaxies as test particles. The mass of the hot gas itself is only about 0.1 of the total mass (Holberg et al., 1973; Lea et al., 1973). We used this argument to indicate that the amount of hot gas in clusters is insufficient to explain the mass paradox in clusters (Einasto et al., 1974b). The luminous mass in member galaxies is only a fraction of the mass of the cluster X-ray emitting gas.

In 1979 a conference was held at Princeton University to discuss scientific programs for the Hubble Space Telescope. I had the luck to be a member of the small delegation from the Soviet Union. After the conference we had the opportunity to visit Harvard Observatory. There I had many discussions with John Huchra on the Harvard redshift survey and on the structure of the cosmic web. Also we met

Riccardo Giacconi, the head of the Einstein X-ray observatory program. At this time there were no doubts that X-ray data give strong support to the high masses of clusters of galaxies.

More recently clusters of galaxies have been observed in X-rays using the ROSAT satellite, and the XMM-Newton and Chandra observatories. The ROSAT satellite was used to compile an all-sky catalog of X-ray clusters and galaxies. More than 1000 clusters up to a redshift ~ 0.5 were cataloged. Dark matter profiles have been determined in a number of cases.

4.2.4 *Gravitational lensing*

Clusters, galaxies and even stars are so massive that their gravity bends and focuses the light from distant galaxies, quasars and stars that lie far behind. There are three classes of gravitational lensing: (1) strong lensing, where there are easily visible distortions such as the formation of Einstein rings, arcs, and multiple images; (2) weak lensing, where the distortions of background objects are much smaller and can only be detected by analysing the shape distortions of a large number of objects; and (3) microlensing, where no shape distortion can be seen, but the amount of light received from a background object changes in time. The background source and the lens may be stars in the Milky Way or in nearby galaxies (M31, Magellanic Clouds).

The strong lensing effect is observed in rich clusters, and allows us to determine the distribution of the gravitating mass in clusters. Massive galaxies can distort images of distant single objects, such as quasars: as a result we observe multiple images of the same quasar. The masses of clusters of galaxies determined using this method confirm the results obtained by the virial theorem and the X-ray data.

Weak lensing allows us to determine the distribution of dark matter in clusters as well as in superclusters.

A fraction of the invisible baryonic matter can lie in small compact objects — brown dwarf stars or Jupiter-like objects. To find the fraction of these objects in the cosmic balance of matter, special studies have been initiated, based on the microlensing effect. Microlensing effects were used to find Massive Compact Halo Objects (MACHOs). MACHOs are small objects such as planets, dead stars (white dwarfs) or brown dwarfs, which emit so little radiation that they are invisible most of the time. A MACHO may be detected when it passes in front of a star and the MACHO's gravity bends the light, causing the star to appear brighter. Several groups have used this method to search for baryonic dark matter. The total mass of these objects forms only a small fraction of the mass of stellar populations observed in galaxies. In other words, MACHOs do not solve the dark matter problem in galaxies.

4.3 Dark matter in galaxies

4.3.1 *The density distribution of dark matter*

Flat rotation curves of galaxies suggest that the radial density distribution in galaxies, including stellar populations, interstellar gas and dark matter, is approximately isothermal: $\rho(r) \sim r^{-2}$. As dark matter is the dominating population, its density profile should also be close to an isothermal sphere. Thus in our models of galaxies we approximated the dark matter population density with a truncated isothermal profile to avoid infinite total mass and infinite density at the center.

In the early 1990's, the results of high-resolution numerical N-body simulations of dark matter halos based on the collisionless CDM model became available. The simulations did not show the core-like smooth behaviour in the inner halos, but were better described by a power-law density distribution, the so-called cusp. Navarro et al. (1997) investigated systematically simulated DM halos for many different sets of cosmological parameters. They found that the whole mass density distribution could be well described by an "universal density profile". This profile, known as the "NFW profile", cannot be applied to the very centre of the halo, since in this case the density would be infinite. Near the centre of the halo the density rises sharply, forming a "cusp".

Bullock et al. (2001) analysed the evolution of profiles of dark halos. They defined the halo concentration parameter $c_{vir} \equiv R_{vir}/r_s$, where R_{vir} is the virial radius of the halo, and r_s is the halo inner radius, where the logarithmic slope of the density profile is -2. The virial radius of the halo of mass M_{vir} is defined as the radius within which the mean density is Δ_{vir} times the mean density. The range of the last quantity is from 178 to 337. The authors fit halo profiles with the NFW profile, and find that the median concentration parameter depends on the redshift: $c_{vir} \propto (1+z)^{-1}$, i.e. with decreasing redshift the concentration parameter increases. Further the authors find that subhalos and halos in dense environments tend to be more concentrated than isolated halos, and that low-mass halos have a larger concentration parameter.

Navarro et al. (2004) proposed another model that fits the density profiles of halos in ΛCDM simulations even better than the NFW model. It was realized by Merritt et al. (2006) that the model advocated by Navarro et al. (2004) had previously been introduced by Einasto (1965, 1968c, 1969a).

The paper by Einasto (1965) was not available in the SAO/NASA Astrophysics Data System (ADS), thus initially this model was often referenced as the Einasto profile without any citation. To help the astronomical community read the original paper we scanned it, and made great efforts to clean the scan — the original was rather unsharp and gray. Now it is available in ADS and has collected 115 citations as of the middle of March 2013. For comparison, the Navarro et al. (1997) paper has over 4000 citations now!!!

Recent studies have shown that the Einasto profile represents the spatial density profiles of dark matter halos rather well (Merritt et al., 2005, 2006; Graham et al., 2006; Ludlow et al., 2010). Ludlow et al. (2011) used Millennium-II simulations as part of the Virgo Consortium of high resolution N-body simulations to investigate the density and pseudo-phase-space density profiles of CDM halos. The Millennium-II simulations is a 10^{10}-particle cosmological simulation of the evolution of dark matter in a 100 h^{-1} Mpc box. The run adopted a standard ΛCDM cosmogony with the same parameters as the Millennium simulation presented by Springel et al. (2005). The authors find that the pseudo-phase-space density profiles are best reproduced by the Einasto profile. The origin of this behaviour is unclear, but its similarity for all halos may reflect a fundamental structural property of DM halos.

A high-resolution 21-cm observation survey of 34 nearby galaxies by Chemin et al. (2011) shows that this profile represents very accurately the density profiles of visible populations. These observations were carried out using The HI Nearby Galaxy Survey (THINGS), see de Blok et al. (2008); de Blok (2010); Oh et al. (2008, 2011); Trachternach et al. (2008) and Walter et al. (2008) for details. Retana-Montenegro et al. (2012) and Salvador-Solé et al. (2012) investigated properties of the Einasto family of density profiles.

In our models of galaxies we have always used the profile (3.1) to represent the density of *visible galactic populations*. Recent studies demonstrate that the dark matter has a density distribution, which is very similar to the distribution of stars in galaxies.

4.3.2 *Distribution of luminous and dark matter in galaxies*

In the last twenty years the study of the distribution of luminous and dark matter has made great progress. Here I shall describe only the most important results of these studies.

The basic problem in the comparative study of the distribution of luminous and dark matter is the decomposition of the total matter distribution into the luminous and dark populations. This problem has been adressed by many authors using various methods. One of the key issues is the amount of dark matter in dwarf galaxies. This problem has been studied, among others, by John Kormendy. He writes in a review paper (Kormendy & Freeman, 2004) that probably these dwarf galaxies formed in an early period of galaxy formation: their central densities

$\varrho_0 \sim (1 + z_{coll})^3$, where z_{coll} is the collapse redshift. The smallest dwarfs formed at least $\Delta z_{coll} \simeq 7$ earlier than the biggest spiral galaxies. The high DM densities of dSphs implies that they are real galaxies formed from primeval density fluctuations. Their central densities are about 100 times larger than DM central densities of giant galaxies. The paucity of stars in these galaxies can probably be explained by supernova winds which blew out most of the remaining gas. High $M/L \approx 100$ ratios of dSph galaxies were confirmed by velocity dispersion measurements of stars. Kormendy & Freeman (2004) found a number of scaling laws between the central density and other quantities (velocity dispersion, absolute magnitude, core radius).

Humphrey et al. (2006); Humphrey & Buote (2010) used Chandra X-ray observatory data to investigate the mass profiles of samples of galaxies, groups and clusters, spanning about 2 orders of magnitude in virial mass. They find that the *total* as well as *DM* mass density distributions can be well represented by a NFW/Einasto profile. This coincidence is remarkable, since the fraction of baryonic matter in the total mass distribution in clusters varies with radius considerably. This "galaxy–halo conspiracy" is similar to that which establishes flat rotation curves in galaxies — the "bulge–halo conspiracy". These coincidences suggest the presence of some sort of interaction between the dominating stellar population (bulge) and the dark matter halo, both on galactic and cluster scales. We note that an analogous relation exists between the mass of the central black hole and the velocity dispersion of the bulge of elliptical galaxies, see Kormendy & Bender (2011); Kormendy et al. (2011) for a recent analysis of this problem.

The "core–cusp problem" has been the subject of many recent studies, based both on observational data as well as on results of very high-resolution numerical simulations. To find the DM-halo density profile de Blok (2010) used a collection of HI rotation curves of dwarf galaxies, which are dominated by dark matter. To get a better resolution near the centre H_α long-slit rotation curves were analysed. These rotation curves indicate the presence of constant-density or mildly cuspy dark matter cores.

Wolf et al. (2010) investigated the mass distribution for dispersion-supported (elliptical) galaxies. For many local spheroidal galaxies redshifts of a large number of individual stars have been determined. This allows one to use conventional Jeans equations to derive the masses of these galaxies. Luminosities are also known, which allows one to find M/L ratios for a wide range of galaxies of different magnitude. The authors derive the dynamical I-band half-light mass-to-luminosity ratio versus the half-light mass in mass interval 10^4 (globular clusters) to 10^{15} (giant elliptical galaxies) Solar masses. Globular clusters are located far from the general trend — evidently they do not contain dark matter. For dark matter dominated systems the M/L ratio has a minimum of about 3 for galaxies of half-light mass 10^{10} Solar masses, for the faintest dwarf and most massive giant galaxies the ratio increases up to a value of about 1000.

Dhar & Williams (2011) analysed the density distribution of a large sample of high-resolution images of elliptical galaxies in the Virgo cluster, using the Hubble Space Telescope and ground-based data which span 10^6 in surface brightness and up to 10^5 in radius down to the resolution limit of the HST. The authors used 2D projections of the spatial (3D) Einasto density profile. All observed galaxies can be fit with 2 or 3 populations with different values of the normalizing and shape parameters of the Einasto model.

In the framework of the Phoenix Project Gao et al. (2012) performed detailed numerical simulations of rich clusters of galaxies. The Phoenix Project follows the design of the Aquarius Project and consists of zoomed-in resimulations of individual galaxy clusters drawn from a cosmologically representative volume. Each cluster is simulated with at least two different numerical resolutions. The highest resolution corresponds to over one billion particles within the cluster virial radius. The Aquarius and Phoenix halos differ by roughly three orders of magnitude in virial mass. The most notable difference is that cluster halos have been assembled more recently and are thus significantly less relaxed than galaxy halos, which leads to decreased regularity. The multimodality of rich clusters of galaxies is well known observationally, see a recent study by Einasto et al. (2012). The density profile of rich clusters is best reproduced by the Einasto profile of various values of the shape parameter (Einasto index).

4.3.3 *Universal rotation curve of galaxies*

Rubin et al. (1985) compared rotation velocities of spiral galaxies of various morphological type and luminosity and found that the shape of the rotation curve depends strongly on the luminosity of the galaxy and the bulge-to-disk ratio. Galaxies of high luminosity have high rotational velocity and high central gradient of the velocity, low-luminosity galaxies have low rotational velocity and low central gradient of velocity. This correlation is almost independent of the morphological type of the galaxy. The infrared absolute magnitude of galaxies, M_H, is strongly correlated with the mass within the isophotal radius, R_{25}, deduced from the 25 mag arcsec^{-2} contour. The infrared mass-to-luminosity ratio, $\mathcal{M}(R_{25})/L_H = 2.1$, is independent of the morphology of galaxies.

Persic & Salucci (1991); Persic et al. (1996) extended the Rubin et al. (1985) study to find the correlation between the shape of rotation curves and luminosities of galaxies. Using a homogeneous sample of about 1100 optical and radio rotation curves and relative surface photometry Persic et al. (1996) investigated the distribution of mass in spiral galaxies over a range of 6 mag out to 1.5–2 optical radii. The authors find that there exists a Universal Rotation Curve (URC) of spiral galaxies. This curve implies a number of scaling properties between the dark (DM) and the luminous (LM) galactic structure parameters: the DM/LM mass ratio scales inversely with the luminosity; the halo core radius is comparable to the galaxy optical radius, but shrinks for low luminosities; the total halo mass scales as $L^{0.5}$. Salucci et al. (2007) continued the URC of spiral galaxies and the corresponding mass distribution out to their virial radius. In low-luminosity galaxies the dark matter dominates at all distances from the galactic center; in high-luminosity galaxies dark matter domination begins in outer regions of galaxies.

Donato et al. (2009) continued the study of the distribution of mass in spiral galaxies. The authors coadded rotation curves of ~1000 spiral galaxies, and performed mass models of individual dwarf irregular and spiral galaxies of late and early types. Their basic finding is that the central projected mass density is constant over a wide range of galaxy masses. They find $\log \varrho_0 r_0 = 2.15 \pm 0.2$, where ϱ_0 is the central density, and r_0 is the core radius of the adopted pseudo-isothermal cored dark matter density profile. The projected mass density is given in units of Solar masses per square parsec. Star clusters of the same luminosity as dwarf galaxies lie far from this relationship, showing a different mechanism of origin, as shown by Gilmore et al. (2007).

Salucci et al. (2012) continued the study of this phenomenon. Observational velocity dispersion profiles are now available for eight Milky Way dwarf spheroidal satellites, mean velocity dispersions range from 5 to 10 km/s. This allows one to calculate mass distribution models. The authors demonstrated that the relationship between the central density and core radius, found by Donato et al. (2009), is valid for galaxies of absolute B magnitude interval from -7 to -22.5, for galaxies of various morphological type from dwarf irregulars and spirals to ellipticals. The authors conclude that *"This result is intriguing, and could point to a common physical process responsible for the formation of cores in galactic halos of all sizes, or to a strong coupling between the DM and luminous matter."*

4.3.4 *The formation of galaxies*

The galactic models discussed so far are static, i.e. the aim of the modelling was to describe the present structure of galaxies. Early studies of the evolution of galaxies considered only the role of gravity in the evolution. Using this approach Eggen et al. (1962) showed that our Galaxy was contracting in its early stage of evolution. Toomre & Toomre (1972) discussed the role of merging in the evolution of galaxies, and suggested that elliptical galaxies are probably remnants of merged spiral galaxies.

Models by Tinsley (1968); Tinsley & Spinrad (1971) considered the physical evolution of galaxies — the change with time of their luminosity, colour, and mass-to-luminosity ratio. My model of the evolution of galaxies (Einasto, 1972b) had a similar aim. In these models the formation and evolution of stars was taken into account to find the evolution of stellar populations.

But none of the earlier models analysed the problem of how galaxies were formed. The problem of the formation of galaxies was considered as a part of the more general theory of structure formation. Peebles (1971a) scenario of hierarchical clustering considers the formation of galaxies as part of the clustering process starting from globular-cluster sized objects. Zeldovich (1970) discussed the evolution of density perturbations assuming an adiabatic nature of the process; galaxy formation was not discussed specifically. In both cases only the role of gravity was taken into account.

Numerical simulations performed in the mid 1970's using the Zeldovich (1970) idea of pancaking showed the formation of a cellular network of high-density regions (Shandarin 1975, private communication). A similar picture was found in the distribution of galaxies (Jõeveer et al., 1977; Jõeveer & Einasto, 1978; Jõeveer et al., 1978). In other words — *the pancaking scenario is actually a scenario of the formation of the structure of the Universe, not of the formation of galaxies*.

Galaxy formation in the framework of the pancake scenario was analysed by Doroshkevich et al. (1978). The authors calculated the temperature, pressure and density in a 'pancake', taking into account also radiative cooling of the gas. Inside the 'pancake' shock fronts and cooling fronts form. The main feature of the formation of galaxies according to this scenario is the presence of three different processes inside the 'pancake'. The essential process is the cooling and fragmentation due to gravitational and thermal instabilities of the thin layer of cooling gas, which leads to the formation of primeval gas clouds. Other processes are the clustering of these primeval clouds to form presently observed galaxies, and the clustering of galaxies to form clusters of galaxies. The main conclusion of the paper is that a protogalaxy has never been an integral gaseous cloud; the initial state of a galaxy was a complex of gas clouds formed within a 'pancake'. The role of dark matter was not studied.

When I understood that dark matter surrounding galaxies is a new population with properties very different from that of all known stellar populations, I tried to clarify its possible role in galaxy formation (Einasto, 1972a, 1974a). But at this time I did not realise that one of the main properties of the dark population, its spatial segregation from known stellar populations, has a deep physical meaning — it must be a non-dissipative population.

The dissipationless character of dark matter was first clearly stated by White & Rees (1978). They investigated the role of dark matter in galaxy formation. The presence of dark matter hints that galaxy formation must be a two-stage process, where galaxies form inside dark matter halos by cooling and fragmentation of gas. Dark matter forms halos of radius and mass about ten times larger than galaxies, thus it must be collisionless, whatever its nature is. The authors suggest that *the segregation of luminous and non-luminous material is incompatible with any theory which tries to build up galaxies and clusters from smaller units in an entirely dissipationless way, since one expects efficient mixing to occur during this process.* The authors write that dark matter could be low-mass stars, remnants of high-mass stars, neutrinos, or black holes which formed before recombination.

The formation of galaxies was discussed during the Study Week on Cosmology and Fundamental Physics in the Vatican in 1981. Faber (1982a) used the White–Rees theory of galaxy formation via core condensation to develop a model for the structure of disk galaxies. She showed that disk parameters such as rotation velocity, radius, and surface brightness scale directly in proportion to the corresponding halo parameters. Scaling relations derived from the model look like the observed Fisher–Tully and radius–luminosity laws. In another talk Faber (1982b) argued that the ellipticals and spheroidal bulges of spiral galaxies are highly condensed systems in which ordinary luminous matter falls deep into the central core of the surrounding non-luminous halo. This model is consistent with the observed scaling laws for elliptical galaxies and the velocity dispersions of spheroidal bulges in spirals. The Hubble sequence can be considered as a sequence of increasing dissipation and central concentration of the luminous matter relative to the surrounding dissipationless halo.

The increasing power of computers allows us to investigate the possible assembly of various populations of galaxies. Sales et al. (2012) studied the origin of disks and spheroids in simulated galaxies similar to the Milky Way, in a series of cosmological gas-dynamical simulations, the Galaxies Intergalactic Medium Calculation (GIMIC), based on the Millennium simulation by Springel et al. (2005). The authors find that *the final morphology of a galaxy results from the combined effects of spin alignment and of hot/cold gas accretion. Disk dominated objects are made of stars formed predominantly* in situ, *and avoid systems where most baryons were accreted cold, or those where spin misalignments are extreme.*

4.3.5 *Modern models of galaxies*

In our first models of the Galaxy, M31 and other nearby galaxies with dark matter coronae only rather rudimentary data on the mass and radius of the dark matter component were available, thus we published our models only in conference proceedings and in small journals in the hope that soon better data will be available (IAU Symposium on the Spiral Structure of Galaxies (Einasto & Rümmel, 1970a), IAU Symposium on External Galaxies and Quasi-Stellar Objects (Einasto & Einasto, 1972a; Einasto & Rümmel, 1972), First European Astronomy Meeting (Einasto, 1974a), Third European Astronomy Meeting (Einasto et al., 1976c,b), Astronomicheskij Circular (Einasto et al., 1978, 1979b)). In these models the M/L ratio of stellar populations improved as new data came in, also parameters of the dark population were improved. In these models we used for all stellar populations the generalised exponential mass distribution, and our standard graphs for various values of the concentration parameter N tabulated by Einasto & Einasto (1972b).

Our young collaborator Urmas Haud developed a new program to automatically adjust the parameters of our multi-component models to get the best fit to all data. In these models all the main galactic populations were present — the core, the bulge, the disk, the stellar halo, the flat population of young stars and gas, and the dark corona. Thus parameters for all these populations must be found. However, the development of this program advanced rather slowly. Only in the late 1980's were we able to publish in the journal "Astronomy & Astrophysics" our first multi-component model of our Galaxy, calculated with the new method (Einasto & Haud, 1989; Haud & Einasto, 1989), later models of the Andromeda galaxy (Tenjes et al., 1994), and some other nearby galaxies (Tenjes et al., 1991, 1998).

In hindsight I think that this long delay with publication in a major journal was a mistake. Actually our models of the Galaxy and M31 (Einasto et al., 1978, 1979b) contained all the essential information, including better data for dark coronas. The publication of these models in the late 1970's in a respectable journal could have better influenced the development of models of galaxies; by the late 1980's and early 1990's it was already too late.

Some authors have constructed two-component models with the bulge and the disk in an attempt to avoid the use of the massive corona (the mass of the ordinary stellar halo consisting of old metal-poor stars is rather small and can be excluded from the mass model). One of these attempts was made by Rohlfs & Kreitschmann (1980). The authors ignored the last measured point in the rotation curve of the galaxy M81, and were able to find a model with only two populations, the bulge and the disk. The disk of the galaxy has a hole in the centre; a similar disk model was presented also by Einasto et al. (1980c). However, using more complete data Tenjes et al. (1998) calculated a multi-component model using the rotation curve of M81 which flattens at large galactocentric distances. Thus a massive corona must be present in this galaxy. Rohlfs & Kreitschmann (1988) constructed a multi-component model of our Galaxy, where a massive corona was added. In this model the bulge has two components, a visible and a dark one. The authors argue that the dark components both in the bulge and the corona are baryonic.

Another attempt to try to avoid the presence of a massive dark halo population was made by Kalnajs (1987). He investigated by numerical experiments the role of the massive halo in stabilising the disk, as suggested by Ostriker & Peebles (1973). He argued that, compared to a bulge, a halo is not very efficient in stabilising the disk. His conclusion was that the stability argument for the presence of a massive halo is not very compelling. The stability of a flat galactic disk was studied earlier by Toomre (1964), who argues that a bulge-type population is needed to stabilise the flat population.

All models where attempts were made to avoid the presence of a massive dark halo have one common aspect — they have as massive a disk as possible. Thus these models are often called "maximum disk" models. In our models we tried to derive the masses of populations as accurately as possible, using all available data and not maximising only one parameter. Instead, we tried to minimise the overall deviations of model parameters from directly observed parameters.

In modern models galaxies are not considered as isolated objects. Actually new data suggest that all giant galaxies have been formed by merging dwarf galaxies. The ongoing merging of giant galaxies can be observed in interacting galaxies, as shown by Toomre & Toomre (1972). In high-resolution simulations of the evolution of the cosmic web the merging of smaller galaxies to form more massive galaxies is well seen.

Recently Tamm et al. (2007) and Tempel et al. (2007) calculated a new model of M31, taking into account the absorption inside the galaxy. The model includes four visible populations: a bulge, a disc, an inner halo and an extended diffuse halo, and a dark matter halo. The authors find that about 40% of the total luminosity is obscured due to the dust. Using chemical evolution models, authors calculated the mass-to-luminosity ratios of the populations. The total intrinsic mass-to-luminosity ratio of the visible matter is $M/L_B = 3.1 - 5.8$ $M_\odot/L_\odot$, and the total mass of visible matter $M_{vis} = (10 - 19) \times 10^{10}$ $M_\odot$. Further the authors use HI and stellar rotation data and stellar velocity dispersions to find a dynamical model, which allows them to calculate more accurately the DM halo density. The authors find that a DM halo having NFW or Einasto profiles gives the best fit with observations.

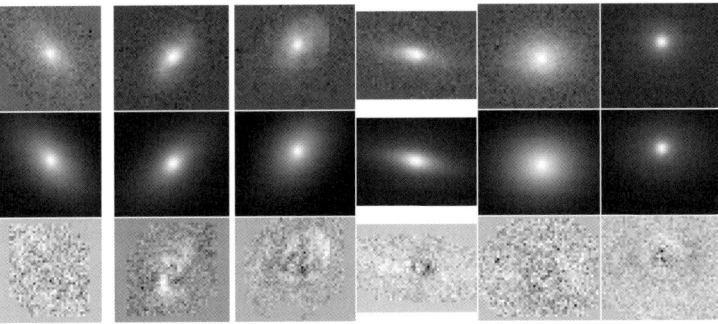

Fig. 4.8 Examples of modelling real SDSS galaxies. Upper row shows the original observations, middle row shows the point-spread-function (PSF)-convolved model galaxies, and lower row shows the residual images (Tempel et al., 2012d).

For the Einasto DM profile, the total mass of M31 is $1.28 \times 10^{12}\ M_\odot$, and the ratio of the DM mass to the visible mass is 10.8.

Presently photometric and redshift data, as well as direct images in $ugriz$ filters are available for almost one million galaxies of the SDSS main galaxy survey. So far this large dataset has been used to construct simple 2-dimensional models of these galaxies. Tempel et al. (2012d) used this dataset to construct 3-dimensional models of SDSS galaxies. They used in models two main populations, the bulge and the disk, applying the Einasto profile with variable shape parameter N. The authors first tested the modelling technique using simulated galaxies. This test shows that the restored integral luminosities and colour indices remain within 0.05 mag and the errors of the luminosities of individual components remain within 0.2 mag. The accuracy of the restored bulge-to-disc ratios is within 40% in most cases. Examples of modelling real SDSS galaxies are shown in Fig. 4.8. As we can see the visual apperence of modelled galaxies is rather close to pictures of actual galaxies, except for a relatively small noise in the residual images. The general balance between bulges and discs is not shifted systematically. Inclination angle estimates are better for disc-dominated galaxies, with the errors remaining below $5°$ for galaxies. In total, 3-D models were found for more than half a million SDSS main sample galaxies.

Tempel et al. (2013) used this sample of 3-D models of SDSS galaxies to investigate the spin alignment of spiral and elliptical/S0 galaxies in filaments. The authors found evidence that the spin axes of bright spiral galaxies have a weak tendency to be aligned parallel to filaments. For elliptical/S0 galaxies, the authors showed that their spin axes are preferentially aligned perpendicular to the host filaments.

4.4 Tartu Observatory in the 1970's

4.4.1 *Computer revolution*

In my life I have experienced the whole computer revolution. When I started my studies in Tartu University, the main computer was a sliding rule. It remained so for almost 25 years. However, sometimes a higher accuracy was needed. One such example was my work on data reduction of the Solar eclipse observations made by Prof. Kipper and his assistants in summer 1945. The reduction of these data was announced as a competition project for students. I observed the same eclipse with my self-made telescope, so for me it was interesting to compare my own modest observations with those made by professional astronomers.

One of the tasks was to calculate the phases of the eclipse for the Tartu Observatory location, since the main goal of observations was the determination of the change of the Solar surface brightness as a function of the distance from the center of the Solar disk towards the limb. For these calculations phases of the eclipse were needed. Input data were available in the Astronomical Almanac, and calculations were needed to find the phases for a particular location. For this task relatively simple formulae of spherical trigonometry were needed. I found all input data, and started computations. Here the accuracy of the slide rule was not sufficient, so I used a Rheinmetall mechanical calculator available in the physics department of the University. The calculations were relatively simple but demanded a lot of time. When calculations were almost completed, I discovered that something was wrong — phases were completely different from expectations. Then I started to look for what is wrong, and discovered that I had taken in my computations the longitude of Tartu Observatory wrongly: I used a positive value for Tartu, forgetting that longitudes to the east are negative. In other words, I found phases for a place in the middle of the Atlantic Ocean!

The deadline to present the paper was already close, thus there was no time to repeat all calculations. But I found a way to use the first part of calculations, thus only the second part must be repeated. The final polishing of the paper was made in a great hurry. For last three days I worked in the Observatory day and night, sleeping on the floor of Rootsmäe's office a few hours per night. I presented the work in time, but was thereafter so tired that I was not able to work or study for a month or so.

However, the main lesson from this study was different. The maximal phase of the eclipse in Tartu was about 0.9, thus the critical region to find the darkening of the Solar limb was not reached. In other words, these observations did not have any scientific meaning, and gave no new information to the understanding of Solar physics. This was a critical lesson. In science only such studies which give answer to an open question are needed.

In 1954 the next Solar eclipse of the same cyclus came about, and the whole staff of the Observatory participated in the observations. This time the zone of the full eclipse crossed Lithuania, and we made an expedition to observe the event. A special camera was used to photograph the Solar corona during the eclipse. However, the sky was partly cloudy, and the corona was not visible. After the expedition Prof. Kipper asked me to do the data analysis, since I already had experience from the earlier eclipse. I refused and declared that our observations have no scientific value, thus it makes no sense to do the data analysis. Kipper took umbrage, and this incident strained our relations for a long time.

In the 1960's Tartu University formed a laboratory for computing. The computer was probably a copy of one of the first American electronic computers; it worked with vacuum tubes. The central computing unit (CPU) was on a rotating magnetic drum which made 100 cycles per second, this determined the speed of the computer. We had access to this computer and did some calculations there.

The first electronic computer the Observatory had was a rather curious one — the programming was done by putting wires into proper places as in old telephone central stations. My wife Liia was one of the programmers, and it took time to learn how to use such methods for programming. But soon we got a better computer where input data were given using punched strips, and later punched cards.

However, our own computers were rather slow, and thus we often used computers installed in Tallinn at the Institute of Cybernetics which had a large computing center. To get a computer of such power was in the 1970's a great problem. Permission from high bureaucrats in Moscow was needed. They discussed whether to give one computer for all three Baltic countries for joint use, or separately for Estonia. The management of the Institute of Cybernetics was able to convince the bureaucrats in Moscow that there were lots of people in Estonia in need of such a large computer, and we could make full use of it. The computer was really big — one full hall was needed. A modest remark: the computing power of this computer was a tiny fraction of the smallest modern notebooks or even mobile phones.

The programming with this computer used punched strips. The actual work was done so. We had several programs running simultaneously. Once a week our Observatory bus made a trip to Tallinn to deliver our strips, and to bring back computer output rolls to fix errors in programs and as well as results sent in the previous week. The polishing of one program took normally several months, so we had always several programs running. All this work was so inconvenient for astronomers that we used a special laboratory for programming. Programmers of the laboratory made the practical programming. My wife Liia also worked in the programming laboratory, and all my programs in the late 1960's and early 1970's were made together with her.

If the results of some program were urgently needed, then we drove together to Tallinn. A course-mate of Liia, Aino Männil, was the deputy head of the computing centre, so we could work during the night when the computer was less used to polish our programs. Aino slept nearby in a camp bed to be ready to help us in case we had problems. So during a few nights we made more progress than in several months using the traditional way of transporting programs back-and-forth to Tallinn. In just such a manner all my results of modelling galaxies and galaxy evolution were obtained.

In the late 1970's our Observatory also got a big computer, not as powerful as the computer in Tallinn, but for those days good enough. The programming was, however, very similar to the earlier time, using punched strips or cards. That was easier — there was no need to send programs to Tallinn, so the process was much more rapid. In the early 1970's our group got several new astronomers, graduates from Tartu University. They learned the programming style in the Observatory quickly, and helped me also when longer and more complicated programming was needed.

In 1976 I attended the IAU General Assembly in France and was able to buy my first pocket computer. Next year I was on a short visit in Germany to attend the IAU Symposium on galaxies. This time it was possible to buy a Texas Instruments *programmable* pocket computer. It had only 100 memory locations for programs and input data, but I was able to write a program to calculate galactic models using the generalised exponential density profile, now called the "Einasto profile". A year later I had a short visit to the USA, for a conference to discuss programs for the Hubble Space Telescope. In a small shop I saw the first personal computer. This was a kit comprising of parts to be assembled by hobbyists, but it was evident that the computing is going towards personal computers.

In 1980 there was an exhibition in Moscow, where for the first time real personal computers were on display. At the time it was extremely difficult to get foreign currency to buy equipment from Western countries. I cannot remember how we managed to buy from this exhibition two computers, Tandy TRS-80 and Apple II. The Tandy computer was used by other people, but I got the Apple II for my personal use. From this moment on I have done almost all my computing with my own computer. We could buy an external floppy disk drive for 8-inch floppies. One floppy disk contained the Fortran compiler, so we could program not only in BASIC but in Fortran too, which was for scientific computing much better. I used my first Apple II computer for many years, and most of the computing needed for my papers in the early 1980's was done with this device.

The same autumn I visited the Institute of Astronomy of Cambridge University, and was rather surprised that no personal computers were used there. Computing was done with a central DEC VAX computer; terminals were in all offices. So I had to learn computing on this machine too. But computing with a personal computer was much more convenient.

I bought my first private personal computer when visiting Germany in 1981, a Commodore VIC-20. But this was not as convenient as the Apple II, so I gave it to my younger collaborators. In 1982 I had a chance to work also with the very first IBM Personal Computer (PC). George Abell had just bought it, and during my visit to Los Angeles I tried it. But at home I continued to work with my Apple.

The personal computing industry developed very rapidly, and in the middle of the 1980's we had to think which computer to buy next. The Apple Lisa was expensive, moreover too few software were available for Apple computers, so we decided to switch to the IBM PC line. During the trip to Nordita in 1987 I was able to spare enough money to buy a PC. The development was so rapid that in my next few visits to the Institute of Astronomy I bought a new PC each time. In 1990 I was visiting ESO, and in the computer shop Vobis I saw a not too expensive notebook. The development of notebooks was also very rapid, thus from this year on in almost every year I sold the old one and bought a new, better notebook model.

In the early 1990's our team got a grant from the Soros Foundation. We used the grant to buy personal computers, so soon all our team members had their own PCs. Also it was possible to buy for the Observatory better computers; in the mid 1980's we got our first UNIX computer. Since then we used for scientific computations first UNIX, then later Linux operating systems. Linux was available also for personal computers, thus most of us switched to Linux. However, there are a lot of programs for non-scientific use, and most of these programs were available only for the Windows operating system. So most of our Linux users had two operatings systems installed, Linux for science, and some version of Windows for everyday life.

About the mid 2000's we discovered again Apple computers. The operating system of all Mac computers is UNIX based, thus all the advantages of Linux are available. But Mac has also programs for almost all everyday applications, many of them the same as for Windows. The user interface of the Mac operating system is much more user friendly than that for Windows. So many of our team members are now using Mac computers. I too have used for about 6 years only Macs — MacBook Pro and MacBook Air for computations. My iPhone is not only for communications, but it contains almost the whole of my music collection, and is a good camera always with me.

4.4.2 Life in the Observatory

In the early seventies the Estonian Academy of Sciences planned to divide the Institute of Physics and Astronomy into two institutes. According to the plan the present Institute was renamed as the Institute of Astrophysics and Atmospheric Physics, and a new Institute of Physics was formed. But here there was a problem. According to rules accepted in the Soviet Academy of Sciences, the number of Doctors of Sciences was of particular importance. In the Soviet system a Doctor of Sciences degree is approximately similar to Doctor Habilitatus in Germany. To get a Doctor of Sciences degree usually about 20 years of hard work was needed, as well as a large number of publications. The number of doctors determined the rank of the institute and thus the size of salaries, and the financing of the institute in general. Since Aksel Kipper was the only Doctor of Sciences among our astronomers, doctoral defences became vital.

We all expected that Grigori Kuzmin should defend the Doctor of Sciences thesis, but he was too reluctant to do this. How to encourage Kuzmin to compile his doctoral thesis? Kuzmin usually achieved his results through a rather brief, yet very intense period of thinking and analysis, but compiling the results into an article took him a lot of time. Several of his results stayed in the drawer because he could not pull himself together to write the article. It was likewise with his doctoral thesis — he could not mobilise himself to write it, arguing that he was more interested in mulling over obscure problems rather than drawing up long volumes that contain nothing new. So what about the thesis?

I discussed the issue with my younger colleagues and we found that his published works were so good that they could be used as chapters of a solid thesis. Having decided this, we announced to Kuzmin that we were going to start putting his papers together to form a Doctor of Sciences thesis. Due to the requirements of that time, this meant re-typing of all papers. After some resisting, Kuzmin agreed. But as he reread the old articles, he come up with new ideas to be used as supplements to the chapters. How this transpired: Kuzmin was up well into the night, writing his additions, which were on the table by the morning for us to somehow make out the garbled handwriting and type them in. Around noon the *Maestro* showed up again and examined the text, adding copious new improvements that we typewrote once more. The process was quite effective and after half a year, a unique piece of work was completed. Most of the additions have unfortunately still not been published, but the defence of the thesis was a great success, and once again, Kuzmin's opponents had to acknowledge the exceptional quality of his work.

Almost all Kuzmin's papers were published in Russian, mostly in Tartu Observatory Publications, some in Soviet journals. So his results were not known to the English speaking astronomical community. About ten years ago one of our young collaborators, Peeter Tenjes, translated Kuzmin's most important papers into English, including his additions to the doctoral thesis. Recently we had a small conference to celebrate the 200th anniversary of the Old Tartu University Observatory. One of the main speakers was Tim de Zeeuw, who gave a review of Kuzmin's work. We discussed the publication of English versions of Kuzmin's papers, either as a special issue of the journal "Baltic Astronomy", or just on arXiv.

After the successful defence of the Kuzmin's doctoral thesis, I decided to use the same method with my own thesis. Most of the chapters comprised of earlier published works. I added several new chapters using my latest results on the evolution of galaxies and the new model of our Galaxy. The defence took place in March 1972, and went equally well. Regrettably, my newer chapters also remained collecting dust. New projects kept me busy, and several pioneering works on the evolution of galaxies are so far not published. Most of my later publications are already written in English, so I am now preparing English versions of these chapters to put onto our website.

By the end of the sixties our team was rather small. However, the studies of the structure of galaxies, including difficulties with the explanation of the dark matter problem, had reached a stage where a need arose to hire young astronomers who could examine the problems more thoroughly. According to our development plans, all the young astronomers would in the 1960's go to work in the field of stellar astrophysics. Now it was time to start the second phase of the plan and to devote more attention to galactic studies. Grigori Kuzmin and I turned to Kipper with the request for permission to start preparing students for studying galaxies. To our surprise, Kipper stated that if we were to present the plan to the board of the Institute, he would oppose it. We were quite astonished — the plan of the development of the observatory was approved by the director himself. But we did not wish to cause frictions, so we decided to work towards achieving the plan, but by going about it discreetly.

We made presentations of the new problems concerning the structure of galaxies to physics students and encouraged them to choose the structure of galaxies as their subject field. Soon we had new young astronomers among us after a long time: Erik Tago, Jaan Vennik, Ants Kaasik, Peeter Tenjes and Peeter Traat. Another problem was in our education. The older generation of astronomers in our group, including me, had a classical astronomical education. But to solve new problems a good knowledge of theoretical physics and cosmology was needed. Thus we started collaboration with Enn Saar and Jaak Jaaniste, who had excellent education in theoretical physics and were already working in the department of theoretical astrophysics.

The study of galaxies soon became our main course, taking precedence over classical problems of stellar statistics and dynamics, performed by Grigori Kuzmin, Heino Eelsalu, and Ülo Veltmann. Mihkel Jõeveer switched to more practical problems of stellar dynamics, in particular to the study of dark matter and the structure and distribution of galaxies.

It was customary to write for the Academy reviews of the results of the past five-years, and to highlight the most significant results. In the mid 1970's again a five-year period was over, and highlights were discussed in the council of the Institute. There was a dispute on whether solving the problem of dark matter could be mentioned among the highlighted results. Kipper opposed this suggestion; according to his opinion our results were too "populistic".

One more episode from the 1970's. At the IAU General Assembly in 1970 in Brighton I met Gerard de Vaucouleurs. We discussed problems of the structure of galaxies — he is one of the best experts in this field. He was interested in our models of galaxies and offered cooperation. Several times in the 1970's he sent me invitations for a visit. At the time it was very difficult to get from Soviet officials permission for such visits, in particular to Western countries. Before the Soviet Academy of Sciences applied for a visa, approval from a number of intermediate instances was needed. The very first instance was our own Institute, thereafter various organs in Tartu, Tallinn and Moscow. Once, when I started the application, director Kipper invited me to his office and suggested I withdraw my application since this could block foreign visits for our stellar astrophysicists. So, the cooperation with Gerard was not realised. Only many years later, when I received an invitation from George Abell to discuss problems of the distribution of rich clusters of galaxies, I had a chance to make a short visit to Austin and to meet Gerard.

These episodes do not diminish Aksel Kipper's role as the founder of the new observatory. He was clearly one of the best scientific organisers in Estonia in the Soviet era. The above-described events show Aksel Kipper's human side — even the best of us have our shortcomings. The way Kipper viewed science was shaped in the thirties, when the most modern branch of astronomy was the physics of stars — their energy sources and evolution. The study of galaxies was not considered as a branch of astrophysics at the time; observational cosmology in the modern meaning had not formed yet. Despite his somewhat skeptical attitude towards the newer directions in cosmology, Kipper never hindered their development. This is how he once expressed his mentality towards the existence of different directions in the observatory: *"Let a hundred flowers bloom, for we cannot foresee, which blossom will come to bear the best fruit."* This is a stance one does not meet everywhere.

Chapter 5

The cosmic web

In this Chapter I shall describe how our team changed our main research goal from galaxies to systems of galaxies. This was a natural extension to our earlier work directed to better understanding of the structure of individual galaxies. This happened in the early 1970's and continued in the 1980's. We had so far little experience in the study of the distribution of galaxies because almost all our attention was directed to galaxies, their populations and the evolution of galaxies. I start the story with a description of the circumstances that brought us to the study of the distribution of galaxies.

5.1 Early studies of spatial distribution of galaxies

As described above, according to the classical cosmological world view, based on the study of the 2-dimensional distribution of galaxies in the sky, most galaxies belong to the general field, and only a relatively small fraction of galaxies is located in clusters. These studies suggested that field galaxies are distributed more-or-less randomly.

One of the first hints of difficulties in the classical paradigm came from the study of the distribution of a homogeneous sample of Sc I galaxies by Rubin et al. (1973, 1976a,b). Rubin compiled an all-sky sample of Sc I and Sc II galaxies in apparent magnitude interval $14.0 \leq m \leq 15.0$. The authors found that the distribution of redshifts of these galaxies is curious. In one large area of the sky redshifts of galaxies cluster around a value 6,400 km s^{-1}, but in another large area redshifts are clustered around a value 4,950 km s^{-1}. Areas of different mean values of redshifts are approximately located in opposite regions of sky, thus Rubin et al. suggested that one possible reason for this anisotropy may be a large motion of the Galaxy and the Local Group with respect to the general field of galaxies.

I discussed with Mihkel Jõeveer these results and we tried to understand the reason for the anisotropy of redshifts. At this time we had already started to investigate the large-scale distribution of galaxies and clusters of galaxies (see the next Section), and had a catalogue of Zwicky near clusters. Our study indicated that Zwicky near clusters are located very inhomogeneously and form large superclusters. The area of the sky where Rubin found lower redshifts of galaxies contains one of the largest nearby superclusters, the Perseus–Pisces supercluster. The area with larger redshifts contains the Coma and Hercules superclusters (Einasto et al., 1975a). These differences in mean redshifts of galaxies could be explained if galaxies also cluster into superclusters, similarly to Zwicky clusters. The number of galaxies with known redshifts was in the mid 1970's still rather small, thus our result was tentative. But it suggested that superclusters are not just clusters of clusters; here both galaxies and clusters of galaxies of various richness form density enhancements.

Chincarini & Rood (1972, 1975, 1976) measured redshifts of galaxies in an area close to the Coma cluster and found that redshifts of galaxies are concentrated around three distinct values: about $1,000 \text{ km s}^{-1}$, $4,000 \text{ km s}^{-1}$, and $7,000 \text{ km s}^{-1}$. Galaxies of the first concentration evidently belong to the Virgo supercluster, and of the third concentration to the Coma supercluster, while the intermediate concentration consists of the N4169 group of galaxies and some other nearby groups. The space between these concentrations does not contain any galaxies in the magnitude interval used for the study, $m_p \leq 15.1$. The Coma supercluster is detected at a radial distance from the center of the Coma cluster of 14.2 degrees; there is no evidence for the existence of a homogeneous field of galaxies between these three concentrations (Chincarini & Rood, 1976).

The study by Chincarini & Rood (1976) is one of the first indications for the presence of voids in the galaxy distribution. A more detailed study of the environment of the Coma superclusters by Gregory & Thompson (1978) has confirmed results by Chincarini and Rood.

5.2 The discovery of the cosmic web

5.2.1 *Zeldovich question*

After my talk at the Caucasus Winter School in 1974 on dark coronae of galaxies Zeldovich turned to me and offered collaboration in the study of the Universe. He was developing a theory of the formation of galaxies (the pancake theory); an alternative whirl theory was suggested by Ozernoi, and a third theory of hierarchical clustering by Peebles. Zeldovich asked for our help in solving the question: Can

Fig. 5.1 Yakov Zeldovich with his wife visiting Estonia, late 1970's (author's photo).

we find some observational evidence which can be used to discriminate between these theories?

Later I heard from my Moscow colleagues that Zeldovich was often interested in understanding new phenomena, and had a habit of finding the best authorities in the new field to learn as much as possible from a collaboration with these people. Moscow is one of the best scientific centres in the world, and there are specialists in practically all fields. Quite often Zeldovich learned the basics of the new field very quickly and was able, either himself or in collaboration with specialists in the particular field, make significant contributions in the new field.

Why he selected our group for this task, I do not know. So far we had no experience in observational cosmology; our work was directed to the understanding of the structure of galaxies. We had theoretical cosmologists in our group (Enn Saar and Jaak Jaaniste), but they also did not have experience in observational cosmology. Thus, initially we had no idea how we can help Zeldovich.

But soon we remembered our previous experience in the study of galactic populations: kinematical and structural properties of populations remember their previous evolution and formation (Rootsmäe, 1961; Eggen et al., 1962). Random velocities of galaxies are of the order of several hundred km/s or less, thus during the whole lifetime of the Universe galaxies have moved from their place of origin only about 1 h^{-1} Mpc (we use the Hubble constant in units $H_0 = 100\,h\,\mathrm{km\,s^{-1}\,Mpc^{-1}}$). In other words — if there exist some regularities in the large-scale distribution of galaxies, these regularities must reflect the conditions in the Universe during the formation of galaxies. Actually we already had some preliminary results: the study of companion galaxies had shown that dwarf galaxies are located almost solely around giant galaxies and form together with giant galaxies groups and clusters of galaxies. In other words — the formation of galaxies occurs in large units, not in isolation. A similar phenomenon is observed in star formation: stars born in star-forming gas clouds which evolve to form stellar associations, groups and clusters, but not as isolated objects (Ambartsumian, 1958).

Thus we had a leading idea on how to solve the problem of galaxy formation: *We have to study the distribution of galaxies on large scales.* Both our galactic astronomy and theoretical cosmology groups participated in this effort.

We started to collect redshift data from all available sources. Our first results from the study of the large-scale distribution of galaxies showed the existence of large superclusters in the Perseus as well as in the Coma and Hercules regions of sky (Einasto et al., 1975a). These superclusters contain several rich Abell clusters, numerous less rich Zwicky clusters, and 'field' galaxies. Both superclusters have a length over 50 h^{-1} Mpc, and form flat systems with axial ratio about 1:5. Thus the form of superclusters is far from a spherically symmetrical shape, and they have several rich clusters as nuclei. The main conclusion of the paper was: superclusters form from a gaseous medium prior to the formation of galaxies or simultaneously in a single process. By random clustering of galaxies it is impossible to form such dense and flat systems as superclusters. We noticed that individual galaxies also have a tendency to cluster at the same regions in space where clusters of galaxies are located.

When we started the galaxy distribution study, our Observatory had only two catalogues of galaxies, the de Vaucouleurs & de Vaucouleurs (1964) Reference catalogue of galaxies, and the Nilson (1973) Uppsala general catalogue of galaxies. In our studies of the nearby groups and dwarf satellite galaxies we made extensive use of these catalogues. The Uppsala catalogue is based on the Zwicky et al. (1968) Catalogue of galaxies and clusters of galaxies, so we needed the original Zwicky catalogues too.

Also we needed data on peculiar galaxies, since our previous experience has shown that active galaxies are good markers of the skeleton of systems of galaxies — they mark the locations of dense regions of space where ordinary galaxies also have a tendency of clustering. We already had the lists of

```
CALIFORNIA  INSTITUTE  OF  TECHNOLOGY
             PASADENA, CALIFORNIA 91109
                  March 19, 1973


Dr. J. Einasto
Tartu Astrophysical Observatory
202444 Tôravere
Estonia
U. S. S. R.

Dear Dr. Einasto,

      Thank you for asking about our Catalogue of Selected
Compact Galaxies and of Post-Eruptive Galaxies. The work
on this catalogue as well as the printing of it was financed
by my wife and myself. In order to break even we therefore
are selling it for 205 Swiss francs per copy. If your
observatory wishes to purchase a copy, please have them send
check for 205 Swiss francs drawn on the Swiss Credit Bank in
Berne and made out in the name of my daughter Margrit Zwicky
to
                  Miss Margrit Zwicky
                  Dorfstrasse 95
                  CH 3073 Guemligen (BE)
                  Switzerland

      The catalogue will be mailed by her postpaid and registered
only after receipt of prepayment.

                              Sincerely yours,

                              F. Zwicky
FZ:h                          F. Zwicky
                              Professor Emeritus of Astrophysics
                              California Institute of Technology
                              and The Hale Observatories
```

Fig. 5.2 Letter by F. Zwicky concerning the purchase of his Catalogue (author's archive).

Markarian galaxies and Vorontsov-Veljaminov interacting galaxies; now we needed the Zwicky Catalogue of Selected Compact Galaxies (Zwicky & Zwicky, 1971), as well as Zwicky catalogues of galaxies and clusters of galaxies. Our Observatory had no money to buy it. So I wrote a letter to Fritz Zwicky asking advice on how to get the catalogues. In his answer Zwicky (see Fig. 5.2) explained that the Compact Galaxies catalogue was published privately, and in order to get it we had to pay 205 Swiss francs to his daughter who lives in Switzerland. Now the problem was how to get the money.

During the IAU General Assembly in Brighton 1970 I met Tarmo Oja, an Uppsala astronomer of Estonian origin. We spent a lot of time together during excursions; I have a photo with him and his wife in the new Greenwich Observatory in Herstmonceux Castle near Brighton. Next time we met was in Athens during the First European Astronomy Meeting. We started to exchange literature — Tarmo subscribed for me "Astronomy and Astrophysics" (Tartu Observatory did not have money for this), and I sent Tarmo Estonian books. Based on this experience I wrote a letter to Tarmo asking whether he could help us and order the Zwicky catalogues for me. He answered very quickly and wrote me that the money had been sent to the address given by Zwicky, and for the Catalogue of galaxies and clusters of galaxies to the California Institue of Technology. All catalogues arrived rather soon.

The Uppsala catalogue contains all data on Zwicky galaxies, but not on Zwicky clusters. With these catalogues we were able to study the 2-dimensional distribution of galaxies of various type in the sky, as well the distribution of groups and clusters of galaxies. Of special interest were Zwicky clusters of distance class "near", since it was the nearby space from where we hoped to get enough redshift data to find the spatial 3-dimensional distribution of *all* astronomically interesting objects from normal and active galaxies to groups and clusters of galaxies. The Estonian Academy of Sciences was able to buy for the Tartu Observatory the Palomar Atlas — paper prints of all photos of the Palomar Observatory Sky Survey made with the large Schmidt telescope. This allowed us to see the structures we found in catalogues as they appear in reality.

A few years later we had the possibility of getting the de Vaucouleurs et al. (1976) Second Reference catalogue of galaxies. We also used a number of other sources such as the Sandage (1978) list of redshifts for 719 bright galaxies, and the Sandage & Tammann (1981) Revised Shapley–Ames catalog of bright galaxies. An additional source was the ZCAT — a compilation of all available galaxy redshifts by John Huchra.

In our work to solve the Zeldovich question we had a close collaboration with his team. In 1975 Sergei Shandarin obtained first results of numerical simulations of the evolution of particles according to the theory of gravitational clustering developed by Zeldovich (1970), see Fig. 5.3. Results of the simulation were published a few years later by Doroshkevich & Shandarin (1978). A figure with the results of these simulations was put on the wall of the Saar's and Jaaniste's office. In this picture a system of high- and low-density regions was seen: high-density regions form compact clumps and are joined by filaments, together they form a cellular network which surrounds large under-dense regions. This simulation was, however, only 2-dimensional and had a qualitative character — we did not know the scale of the network seen in the simulation.

Our challenge was to find out whether the real distribution of galaxies shows some similarity with the theoretical picture. We wanted to understand the *actual*

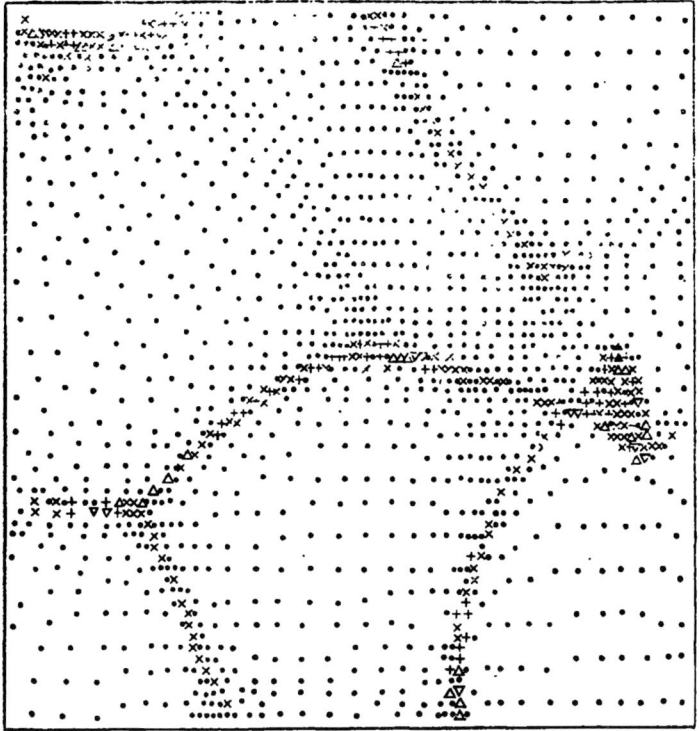

Fig. 5.3   Distribution of particles in simulations (Shandarin 1975, private communication), Fig. 7 of Doroshkevich & Shandarin (1978) and Fig. 6 of Einasto et al. (1980a)).

distribution of galaxies and systems of galaxies, and to compare the real distribution with model distributions of all theoretical scenarios. We did this, step by step, and our final conclusion was that none of the theoretical scenarios suggested so far meets all observational constraints. But this was a long way. Our immediate goal was to compare the actual distribution with the prediction in hand — the Zeldovich pancake scenario.

If we identify rich knots in the simulated structure with superclusters, then results of simulations suggest that superclusters must be joined by galaxy or cluster chains to a connected network. To identify these chains it is needed to study *the global distribution of galaxies and clusters on large scales*, not just the local environment of superclusters. The Peebles scenario of galaxy formation predicted a more-or-less random distribution of galaxies; the Ozernoi scenario did not have any prediction on the distribution of galaxies. Thus we hoped that the large-scale distribution of galaxies and systems of galaxies contains the key information to discriminate between basic scenarios of galaxy and structure formation.

To find the global distribution of galaxies and systems of galaxies we used several methods. Jaaniste and Saar suggested studying the distribution of nearby Zwicky clusters. First of all, the Zwicky catalogue on near clusters is *complete*, and many bright galaxies of nearby Zwicky clusters had at this time measured redshifts, so we hoped to determine the distribution of clusters, and to find some regularities there. To see the distribution better we built in the office of Saar and Jaaniste a 3-dimensional model from plastic balls. Some regularity was evident: there were several clusters of Zwicky clusters — superclusters — one of them in the Perseus region. But too many clusters had no galaxies with measured redshifts, so it was difficult to get an overall picture.

Mihkel Jõeveer found a simple method how to estimate mean redshifts of all near clusters. Most near clusters contained enough member-galaxies to find the cluster luminosity function using apparent magnitudes. By comparing these functions with similar functions for clusters with known redshift it is easy to estimate the redshifts of all clusters. The sample of Zwicky near clusters was used in our preprint by Jõeveer et al. (1977), and by Einasto et al. (1980a) (Figs. 5.7, 5.8 and 5.9) and by Tago et al. (1984, 1986).

Our young collaborator Erik Tago started to search nearby systems of galaxies using prints of the Palomar Sky Survey. His first goal was to look for nearby galaxy systems in the Coma and the Virgo supercluster region. We all followed his search with curiosity. Soon a poor cluster was found, then the next one. Later, when Zwicky catalogues of galaxies were available, we discovered that all these clusters Erik had found are listed as Zwicky 'near' clusters. Initially there were no redshifts for these galaxies. As our redshift compilation improved Erik was able to find distances to these clusters. The clusters formed a chain though a large void between the Virgo, the Coma and the Hercules superclusters. Tago et al. (1986) argued that this string of galaxies can be an evidence for a Lagrangian Singularity in the Zeldovich pancake scenario.

A third approach was applied by Mihkel Jõeveer. He used wedge-diagrams, invented just when we started our study. He applied two ideas to understand the spatial structure of the distribution of galaxies and systems of galaxies. Most important was his idea to investigate the distribution not only of ordinary galaxies, but all other interesting astronomical objects — peculiar galaxies, such as Markarian galaxies, radio galaxies, galaxies from the Zwicky list of compact galaxies, and of various systems of galaxies from groups to clusters. His second idea was: he made a number of relatively thin wedge diagrams in sequence, and plotted in the same diagram ordinary and active galaxies, as well as groups and clusters of galaxies. By comparing neighbouring wedges in declination and in right ascension it is possible to understand the three-dimensional distribution of objects.

In these diagrams a regularity was clearly seen: *galaxies and galaxy systems populate identical regions, and the space between these regions is empty*. Ordinary galaxies are mostly located along a network of strings or filaments. Groups and clusters of galaxies, as well as active (Markarian and radio galaxies) form the skeleton of the network. Galaxies lie close to the skeleton, or form bridges between elements of the skeleton. After this success we concentrated our efforts to the study using wedge-diagrams. Most attention was directed to the Perseus–Pisces supercluster, well seen also in our model with balls as a cluster of Zwicky clusters. In this region there are almost no foreground galaxies as shown already by Rubin et al. (1973), thus we see the structure of the supercluster very clearly.

When we had our first results using both Zwicky clusters and wedge diagrams with all objects, we immediately understood that distribution of galaxies and of systems of galaxies tells us something principally new about the formation and evolution of the Universe — *the Universe has structure, it is not structureless as believed so far*. We had again the feeling that we have reached a tip of a mountain, and behind the peak there is a completely new and unexplored landscape. Also we understood that whatever we did next, we would be the first to do it — it was a completely new area of research.

To our great surprise and joy slices with galaxies and clusters were quite similar to the theoretical picture predicted by Zeldovich. We made wedge-diagrams for the full sky, for three declination wedges they are shown in Figs. 5.4, 5.5 and 5.6. Wedge diagrams were shown in the Tallinn symposium, Fig. 5.4 was published in the symposium proceedings by Jõeveer & Einasto (1978). We also made a very detailed analysis of the Perseus supercluster region and made pictures of the distribution of Zwicky clusters of the class "near" in the Perseus region of sky, see Figs. 5.7, 5.8 and 5.9 for clusters in three redshift intervals. The distribution of Zwicky clusters in the Perseus region was shown in our preprint by Jõeveer et al. (1977), and published in detail by Einasto et al. (1980a). Similar distributions were found for the Coma region of sky, published a few years later by Tago et al. (1984).

5.2.2 *The Tallinn symposium on large scale structure of the Universe*

Already in 1975, after the Tbilisi Meeting, we discussed with Zeldovich the possibility of organising a real international conference devoted solely to cosmology. Due to the Soviet bureaucratic system it was extremely difficult for Soviet astronomers to attend international conferences in Western countries; thus the only possibility of having better contact between Soviet and Western cosmologists was

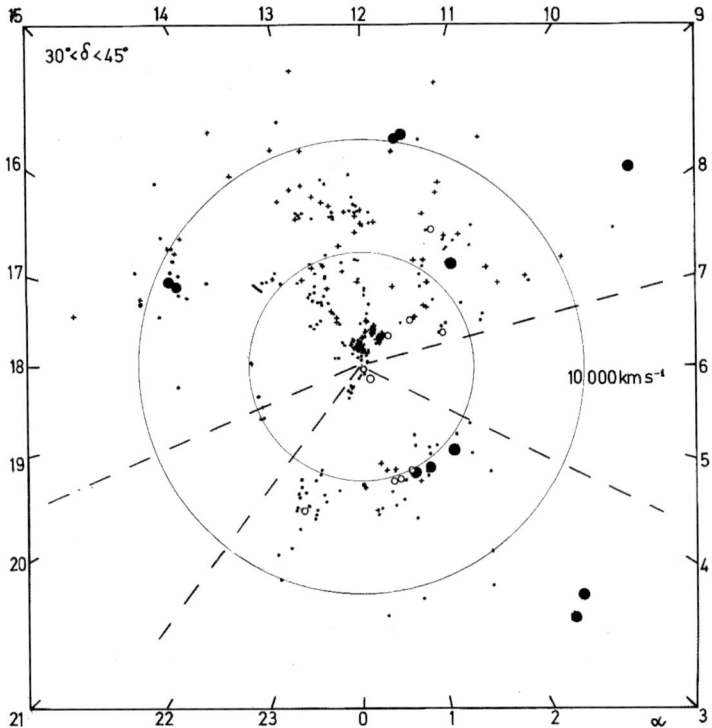

Fig. 5.4 Wedge diagram for the 30°–45° declination zone. Filled circles show rich clusters of galaxies, open circles — groups, dots — galaxies, crosses — Markarian galaxies. Clusters at RA about 2 h belong to the main chain of clusters and galaxies of the Perseus–Pisces supercluster; galaxies and clusters near the center at RA about 12 h are part of the Local supercluster, and galaxies and clusters at redshift about 7,000 km/s and RA between 10 h and 13 h belong to the Coma supercluster. Note the complete absence of galaxies in front of the Perseus–Pisces supercluster, and galaxy chains leading from the Local supercluster towards the Coma supercluster (Jõeveer & Einasto, 1978).

to hold the conference within the Soviet Union or in nearby "Socialist" countries. Actually a conference on cosmology was recently held in Krakow, Poland, during the IAU General Assembly, where Zeldovich had a talk on his scenario of structure formation (Doroshkevich et al., 1974).

Zeldovich suggested holding the symposium this time in Estonia — here the organisation was much more easier than in Moscow. The only suitable place was our capital Tallinn, which had conference facilities, and was open to foreign vistors, whereas Tartu was at this time a closed city due to its proximity to a large military airfield. Initially I suggested having a discussion on dark matter, but in Zeldovich's opinion this topic was still too controversial, and it would be better

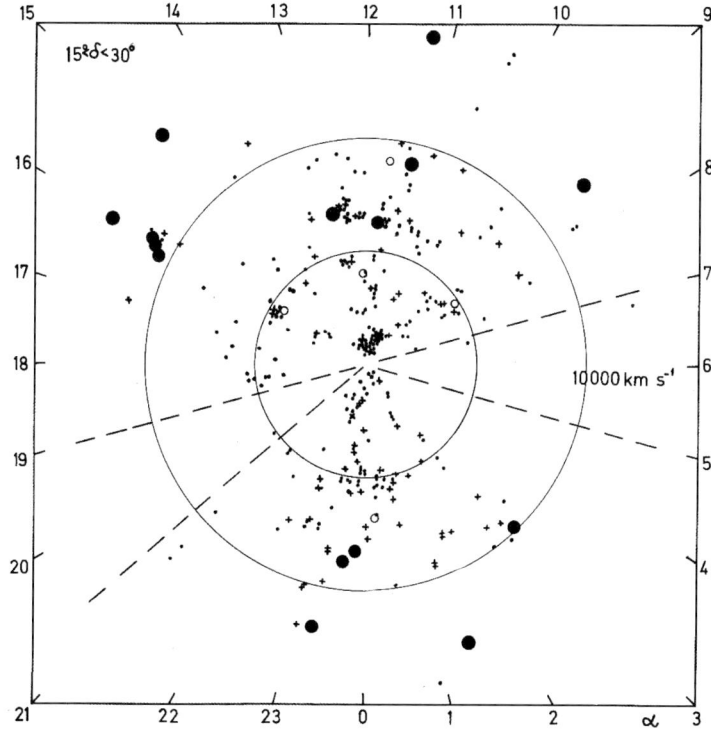

Fig. 5.5 Wedge diagram for the declination zone 15°–30°. Galaxies and clusters near the center at RA about 12 h belong to the Local supercluster; two rich clusters at redshift 7,000 km/s and RA about 12 h are the main clusters of the Coma supercluster, galaxies at redshift 5,000 and RA about 23 h belong to the Perseus–Pisces supercluster. Note the galaxy chains leading from the Local supercluster towards the Coma and the Perseus–Pisces superclusters (Jõeveer & Einasto, 1978).

to have a more neutral topic. After some discussion we decided to devote it to "Large Scale Structure of the Universe". At this time we had no idea what this term actually means, I had one example of a planned IAU symposium on "Large Scale Characteristics of the Galaxy", to be held in 1978 in College Park, USA.

To discuss the program of the planned symposium the scientific organising committee met during the IAU General Assembly in Grenoble 1976. I was the sole representative from the Soviet Union, and we discussed both the program and the candidate for the chairman of the SOC. Most members of SOC were from Western countries and wanted the chairman to also be from a Western country. So I suggested Malcolm Longair. He had spent a long period in Moscow and was well familiar with the work of Moscow theorists, and Zeldovich fully trusted him. This

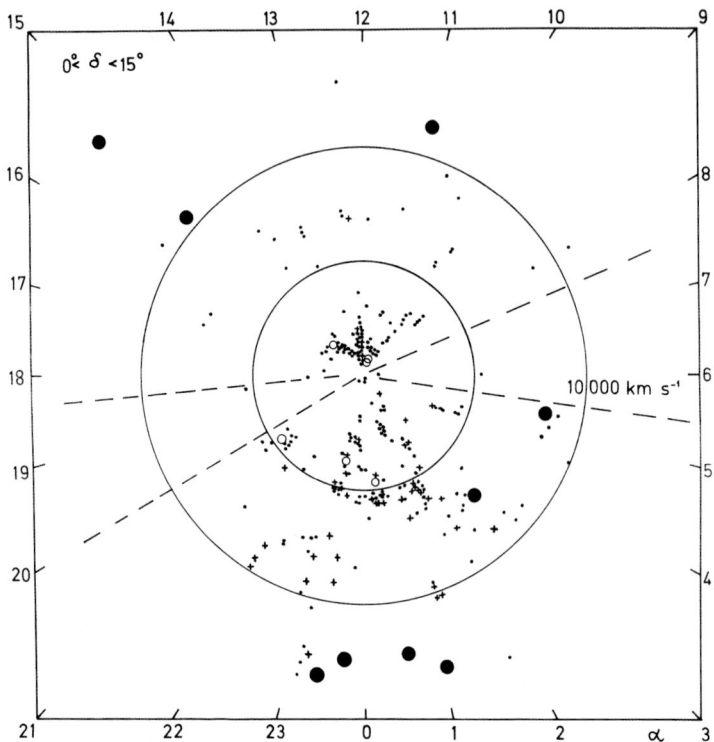

Fig. 5.6 Wedge diagram for the declination zone $0°$–$15°$. Galaxies and groups at redshift 5,000 between RA 23 h and 1 h belong to a supercluster with Abell cluster A194 as the main cluster, see Fig. 5.7. Clusters at redshift about 13,000 km/s at RA between 23 h and 1 h belong to a system of superclusters at the far side of a large void, seen in Figs. 5.7 and 5.9. Note the galaxy chains leading from the Local supercluster towards the A194 supercluster (Jõeveer & Einasto, 1978).

suggestion was accepted and, as time has shown, was very good — he was really a very effective leader.

In preparations for the symposium we had two goals in mind. The first one was scientific. In 1976 we already had preliminary results of our own study of the distribution of galaxies and clusters, and knew that only a discussion of *global* properties of the distribution can answer basic theoretical questions concerning the formation and evolution of the Universe. Thus we suggested discussing all major global aspects of the observational and theoretical cosmology. At first glance this seems to be too broad a scope, but, as results of the symposium showed, this broad scope was needed to understand major theoretical problems on the evolution of the Universe.

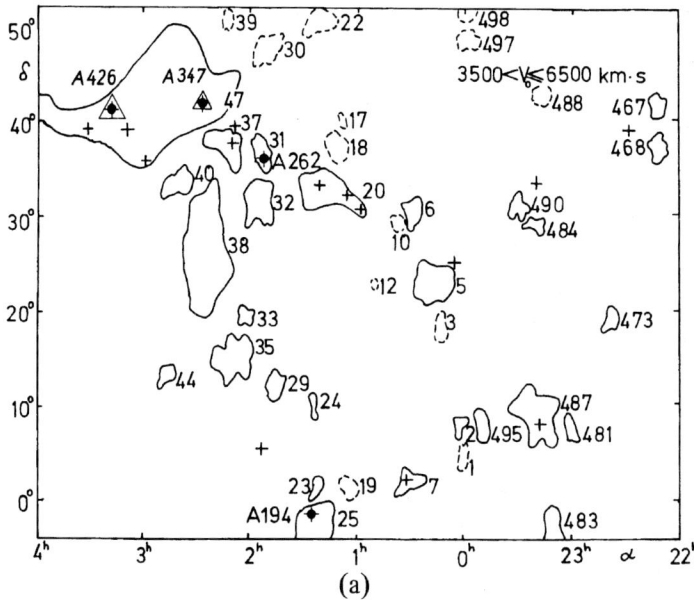

Fig. 5.7 The distribution of Zwicky clusters of the distance class 'near' in the Perseus area of the sky in redshift interval $3,500 \leq V_0 < 6,500$ km/s. Abell clusters A426, A347, A262 and Zwicky clusters 37, 31, 20, 10, 6, 5 form the main chain of clusters of the Perseus–Pisces supercluster. The Northern chain is formed by Zwicky clusters 30, 22, 498, 497, the Southern chain by Zwicky clusters 40, 38, 33, 35, 29, 24; this chain forms a bridge towards the supercluster which surrounds the A194 cluster (Jõeveer et al., 1977; Einasto et al., 1980b).

Our second goal was of a completely different nature. I had attended previously several international astronomical meetings elsewhere in the Soviet Union, and had been astonished by the rather bad organisation. Soviet scientists wanted Russian to be a major scientific language, and quite often talks by Soviet astronomers were in Russian with no adequate English translation. Also it was difficult to follow discussions because of the language conflict. Other aspects of local organisation were also poor — meals, accommodation, cultural programs. Thus we had a very special goal — to prepare the symposium on a really high international level.

First of all, we assembled our Technical Organising Committee (TOC) using almost all active astronomers of Tartu Observatory, not only of our cosmology team. This Committee worked completely independently of the Scientific Organising Committee, and had full freedom and possibilities for any initiative. Among us we called the TOC "Glavsympstroy" to parody the Soviet style abbreviations of long names of organisations. There were people responsible for auditorium, for

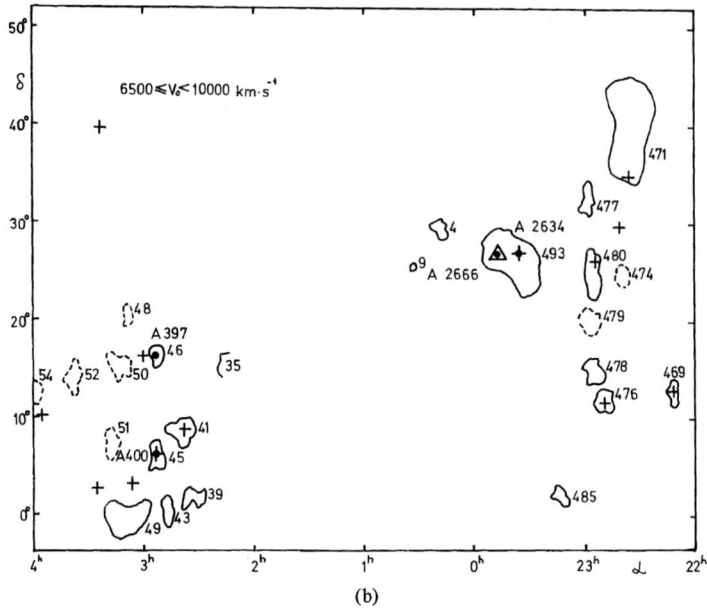

Fig. 5.8 The distribution of Zwicky clusters of the distance class 'near' in the Perseus area of the sky in redshift interval $6,500 \leq V_0 < 10,000$ km/s. Clusters A397, A400 and numerous Zwicky clusters form a cluster system at R.A. $\alpha \sim 3^h$, clusters A2634, A2666 and surrounding Zwicky clusters form another cluster system at R.A. $\alpha \sim 23^h$. These cluster systems surround a large void between them, the Perseus–Pisces supercluster in the foreground, and another supercluster in the background, seen in the next redshift interval (Jõeveer et al., 1977; Einasto et al., 1980b).

equipment, for distribution of the guests to hotels, for the cultural program, for English translation etc. Some of the TOC members had good relations with the Moscow circles, so we were able to hire the best synchronous translators from Moscow, really good professionals. This allowed us to have all talks and discussions in English; the translation was so good that even people with bad understanding of English could follow talks and discussions.

Some members of the TOC knew Estonian people active in music, so it was possible to organise for the symposium a special concert in the historical Tallinn Rathaus (Town Hall) where a young conductor Tõnu Kaljuste gave a concert of classical music conducting the recently formed Chamber Choir. In the following years this choir has given concerts all over the world and is counted as one of the very best. In 1977 the choir was not yet internationally known, and astronomers were surprised to hear a concert at this high level.

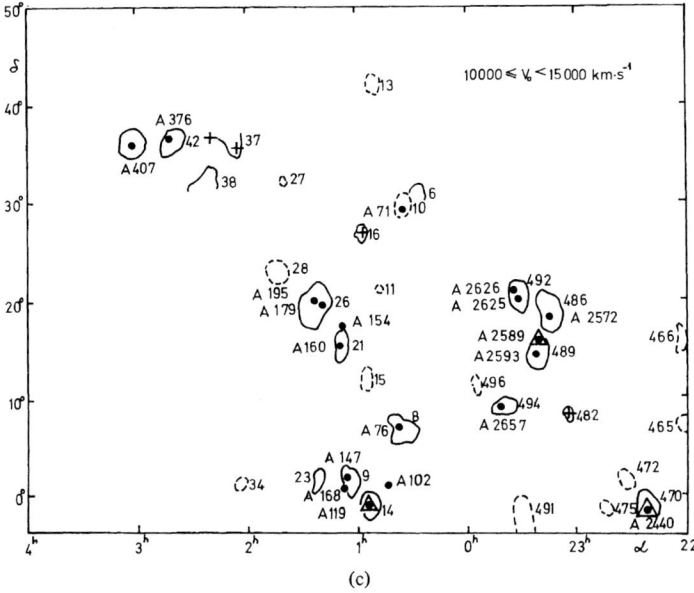

Fig. 5.9 The distribution of Zwicky clusters of the distance class 'near' in the Perseus area of the sky in redshift interval $10,000 \leq V_0 < 15,000$ km/s. Numerous Abell and Zwicky clusters form in this region several rich superclusters. Together with superclusters seen in lower redshift regions these superclusters surround a large void (Jõeveer et al., 1977; Einasto et al., 1980b).

The symposium was held in September 1977 with reception, cultural program, and banquet as normal for an international meeting. All talks and discussions were tape-recorded, and discussion sheets edited and typed by our secretaries, who sometimes worked the full night to prepare the typed text for further editing. This allowed Malcolm to take with him the full set of typed discussions. To help final editing a copy of all tapes was given to Malcolm. All this helped to prepare the symposium proceedings. This symposium was of such interest to the Russian speaking community that a full Russian translation was soon prepared and printed.

Years later when visiting other observatories some astronomers told me that the Tallinn conference was one of the best they ever have attended. Quite recently, I met Hugo van Woerden in Groningen. We discussed problems of common interest, and also the development of the understanding the structure of the Universe. He remembered the Tallinn conference and told me that he and other participants were impressed seeing the enthusiasm that *"radiated from eyes of all members of the local organising committee"*.

We had wanted to show that Estonia is not the same as the rest of the Soviet Union. We had wanted to show that we have our own culture and traditions, and we belong to the international community. We achieved our goal.

5.2.3 *Superclusters, filaments and voids*

As the conference progressed, it became clear that the structure of superclusters was its crux. The first speaker who presented new data on the three-dimensional distribution of galaxies was Brent Tully (Tully & Fisher, 1978), a well-known observer who had used radio observations to determine the distances to the galaxies in the Local Supercluster. The core of his presentation was a film of the Local Supercluster. To obtain a spatial image of the supercluster, he used the simple trick of making the image rotate with the use of a computer, which created a three-dimensional illusion. This method has later been frequently used by observers of superclusters, as well as by theorists to illustrate the distribution of particles in N-body simulations. The film showed that the Local Supercluster consists of a number of chains of galaxies which branch off from the supercluster's central cluster in the Virgo constellation. No galaxies could be seen in the space between the chains. This image strongly resembled the one we had acquired of the Perseus supercluster. The structure of the Local Supercluster had been studied for years by the renowned American astronomer Gerard de Vaucouleurs, who also participated in the conference. He implemented supergalactic coordinates which were oriented to follow the Local Supercluster's denser flat part.

Then it was our turn (Jõeveer & Einasto, 1978). We had been studying the structure of the Perseus–Pisces Supercluster and its surroundings closely, so we had a good understanding of the structure of the Perseus–Pisces and the Local superclusters, as well as of the global network of superclusters and galaxy chains/filaments. The galaxies of the Perseus–Pisces Supercluster form a long chain in which clusters and groups of galaxies are embedded as pearls, see Fig. 5.7. The supercluster's distance from us is about $50\,h^{-1}$ Mpc. The main chain of clusters and groups lies nearly perpendicular to the line of sight. Especially characteristic of the chain is its thickness — it is very narrow, as thick as the diameter of the clusters. There are no galaxies either up or downwards of the chain (as seen from our viewpoint), as well as in front of and behind the chain. Thus the chain is an elongated, essentially one-dimensional formation that is surrounded by a void on each side. Its total length is about hundred megaparsecs. At the one end of the chain is the central cluster of the Perseus–Pisces Supercluster (Abell cluster A426, see Fig. 5.7). Further continuation of the supercluster cannot be observed for it is hindered by the absorption effect of the Milky Way. The other end of the chain reaches the next supercluster, which is located somewhat closer to us (Zwicky clusters 481, 487, seen in Fig. 5.7). Another chain of clusters revolve around a large void behind the Perseus Supercluster and form the next system of superclusters behind it at about $130\,h^{-1}$ Mpc, see Figs. 5.8 and 5.9.

But that was not all of the information about the chain of clusters. Comparison of adjacent slices showed that chains of galaxies and systems of galaxies (groups and clusters) form an almost continuous network. Here and there in the network there are denser regions with more clusters in an aggregate, such gaskets can be considered as superclusters of galaxies. The superclusters in turn are branched; in addition to clusters, there are numerous galaxies which are not randomly scattered but also located as chains. The structure of the Perseus–Pisces supercluster was rather similar to the structure of the Local supercluster, consisting of numerous galaxy filaments, as seen from the movie by Brent Tully, only the richness of knots in chains is different. Chains of the Local supercluster are less rich and contain no clusters, whereas chains of the Perseus-Pisces supercluster contain numerous groups and clusters, in addition to the main cluster A426.

Jõeveer et al. (1978) estimated also the filling factor of the Universe covered by superclusters of galaxies and other filled regions (groups outside superclusters). The data indicated that superclusters fill only about 4 per cent of the total space; the remaining 96 per cent of space forms voids between superclusters. Since voids have been found to exists within superclusters, the filling factor of the Universe with systems of galaxies can be even smaller.

In Figures 5.7, 5.8 and 5.9 we presented the distribution of "near" Zwicky clusters in the sky in the Perseus region in three distance classes. Clusters are numbered according to the Uppsala General Catalogue, Abell clusters are also shown. Cluster contours have been drawn according to Zwicky et al. (1968). Contours of clusters with measured redshifts are drawn by solid lines, clusters with estimated redshifts by dotted lines.

What is especially important is the fact that the Zwicky "near" cluster catalogue is *complete* — it contains all clusters Zwicky had discovered in this distance interval. Thus there are no selection effects which could distort the distribution. In our first distance interval (3,500–6,500 km s^{-1}) we see the filamentary Perseus–Pisces supercluster, in the second distance interval (6,500–10,000 km s^{-1}) the large void behind this supercluster and clusters surrounding the void, in the third interval (10,000–15,000 km s^{-1}) there is a system of several rich superclusters at mean distance $130\,h^{-1}$ Mpc, which surrounds the far side of the void of diameter $\sim 70\,h^{-1}$ Mpc.

These figures were demonstrated during the Symposium and were included in the Preprint (Jõeveer et al., 1977), but published in the Monthly Notices only in our paper by Einasto et al. (1980a). Later analysis by Einasto et al. (1994b, 1997b, 2001) has shown that most Abell clusters seen in Fig. 5.9 belong to a complex of superclusters at the far side of the void.

A similar analysis was made a few years later by Tago et al. (1984) in the Coma supercluster and its large-scale environment. The distribution of Zwicky clusters of distance class "near" is shown at various redshift intervals in Figs. 1 and 2 of Tago et al. (1984).

The network of superclusters, the chains that form them and the voids in between were comparable to the theoretical model found in the Zeldovich group. To designate this network, we used the term "the cellular structure of the Universe". In this terminology a "cell" is a low-density region surrounded from all sides by rich clusters and superclusters. Later we used the term "supervoid" to designate supercluster-defined voids (Lindner et al., 1995). Our analysis presented during the Tallinn Symposium showed that cells are not empty — they are crossed by filaments consisting of galaxies and groups of galaxies (poor Zwicky clusters), as seen in Figs. 5.4, 5.5, 5.6, 5.7, 5.8, and 5.9. This means that chains/filaments are of two types: in superclusters they are rich and contain rich clusters and groups, while in cell interiors they are poor and contain only galaxies or poor groups/clusters.

The presence of holes (voids) in the distribution of galaxies was reported also by other groups: by Tifft & Gregory (1978), and Tarenghi et al. (1978) in the Coma and Hercules superclusters, respectively. A few years later a large void in Bootes was discovered by Kirshner et al. (1981), similar in size to the void described above behind the Perseus–Pisces supercluster. The Bootes void in our terminology is the interior of a cell, i.e. a supervoid.

Theoretical interpretation of the observed cellular structure was discussed at the symposium by Zeldovich (1978). As noted by Longair (1978) in his concluding remarks, *"the discovery of the filamentary character of the distribution of galaxies, similar to a lace-tablecloth, and the overall cellular picture of the large-scale distribution was the most exciting result presented at this symposium"*. These results demonstrated that the pancake scenario by Zeldovich (1970) has many advantages over other rival scenarios. The term "Large Scale Structure of the Universe" got its present meaning.

However, we noticed also some differences between the observed distribution of galaxies and clusters and the Zeldovich scenario as shown in Fig. 5.3. More on this in Chapter 7.

So far the general perception among astronomers had been that galaxies and clusters are located in space almost randomly. This view was based on the apparent distribution of galaxies and clusters, where voids and sharp features like cluster chains are not visible. Before the Tallinn Symposium I sent the preprint of our

Fig. 5.10   Jim Peebles explaining to Scott Tremain secrets of structure formation, IAU Tallinn Symposium 1977 (author's archive).

results (Jõeveer et al., 1977) among others to Jim Peebles. I got a rapid reply with high-quality photos of the Lick survey and a computer generated 2-D distribution of galaxies using the clustering algorithm by Soneira & Peebles (1978). In the accompanying letter Jim asked: "Can you find filaments in the real and simulated surveys?" In the introduction of the paper Soneira & Peebles (1978) argue: *We know that the eye does tend to judge in a biased way — for example, one readily picks out "chains" of points in a uniform random distribution'*. Jim and his collaborators have worked very hard to investigate the Lick galaxy counts and other catalogues of galaxies using two-dimensional data and analysis. Our analysis suggests that galaxy filaments are clearly seen only in the three-dimensional distribution of galaxies.

Anyway, the attitude of the astronomical community to our results was rather skeptical, and we had difficulties in publishing the detailed analysis, which was distributed before the Symposium as a preprint. The referee argued that our results are not sufficiently reasoned and demanded that we exclude arguments on the similarity of the present distribution of galaxies to the distribution at the early epoch at the formation of the structure, as well as the need to study the global distribution of galaxies. Also the referee demanded we exclude a detailed theoretical interpretation of the continuous cosmic web which joins all large-scale structures to a single network. We had to make changes in the text many times. Finally the paper was published with an inappreciable title "Spatial distribution of galaxies and of clusters of galaxies in the southern galactic hemisphere" (Jõeveer et al., 1978).

Our experience confirms the opinion of Ernst Öpik (1977), who writes: *"From my long experience, both with my own papers, and with those sent to me for refereeing, I have a feeling that the "recognized" journals usually accept without difficulty papers with a middle-of-the-road content, useful contributions to research which already has established itself or accumulations of additional new material. As to pioneering work, papers of this kind often are running the risk of rejection or of excessive curtailment."*

The discovery of voids in the galaxy distribution was recently discussed by Thompson & Gregory (2011). The authors give a fairly complete story of the history of the discovery of voids. I discussed the story with Laird privately via e-mail. Their paper in a refereed journal (Gregory & Thompson, 1978) appeared earlier than ours (Jõeveer et al., 1978). On the other side, we made a preprint (Jõeveer et al., 1977) earlier, and distributed the preprint to participants of the Tallinn symposium. Our paper also arrived at the journal office earlier, but much more time was needed to satisfy all the suggestions of the referee. However the timing is not that important. More important is that we came to the same conclusion on the presence of voids completely independently. To be accurate, the presence of voids in galaxy distribution was first described by Chincarini & Rood (1976).

Gregory & Thompson (1978) used their own observations with limiting magnitude $m_p < 15$ in a 260 degree^2 region in the Coma supercluster, the largest possible area around the Coma cluster which could be studied with available equipment in a reasonable timescale. Similar relatively small regions were studied by Tifft & Gregory (1978) and Tarenghi et al. (1978).

What was new in our papers was the complete description of the cosmic web with superclusters, galaxy chains joining superclusters to a connected network, and voids between them. We studied the whole Northern hemisphere from equator up to declination $\sim$50 degrees. If we exclude regions close to the Milky Way zone of avoidance this makes about one quarter of the whole sky, i.e. about 10 thousand square degrees. Our main goal was *to find basic properties of the whole cosmic network*. We found that different objects — clusters and groups of galaxies, radio and Markarian galaxies — populate identical regions, and thus can be used as markers of the skeleton of the overall network. For these objects almost complete redshift data were available up to redshift $\sim$0.05, i.e. distance 150 h^{-1} Mpc. Redshift samples for galaxies were incomplete, but all available data showed that galaxies have a tendency to cluster in regions marked with clusters and active galaxies. Of course, dwarf galaxies could be distributed more evenly, and fill the voids between galaxy clusters and chains. However, our previous studies of the distribution of dwarf galaxies had shown that dwarf galaxies have a strong tendency to cluster around giant galaxies and to form loose groups around them (Einasto et al., 1974c, 1975b). Thus our approach and that of Gregory & Thompson (1978) were complementary. The advantage of studying the global distribution is that in this case it is possible to investigate properties of *the whole cosmic web*, not only the environment of superclusters.

To summarise: all essential properties of the cosmic web were discussed by our team at the IAU Tallinn Symposium in 1977 (Jõeveer et al., 1977; Jõeveer & Einasto, 1978), and in subsequent papers by Jõeveer et al. (1978) and Einasto et al. (1980a,b). The basic properties of the web are the following:

(1) The Universe is not structureless but forms the cosmic web as we call it now;
(2) Superclusters consist of chains/filaments of galaxies, groups and clusters of galaxies;
(3) Galaxy and group/cluster filaments form bridges between superclusters;
(4) Chains of galaxies and clusters are essentially one-dimensional, clusters in chains are often elongated along the chain;
(5) The space between galaxy and cluster filaments is almost devoid of galaxies, and form holes/voids;
(6) Voids defined by galaxy filaments have diameters $10\ldots30\,h^{-1}$ Mpc, voids defined by rich clusters and superclusters have diameters up to $\sim 75\,h^{-1}$ Mpc;
(7) Superclusters occupy about 4% of the total space, the remaining 96% comprises of voids;
(8) The filamentary character of galaxy distribution cannot be formed by clustering of already-formed galaxies — galaxies can form only inside the filaments made of pre-galactic matter, after the dissipation of kinetic energy perpendicular to the filament;
(9) The comparison of the observed structure with results of numerical simulations indicates the presence of non-clustered pre-galactic matter in voids — gravity cannot evacuate voids completely; galaxies form only in regions of enhanced density (see Chapter 7 for discussion of this property of the web);
(10) Observed large-scale distribution of matter is the remnant of singularities in the initial perturbation field.

However, further development of the cosmic web concept was influenced by the East–West controversy during the Cold War. Fairall (1998) characterises the reception of the concept of the cosmic web by the Tartu team with the following words: *"News of the claim travelled widely, but converts outside the Soviet Union were few; the idea seemed to overthrow all that was understood in the West about clustering. It was seen to support the Soviet theoretical cosmological view — in particular that of Zeldovich and colleagues in Moscow — involving 'pancaking', rather than the American views of galaxy formation, at a time when 'cold war' rivalry flourished."*

The reception of the concept of the cosmic web was partly hindered by the suspicion that our data are incomplete and thus biased, as noted by Joe Silk in his interview in Lightman & Braver (1992). This shows that readers have not noticed one of the basic aspects of our study: to find the skeleton of the cosmic web we used data not only for galaxies, but also for near clusters of galaxies and active galaxies — these data were *complete*. Just the combination of various markers of the cosmic web allowed us to find almost all essential features of the web already in the late 1970's, much earlier than similar results were obtained from complete redshift data covering large contiguous regions on sky.

Soon we understood that the structure of the cosmic web contains information not only on the scenario of galaxy formation, but also on the nature of the dark matter — the basic constituent of the matter in the Universe. Further development of our understanding of the nature of the dark matter and of the structure of the cosmic web occurred hand in hand. It shall be described in the following sections.

5.3 Tartu Observatory in the early 1980's

5.3.1 *Southern base of Tartu Observatory*

All our results on the distribution of galaxies and the structure of the Universe have been achieved by analysing observations that had been made elsewhere. In the 1970's we attempted to measure radial velocities of galaxies using our 1.5-m telescope and later, in cooperation with the Byurakan Observatory, using their 2.6-m telescope. Results of these radial velocity measurements were published by Vennik et al. (1982); Vennik & Kaasik (1982). In Tartu Observatory we got redshifts for 62 galaxies, in Byurakan for 34 galaxies, located in groups of galaxies. However, the climate both in Estonia and Armenia does not allow a statistically adequate sample of galaxies to be observed.

After the Tallinn symposium in 1977 I understood that for cosmological studies an observatory in a very good climatical region is needed. To maintain such an observatory the collective efforts of many scientific centres are needed. Examples of such joint observatories are the European Southern Observatory, the National Optical Astronomy Observatory in the USA and similar observatories in Japan, China and elsewhere.

To investigate the possibility of the creation of a joint astronomical observatory in the USSR I visited in autumn 1977 Byurakan Observatory and discussed the idea with Viktor Ambartsumian, the most influential astronomer of the USSR. We both agreed that the creation of a joint astrophysical observatory would be very important for the further development of astronomy. Ambartsumian wrote a letter to the vice-president of the USSR Academy of Sciences with the suggestion to consider the possibility of creating a joint astrophysical observatory. With this letter all doors were open to start actual preparations.

Next year we started conducting expeditions to Central Asia to find a possible location for the joint observatory, because observing conditions within the Soviet Union were the best in the desert mountain regions of Central Asia. First I visited with some of my collaborators all Central Asian republics of the USSR, where local academies or universities already had their observatories. Soon we understood that the best location is in the Maidanak mountain region of the Uzbek SSR, situated a few hundred kilometers south of Samarkand. Here several observatories already had their Southern bases: the Moscow and the Leningrad universities and the Lithuanian Academy of Sciences. The major advantage of the Maidanak mountain was the availability of roads to the mountain, built for a military observatory a few kilometers away from the university observatories.

It soon became evident that it is not easy to create the joint observatory, because everyone wanted to have its own observatory. Thus we soon changed our initial plans and started to prepare our own Southern observational base. Our preparations to erect a Southern base were otherwise successful, but financing the construction of the telescope was a weak point. Finally we found a suitable source: in the late 1980's it was possible to get a subsidy (similar to grants at present time) from the central sources in Moscow. By explaining our request with the need to study the structure of the Universe on large scales, we received subsidies for several years. We spent them on the engineering of the telescope in the laboratory of the Leningrad University; we also bought a 2.3-m diameter mirror blank for the telescope, which was perfect for our needs.

During the preparations to build our Southern observational base I often visited Byurakan Observatory to discuss with Viktor Ambartsumian the progress of the project. One of my last visits was after the Armenian earthquake in 1988. When scientific discussions were over, we started to speak on other interesting events, and Ambartsumian told me an anecdote. *During the earthquake many people were buried under beton-blocks used in building standard houses at the time. When rescue workers found one man alive under the block, his first question was: "is Armenia already free". When he heard the answer that not yet, he asked to put him again under the block and take out when the country is free.* It was the time when

Fig. 5.11 Discussing with Viktor Ambartsumian the development of astronomy in a visit to Byurakan Observatory (author's photo).

Soviet republics had started their independence movements, and Ambartsumian was one of the leaders of the Armenian freedom movement.

Ambartsumian also told me the history of Armenia. Armenians are very proud of their culture and history. Armenia is located very close to the middle of the fertile crescent — the region of the development of the earliest human civilizations. The Armenian civilisation is one of the oldest ancient civilisations which is still preserved today. The Armenian state was formed more than 3 thousand years ago. Armenia was the first sovereign nation to accept Christianity as a state religion in 301. The summer residence of the head of the Armenian Apostolic Church, the Catholicos of All Armenians, is located near the Byurakan Observatory in the Southern hill of the Aragaz Mountain. In my visits to Byurakan I often made long walks to the Catholicos residence. For many centuries in the middle ages Armenia was divided between the Ottoman and Russian Empires. It regained for a short period its independence in 1918, before it was permanently restored in 1991. The most tragic event in Armenian history was the Armenian Genocide during World War I. Together with my colleagues from the Byurakan Observatory I visited the Memorial of the Genocide.

Business in Central Asia was rather bumbling, however, and when communicating with the outside world got easier at the end of the 1980's, we started to investigate whether we could place the telescope on the Canary Islands instead, which already hosted several telescopes stationed there by Western European countries. These plans had to be rescinded when the Soviet Union was abolished and Estonia became independent. We could not find the necessary funds for the construction of the telescope under the new circumstances — Estonia did not have the means and we were unable to obtain such a large sum from elsewhere. Thus we decided to cancel our efforts to build our own Southern telescope.

The development of astronomy in recent years has shown that there was a real need for a telescope constructed for the purpose of studying cosmological problems. There is a project underway between the astronomers of the USA, Japan, Germany, United Kington, Korea and several other countries — the Sloan Digital Sky Survey (SDSS). However, it is far more complex than we had anticipated when starting our telescope project. The originally planned SDSS I survey and its first extension SDSS II are finished, the new extension SDSS III is now underway, and there are many countries participating. Such projects are only feasible for transnational cooperations. Finland, incidentally, is taking part in a joint venture, the Nordic Optical Telescope, the telescope itself being situated in La Palma, Canarias Island.

As for small observatories akin to Tartu Observatory, the only option is to use observational data received from major joint observatories. If we have the means to process the observational data which has nowadays proven to be very large-scale, we can compete with other observatories quite successfully. Contemporary personal computers already have enough power for highly advanced scientific tasks, while the prices have settled to levels that are also affordable in Estonian conditions.

5.3.2 *Studies of ancient astronomy*

In addition to the philosophical seminars, ancient astronomy studies and conferences also turned out to be very fruitful. They were organised by Heino Eelsalu, who had a keen interest in ancient astronomy and the history of astronomy. Papers in this field were mostly published in local journals, so Eelsalu learned all the local languages of the Baltic region, starting with Latvian, our southern neighbours language, and ending with all Scandinavian languages, in addition to the major languages English, German, French and Russian. He started several series of publication of this topic: "Rara Astronomica in Estonia", "Archivalia Cosmographica" etc. as Teated (Communications) of Tartu Observatory.

Eelsalu found that rock art in Karelia and ancient Estonian folk songs *regilaul* contain important astronomical and historical information. So he initiated the formation of an *Estonian Society of Prehistoric Art* in the 1980's. The Society made expeditions to Karelia, to study rock art from Lake Onega that date to 4,000–2,000 B.C. Many of our astronomers participated in these expeditions, including my daughter Maret. In these studies we had a very close collaboration with one of the best specialists of Estonian ancient culture, Mikk Sarv. Mikk writes on the meaning of *regilaul* as follows.[1]

"The term 'regilaul' seems to derive from the word regi (sleigh) that denotes the oldest means of transport. A sleigh would take one across a boggy patch of land, even in summertime, when a carriage would be hopelessly stuck. In the winter, a sleigh would carry you straight across swamps and lakes and along rivers and winter roads. Sleighs would help traders to cover thousands of kilometres on their journeys and to connect people from coast to coast. Sleighs leave behind them two trails running towards the horizon on the plain fields. These trails are always next to each other but they can never meet. A song would help to make long journeys on a sleigh shorter, keeping the travellers awake and lifting their spirits.

Estonian regilaul is an ancient and powerful means of communication for a headstrong nation with the rest of the world. To use it one needs to embrace timelessness and to picture in one's soul the two trails behind a sleigh heading towards the horizon. Next to each verse there is another one that takes you in the same direction in a slightly different manner. Each verse moves along its trail carried forward by words beginning with and containing similar sounds. Singers cast ancient myths, fairy tales and shaman journeys into the form of regilaul and handed them down from one generation to the next. At the same time, regilaul as a form of art is also a highly sophisticated means of expression. Skilled singers can transform ordinary speech into regilaul and perform it at any time. This used to be common practice during festive rituals, such as weddings, or during incantations and swearing-in ceremonies.

A song festival with choirs from all corners of Estonia was held in Tartu in 1869 to celebrate the abolition of serfdom. Since then song festivals have been held every five years as a tribute to the idea of freedom. In 1989, one-fifth of the Estonian population gathered on the song festival grounds in Tallinn to sing their

---

[1] http://www.estinst.ee/publications/estonianculture/I_MMIII/sarv.html

country free from occupation. The Estonian regilaul has had a central role in all these instances of communal singing. If we take the very literal meaning of the word 'laulupidu' (song festival) then it means "to hold a song" (laulu pidama): i.e. to stop all other activities so that people can devote themselves to the song. For us Estonians, singing can be used to change the world."

I had once a personal experience of a spontaneous singing of a 'regilaul' at our family gathering at my country-home Egeri in summer 1999. One of my relatives, the wife of the son of my brother Rein, was from the South-Estonian region, where the 'regilaul' tradition is still alive. During the family gathering she started to sing, and I was able to record the event (see the movie in the website which accompanies this book).

Another small remark of my own experience with song festivals. I have participated in several festivals. I had a very special experience at the festival in 1969, which celebrated 100 years since the first festival. I was there with Maret who was at the time 10 years old. On the last day of the festival we were already rather tired and had a 200 km drive to return home, so we left the festival and listened to it on our car radio. When the official program was over the joint choir started to sing spontaneously "Land of my Fathers, Land That I Love" (Mu isamaa on minu arm). Then something extraordinary happened: the whole festival public, more than a hundred thousand people, stood up and joined the choir. The people had tears in their eyes. This was the first solemn but very powerful peaceful demonstration of our idea of freedom. This song became an unofficial anthem for the Estonians during the occupation years. It took more than twenty years before real freedom was achieved, but this was the beginning. For many Estonians, including me, this song is the true anthem due to its role during the Soviet period, and its solemn melody and lyrics.

Eelsalu initiated a series of interdisciplinary conferences devoted to paleoastronomy and related topics. One of these conferences was held at the Viljandi theater on April 21, 1983. I remember very well this conference. The theatre hall was heavily crowded by Tartu University students, schoolchildren from Tartu and Viljandi, and leaders of the Estonian culture. Scientific presentations were combined with recitation performances on ancient religion, accompanied by folk music, performed by Mikk Sarv and his ensemble. At that time there were practically no other events supportive of patriotic sentiments. The atmosphere was very similar to the song festivals in 1989, where Estonia people sang the country free.

Chapter 6

The nature of dark matter

By the end of the 1970's most objections against the dark matter hypothesis had been rejected. In particular, luminous populations of galaxies were found to have lower mass-to-luminosity ratios than expected previously, thus the need of extra dark matter both in galaxies and clusters became even stronger. However, there remained three problems:

- It was not clear how to explain the Big Bang nucleosynthesis constraint on the low density of matter, and the smoothness of the Hubble flow — the main argument in favour of the classical cosmological paradigm.
- If the massive halo (corona) is neither stellar nor gaseous, of what stuff is it made of?
- And a more general question: in Nature everything has its purpose. If 90% of matter is dark, then there must be a reason for its presence. What is the role of dark matter in the history of the Universe?

First I shall discuss baryons as dark matter candidates.

6.1 Baryonic dark matter

6.1.1 *Early discussions on the nature of dark matter*

The nature of dark matter has been discussed from the very beginning of the discovery of the possible presence of dark matter in clusters and galaxies. It is clear that initially only baryonic dark matter candidates were considered.

Zwicky (1937) writes: *"We must know how much dark matter is incorporated in nebulae in the form of cool and cold stars, macroscopic and microscopic solid bodies, and gases"*. Oort (1940) investigated the structure of the galaxy NGC 3115 and found that on the periphery the local mass-to-luminosity ratio is very high, of the order of 250 in Solar units. He notes: *"The spectrum of the nebula shows the characteristics of G-type dwarfs. Since M/L cannot be much larger than 1 for such stars, the remainder must mainly consist either of extremely faint dwarfs having an average mass to light of about 200, or else of interstellar gas or dust".*

The study by Kahn & Woltjer (1959) of the dynamics of the Local Group of galaxies is devoted mainly to the discussion of the possible nature of the matter which makes the system dynamically stable. The authors consider stars as possible candidates of the invisible matter and find that this is unlikely. As a possible candidate of the intergalactic matter they suggest ionised hydrogen (and helium). They find that the temperature of the gas should be about 5×10^5 degrees. The cooling time for such gas is of the order of 10^{10} years, thus the whole system is stable.

6.1.2 Stellar or gaseous dark coronae

When I started the modelling of the Galaxy and M31, my basic goal was to represent in the model all known stellar populations as well as possible. Our previous experience had shown that the best function to represent the mass and light distribution of individual populations is the generalised exponential function with variable shape parameter N. Using photometric data it was relatively easy to find the shape parameter, as well as scaling parameters defining the effective radius and the total luminosity of the population. To find the mass distribution from these data, only one additional parameter was needed: the mass-to-luminosity ratio of the population. This was the critical step in the model construction. As described in the previous Chapters I used various methods to calibrate the M/L values of galactic populations.

In my Athens talk in September 1972 I considered the possible nature of the corona, and found that neither stars nor neutral gas can be considered as possible candidates for the coronal matter, see Ch. 4. In the Caucasus Winter School in January 1974 the problem of the nature of the corona was at the center of discussions. I listed my arguments against the stellar nature of galactic coronae.

Physical and kinematical properties of stellar populations depend almost continuously on the age of the population. The oldest have the lowest metallicity and M/L-ratio, and there is no place to put the new population into this sequence. The continuity of stellar populations of various age is reflected also in their kinematic characteristics, such as the velocity dispersion and galactocentric centroid velocity, expressed in the Strömberg diagram, see Fig. 6.1. Here open circles mark metal-poor populations, while dots mark populations with normal metal abundance. Populations were chosen using various physical parameters — variables of different type, star clusters of various age and metallicity etc. We see that there

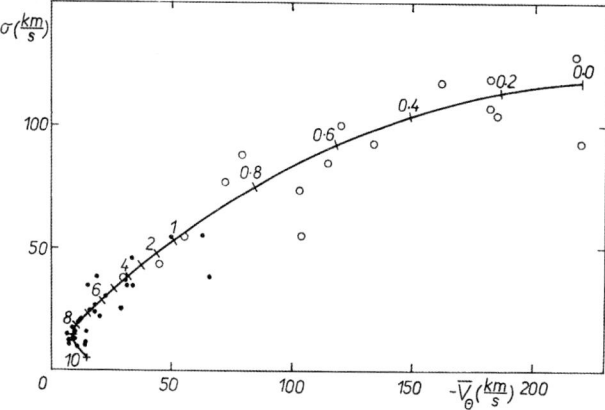

Fig. 6.1 The Strömberg diagram for populations. In the horizontal axis we show the heliocentric centroid velocity of the population in the direction of the Galactic rotation; in the vertical axis we plot the mean velocity dispersion $\sigma = 1/3\sqrt{\sigma_R^2 + \sigma_\theta^2 + \sigma_z^2}$. Open circles are for metal-poor populations, dots for populations with normal metal abundance. The numbers give the birth-dates in 10^9 years starting from the formation of the oldest populations, assuming for the age of the Galaxy 10^{10} years (Einasto, 1974b).

exists an almost continuous sequence of populations. Populations in our galactic models are actually representatives of populations which have a certain range of ages and metallicities. Similar sequences exist between kinematical and physical parameters, such as colour, M/L-ratio etc.

The dark population is almost spherical and thus non-rotating. It has a much larger radius than all known stellar populations. Thus, in order to be in equilibrium in the Galactic gravitational potential, its objects must have much higher velocity dispersion than all known stellar populations. For this reason in the Strömberg diagram the dark population is located far away from all known populations above the diagram seen in Fig. 6.1. The conventional halo population, consisting of old metal-poor stars, has the highest velocity dispersion. This population has the lowest M/L-ratio among known populations. In contrast, the new dark population must have a very high M/L-ratio.

This simple comparison of properties of known populations with those of the dark corona shows that the kinematics, the spatial distribution, and physical properties of the dark population are completely different from properties of known populations, and that there exist no intermediate ones which could form a link between known populations and the dark corona.

This isolated location in the Strömberg diagram is an indirect argument against the stellar origin of the new population. The M/L value and the spatial distribution of the dark population differ greatly from respective properties of known stellar populations, thus it must have been formed much earlier than all known populations to form the gap in relations between various physical, kinematical and spatial structure parameters. The total mass of the new population exceeds the masses of known populations by an order of magnitude, thus we have a problem: How to transform at an early stage of the evolution of the Universe most of the gas into these invisible stars? It is known that star formation is a very inefficient process: in a star-forming gaseous nebula only about 1% of matter transforms to stars (as noted already in Chapter 3).

This topic was discussed in the 1975 Tallinn conference on dark matter. If the dark halo is of stellar origin, then these stars form an extended population around galaxies. From hydrodynamical considerations it follows that coronal stars must have much higher velocity dispersion than the stars belonging to the ordinary halo. No fast-moving stars as possible candidates for a stellar dark halo were found (Jaaniste & Saar, 1975). The publication story of the last paper is interesting. First the authors submitted the paper to "Astrophysics and Space Science", but the editor Prof. Salomon Pikelner rejected the paper with justification: *"you already have a paper on dark matter"* (Chernin et al., 1976). His attitude to the dark matter subject was characteristic for the majority of the astronomical community: that this problem is not a topical one. He could not imagine that in years to come tens of thousands of papers will be written on this subject!

It is interesting to note that many astronomers did not understand the difference between the global and local dark matter even many years after their different nature was noticed and widely discussed. For instance, at the 1977 Tallinn IAU Symposium on Large Scale Structure of the Universe I had a short communication on the discrepancy of M/L-ratios (Einasto, 1978). In the discussion Jerry Ostriker argued that this is not a real discrepancy since a *measurement* is divided by an *assumption*. Actually I compared global measured M/L-ratios with local measured M/L-ratios using different data — global mass and light distribution from rotation curves and photometric profiles of galaxies, and local M/L-ratios found for star clusters, and in some cases for individual stellar populations. Jerry was supported by Beatrice Tinsley who argued that most stellar mass is contained in invisible dwarf stars (yes, but not so much). Both Jerry and Beatrice have apparently not studied properties of stellar populations in such detail, and did not appreciate the need for *independent dynamical calibration* of stellar population M/L-ratios, which is the central problem when we discuss the possible stellar nature of dark matter. In contrast, the Faber & Gallagher (1979) review of masses of galaxies is specifically addressed to the analysis of M/L-ratios of galaxies in general, and also of galactic populations.

Gaseous coronae of galaxies and clusters were discussed in the early 1970's by Field (1972), Silk (1974), Komberg & Novikov (1975) and others. The conclusion was that gaseous coronae of galaxies and clusters cannot consist of neutral gas since the intergalactic hot gas would ionise the coronal gas. My own argument was that neutral gas would be seen through its 21-cm line, but such corona is not observed. The preliminary conclusion following this argumentation was that coronae can consist of hot gas, a suggestion made already by Kahn & Woltjer (1959) and Einasto (1974a). However, early observations by the UHURU X-ray satellite suggested that the mass of the hot gas is not sufficient to stabilise clusters of galaxies, see the discussion in the previous Chapter. I used these data as an indication that the presence of hot gas in clusters does not solve the mass discrepancy observed in clusters of galaxies (Einasto et al., 1974b). However, the morphological segregation of companion galaxies suggests that at least part of the coronal matter must be gaseous (Einasto et al., 1974c).

Modern data confirmed that a fraction of the coronal matter around galaxies and in groups and clusters of galaxies consists indeed of X-ray emitting hot gas, but the amount of this gas is not sufficient to explain the flat rotation curves of galaxies and the masses of clusters of galaxies.

6.1.3 *Nucleosynthesis constraints of baryonic matter*

According to the Big Bang model, the Universe began in an extremely hot and dense state. For the first second it was so hot that atomic nuclei could not form — space was filled with a hot soup of protons, neutrons, electrons, photons and other short-lived particles. Occasionally a proton and a neutron collided and stuck together to form a nucleus of deuterium (a heavy isotope of hydrogen), but at such high temperatures they were broken immediately by high-energy photons. When the Universe cooled off, these high-energy photons became rare enough that it became possible for deuterium to survive. These deuterium nuclei could keep sticking to more protons and neutrons, forming nuclei of helium and other light elements. This process of element-formation is called "Big Bang nucleosynthesis", suggested by Alpher et al. (1948). The denser the proton and neutron "gas" is at this time, the more light elements will be formed. As the Universe expands, the density of protons and neutrons decreases. Neutrons are unstable unless they are bound up inside a nucleus. After a few minutes the free neutrons will be gone and nucleosynthesis will stop. The relationship between the expansion rate of the Universe and the density of protons and neutrons (the baryonic matter density) determines how much of each of these light elements are formed in the early Universe.

According to nucleosynthesis data baryonic matter makes up 0.04 of the critical cosmological density, assuming a Hubble constant $h \sim 0.7$. Only a small fraction, less than 10%, of the baryonic matter is condensed to visible stars, planets and other compact objects. Most of the baryonic matter is in the intergalactic matter (Warm-Hot-Intergalactic-Medium, Lyman$_\alpha$-forest), some of it concentrated also in hot X-ray coronae of galaxies and clusters.

After my talk in the Caucasus Winter School and publications of papers by Einasto et al. (1974b) and Ostriker et al. (1974), Zeldovich started to think about the discrepancy between new density estimates and the nucleosynthesis data. He writes (Zeldovich, 1975): *"A major problem is now ripe for solution: the conflict between the mean density of matter in the Universe, and the interstellar deuterium abundance. This deuterium is believed to be primordial, having been produced by nucleosynthesis during the few seconds immediately after the singularity."* As a solution he suggested a cosmological model with strong inhomogeneities in the baryon density near the singularity, where most of the volume has low density as suggested by nucleosynthesis data, and the rest has higher density as suggested by new density estimates. This model is rather speculative and Zeldovich soon abandoned it.

The results of early discussions on the nature of dark halos were inconclusive — no appropriate candidate was found. For many astronomers this was an argument against the presence of dark halos.

6.2 Non-baryonic dark matter

6.2.1 *Cosmic microwave background radiation*

According to the current understanding, the Universe began with a Big Bang and was initially very hot. It expanded rapidly and cooled, and at a certain epoch (about 300 hundred thousand years after the Big Bang) was cool enough for atoms to recombine. The effective temperature of this radiation drops as the Universe expands. Alpher et al. (1948) predicted that the radiation from this epoch should be still present. The authors predicted that the present temperature of this cosmic microwave background (CMB) radiation should be approximately 5 degrees Kelvin. The possibility of detecting this radiation was suggested by Doroshkevich & Novikov (1964).

The CMB radiation was actually detected by the American radio astronomers Penzias & Wilson (1965). They worked at Bell Labs in New Jersey with ultrasensitive cryogenic microwave receivers for radio astronomy observations. Their goal was to eliminate all possible terrestrial sources of noise. Their finding was,

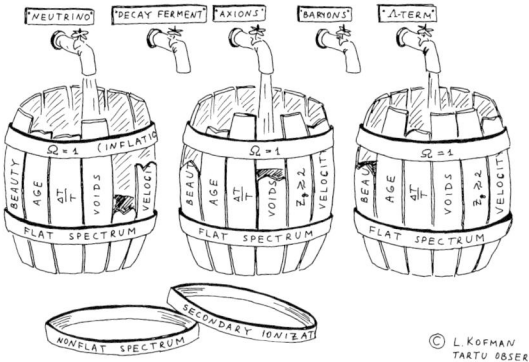

Fig. 6.2 The barrel diagram — principal models of the formation of structure in the Universe using different candidates of dark matter are shown as barrels. There are three main candidates for dark matter: neutrino, axion (and other cold particles), and the cosmological constant. The barrels are hooped together by two principal assumptions, $\Omega = 1$, and a flat spectrum of initial perturbations. If these assumption do not work, there are some hoops in reserve: a non-flat spectrum and secondary ionization. Various observational tests are expressed as staves. The height of a stave indicates the degree of accordance of model with this particular test. One test is the beauty or internal harmony of the model. The level of the liquid in the barrel is equal to the height of the shortest stave, which determines the degree of acceptance of the model. If necessary, a cocktail from several liquids can be made, or some ferment as neutrino decay is added. Idea and artwork by L. Kofman (Einasto et al., 1987).

after eliminating all possible known sources, a certain noise remained. This noise was identified by Dicke et al. (1965) as the CMB radiation. The temperature of the radiation is 2.7 K, and the spectrum peaks in the microwave range, corresponding to a 1.9 mm wavelength.

The discovery of the CMB radiation is a strong test of the Big Bang model of the Universe. CMB radiation contains a lot of information on the structure of the Universe at early times. CMB observations give the strongest argument against the Steady State theory of the formation of the Universe. The CMB radiation has a black-body spectrum with deviations of the order of 3×10^{-5} — this is the most exact black body known (Mather et al., 1990, 1994). This accuracy can only be explained if the Universe was hot at early times as assumed in the Big Bang model.

6.2.2 *Fluctuations of the CMB radiation*

The detection of the CMB radiation was a very important observation which cast doubts on baryonic matter as the dark matter candidate. Initially the Universe was very hot, the gas was ionised, and all density and temperature fluctuations of the primordial soup were damped by very intense radiation. But as the Universe expanded, the gas cooled, and at a certain epoch called recombination the gas became neutral. From this time on, density fluctuations in the gas had a chance to grow by gravitational instability. Matter is attracted to the regions where the density is higher, and it flows away from low-density regions. But gravitational clustering is a very slow process. Calculations showed that density fluctuations are of the same order as temperature fluctuations. Thus astronomers started to search for temperature fluctuations in the CMB radiation. None were found. As the accuracy of measurement increased, lower and lower upper limits for the amplitude of CMB fluctuations were obtained.

Let me recall one moment in our search for understanding of cosmic evolution in the late 1970's. At the Tallinn symposium Parijskij (1978) made a report on his search for temperature fluctuations of CMB radiation with RATAN-600, the largest and most sensitive radio telescope at the time. No temperature fluctuations were found. The upper limit was about 10^{-4} of the mean temperature. After the talk Zeldovich discussed these results with Parijskij and expressed his opinion that something must be wrong in his observations. Theoretical calculations show that at the epoch of recombination the density (and temperature) fluctuations must have an amplitude of the order of 10^{-3}, otherwise structure cannot form, since the gravitational instability that is responsible for the growth of the amplitude of fluctuations works very slowly in an expanding Universe — the amplitude of fluctuations grows linearly with the expansion factor $a = 1/(1+z)$, and the redshift of CMB radiation is $z_{CMB} \approx 1000$.

What was actually wrong was our understanding of the nature of dark matter.

Fluctuations of the CMB were finally measured by the COBE satellite by Smoot et al. (1992) and Bennett et al. (1996). Four-year COBE measurements yielded for the amplitude of temperature fluctuations 15.3 ± 3 μK, and for the power law spectral index $n = 1.2 \pm 0.3$. Modern WMAP and Planck satellite data give even more accurate values of these parameters. Seven-year observations with WMAP satellite yield for the spectral index of fluctuations a value $n = 0.968 \pm 0.012$ Komatsu et al. (2011). The angular power spectrum of CMB temperature fluctuations has the first maximum at wavenumber $l = 200$, which suggests that the total matter/energy density of the Universe is equal to the critical density (Bennett et al., 2003).

6.2.3 Neutrinos as dark matter candidates

Already in the 1970's suggestions were made that some sort of non-baryonic elementary particles, such as massive neutrinos, may serve as candidates for dark matter particles. There were several reasons to search for non-baryonic particles as a dark matter candidate. First of all, no baryonic matter candidate fit the observational data. Second, the total amount of matter is of the order of 0.2–0.3 in units of the critical cosmological density, while the nucleosynthesis constraints suggest that the amount of baryonic matter cannot be higher than about 0.04 of the critical density.

The only known non-baryonic particle was the neutrino, thus it was natural that first neutrinos were considered as dark matter particle candidates. Szalay & Marx (1976) considered neutrinos as dark matter candidates using as argument the total density of matter and the density of known baryonic matter. A similar suggestion was made by Rees (1977). An experimental study by Lubimov et al. (1980) suggested that electronic neutrinos might have finite rest mass on the order of $m_\nu \approx 30\,\text{eV}$. Bisnovatyi-Kogan & Novikov (1980) used this estimate to show that if this mass estimate is correct, the total mass density due to neutrinos is close to the critical density, and neutrinos could be dark matter particles. Chernin (1981) showed that, if dark matter is non-baryonic, then this helps to explain the paradox of small temperature fluctuations of the cosmic microwave background radiation. Density perturbations of non-baryonic dark matter already start growing during the radiation-dominated era, whereas the growth of baryonic matter is damped by radiation. If non-baryonic dark matter dominates dynamically, the total density perturbation can have an amplitude of the order 10^{-3} at the recombination epoch, which is needed for the formation of the observed structure of the Universe.

This problem was discussed at a conference in Tallinn in April 1981. Here all prominent Soviet cosmologists and particle physicists participated. The central problem was the nature of the dark matter. In the conference banquet Zeldovich gave an enthusiastic speech: *"Observers work hard in sleepless nights to collect data; theorists interpret observations, are often in error, correct their errors and try again; and there are only very rare moments of clarification. Today it is one of such rare moments when we have a holy feeling of understanding secrets of the Nature."* Non-baryonic dark matter is needed to start structure formation early enough. The non-baryonic nature of dark matter explains the role of dark matter in the evolution of the Universe, as well as the discrepancy between the total cosmological density of matter and the density of baryonic matter, as found from the nucleosynthesis constraint.

The non-baryonic dark matter and the relationship between particle physics and cosmology were discussed in detail during the "Study Week on Cosmology and Fundamental Physics" in the Vatican, September 28–October 2, 1981 (Brueck et al., 1982). The concept of non-baryonic dark matter was generally accepted for the same reasons as discussed in Tallinn in April. Joe Silk (1982) discussed in detail fundamental tests of galaxy formation theories, based on adiabatic and isothermal scenarios. He also came to the conclusion that dark matter must be non-baryonic for the same reasons as mentioned above. In addition to neutrinos he considered

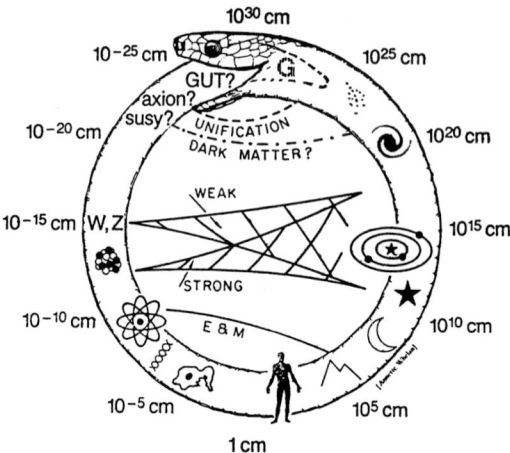

Fig. 6.3 The modified Uroborus. There are links between the microworld of particles (left), and the macroworld of cosmos (right) (Primack, 1984).

photinos as one of the possible candidate for the dark matter, see the next section. Joe concludes his analysis as follows: *"It seems that the large-scale structure of the Universe is intimately related to its microscopic structure on elementary particle scales. This is perhaps not surprising if one recalls that it is the initial seed of fluctuations at the Planck epoch that are likely to determine the asymptotic growth of irregularities in the expanding Universe."*

These two conferences mark probably the birth of astroparticle physics. Cosmologists and particle physicists understood that properties of the micro-world and macro-world are intimately related. The relationship between the micro- and macro-world can be expressed by the modified Uroborus-symbol, as shown in Fig. 6.3.

Uroborus is an ancient symbol depicting a serpent or dragon eating its own tail. It symbolizes the recycling and renewal of the Universe. Primack (1984) and Rees (2000) noticed that the Uroborus symbol can be used to illustrate the links between the micro-world, shown in the left part of the Figure, and the macro-world in the right part. Let us put the human at the center (lowest part) of the serpent near the scale 1 cm. There are left–right connections across it: medium small to medium large, even smaller to even larger. Properties of atoms determine properties of objects on the Earth, properties of nuclei of atoms determine properties of the Sun and stars. Dark matter holds together galaxies and systems of galaxies. The inflation model suggests that properties of the whole Universe depend on interactions on the grand unification scale (Primack, 1984).

Now, finally, the presence of dark matter was accepted by leading theorists. The search of dark matter can be illustrated with the words of Sherlock Holmes *"When you have eliminated the impossible, whatever remains, however improbable, must be the truth"* (cited by Binney & Tremaine (1987)).

This was, however, not the end of the story. The neutrino-dominated or hot dark matter generates almost no fine structure of the Universe, as shall be discussed in the next Chapter. Thus some other solution had to be found.

6.2.4 *Cold dark matter*

In the early 1980's difficulties with baryonic dark matter were well known and non-baryonic dark matter was seriously considered. Also difficulties with neutrino based dark matter were known, thus dissipationless particles heavier than neutrinos were suggested by Blumenthal et al. (1982), Bond et al. (1982), and Peebles (1982). Here hypothetical particles like axions, gravitinos or photinos play the role of dark matter.

Primack & Blumenthal (1984) discussed arguments that dark matter is not baryonic, based on the deuterium abundance, and the absence of small-scale fluctuations in the microwave background radiation. The authors suggested the following classification of elementary particles as dark matter candidates: hot, if free streaming erases all but supercluster-scale fluctuations; warm, if free streaming erases fluctuations smaller than galaxies; and cold, if free streaming is unimportant. Hot particles are light (~ 100 eV) and remain relativistic until just before recombination. Warm particles are 10–100 times heavier and thus become non-relativistic sooner. Cold dark matter candidates are either very heavy particles that become non-relativistic very early (gravitinos, photinos), or are particles that have almost zero peculiar velocity (axions) (Faber, 1984).

The nature of dark matter can be checked also by numerical simulations of the structure evolution based on various assumptions on the form of the power spectrum of density perturbations for different species of the non-baryonic dark matter, and by comparison of simulations with the observed large-scale distribution of galaxies.

With these simulations the dark matter problem was closely related to the problem of the structure of the Universe on large scales, i.e. the structure of the cosmic web.

Numerical simulations made by the Zeldovich team in the 1970's were based on the assumption that small-scale density perturbations were damped, due to the interaction of matter and radiation during the radiation dominated era of cosmic history. Such a perturbation spectrum is essentially equivalent to the perturbation spectrum of neutrino-dominated dark matter due to the high speed of neutrinos which smear out small-scale perturbations. As shall be discussed in the next Chapter, the neutrino-dominated model has several difficulties: the absence of the fine structure of the cosmic web, and the late formation of superclusters (Frenk et al., 1983; Melott et al., 1983).

One of the first to simulate the formation of the cosmic web using both neutrino-dominated (Hot Dark Matter) and axion-dominated (Cold Dark Matter) dark matter was Adrian Melott. In summer 1983 Adrian visited Moscow to discuss his models with the Zeldovich team. We were also interested in analysing his models, so we invited him to Tallinn. He came together with Anatoly Klypin and Sergei Shandarin. Our first goal was to compare both models using our connectivity and multiplicity criteria. The first attempt failed, as our computers were too slow to analyse such large datasets as given by models with a resolution 32^3 particles. Then Enn asked us to pause for a few days, and he will try to elaborate a new program for the analysis which would admit the use of large datasets.

Enn invented a different algorithm to calculate the connectivity of a sample of particles, which was several hundred times faster than the previous one (among us we called the program for its exceptional efficiency WW — Wunderwaffe, to parody Nazi-Germany attempts to create wonder-weapons). So, a few days later we continued our analysis. Results of the analysis showed that the CDM model fits both the connectivity as well as the multiplicity criteria. Thus the CDM model is to be taken seriously. The main body of the paper by Melott et al. (1983) was already written, but not finished, because we did not yet have results of the connectivity and multiplicity tests. In a few days these tests were completed with the new program. The paper was finished and sent to the publisher rapidly. The paper appeared at the end of 1983, just in time. This paper is probably the first one where the advantages of the CDM model were discussed using several quantitative tests.

In the CDM scenario the structure formation starts at an early epoch, and superclusters consist of a network of small galaxy filaments, similar to the observed distribution of galaxies. Presently, the CDM model with some modifications (the cosmological constant or Λ term was added) is the generally accepted model of structure evolution. The properties of the Cold Dark Matter model were analysed in detail in the paper by Blumenthal et al. (1984).

In 1988 Joel Primack visited Tartu Observatory, and we discussed among other topics the formation of the CDM concept. He told me the story of the Blumenthal et al. (1984) paper (another description of the formation of the CDM concept is given in the interview by Sandy Faber to the American Institute of Physics). Sandy and Joel met during the Vatican Study Week in 1981. Joel is a theoretical physicist while Sandy is an exceptionally talented observer with good theoretical background. So they started to think about how to explain the formation of galaxies and the structure of the Universe from tiny fluctuations of the primordial hot gas. Step-by-step they understood that in order to explain the smallness of CMB fluctuations it is needed to assume that dark matter particles must be not only non-baryonic, but heavier than neutrinos, which allows an earlier start to the growth of density fluctuations. The terms Hot, Warm and Cold Dark Matter were suggested by Joel. The Blumenthal et al. paper is written very clearly. With the acceptance of the CDM model the modern period of the study of dark matter begins.

6.2.5 Dark matter in dwarf galaxies

The importance of the possible presence of dark matter in dwarf galaxies was clear already in the late 1970's. In those years Vera Rubin measured rotation curves of bright galaxies. I wrote a letter to Vera asking whether it would be possible to measure rotation curves also for dwarf galaxies. We met during the IAU General Assembly in 1997 and discussed the issue. She remembered my letter and answered that it was very difficult to measure rotation curves of dwarf galaxies using the equipment she had.

Dwarf spiral and irregular galaxies contain hydrogen gas, and using radio observations of the 21-cm line it is possible to derive rotation curves for faint galaxies. Actually the Bosma (1978) thesis contains rotation data for a number of dwarf galaxies with flat rotation curves at the level between 50 and 150 km/s. Thus his data strongly suggest the presence of dark matter in dwarf galaxies.

Faber & Lin (1983) calculated masses and mass-to-luminosity ratios for dwarf spheroidal satellites of our Galaxy using the tidal limit theory. The mean mass-to-luminosity ratios are about one order of magnitude larger than those of globular clusters. Thus dwarf spheroidal galaxies contain large amounts of non-luminous matter and resemble in this regard bigger galaxies. In contrast, globular clusters of similar luminosity have low M/L-ratios and contain no dark matter.

Lin & Faber (1983) discussed the implications of non-luminous matter in dwarf spheroidal galaxies. They showed that phase-space constraints in dwarf galaxies sets a lower limit of several hundred eV on particle mass, if the dark matter consists of noninteracting fermions. This limit rules out neutrinos as dark matter particles.

6.2.6 Missing satellite problem and warm dark matter

As the power of computers improved larger and more detailed simulations were performed. This allowed the study of the evolution of fine structure in dark matter halos and their substructure represented by subhalos. The best possibility to compare results of these calculations with observations are satellites of nearby galaxies, especially satellites of our own Galaxy. A detailed comparison of the inner structure of halos with the structure and abundance of satellites of our Galaxy was made by Klypin et al. (1999). The authors find that the ΛCDM model predicts that there should be a remarkably large number of DM satellites with circular velocities $V_{circ} \approx 10 - 20$ km s^{-1} orbiting our galaxy, approximately a factor of 5 more than the number of satellites actually observed in the vicinity of the Milky Way or Andromeda galaxy. Thus these calculations raised the question: Where are the missing satellites?

This result may be explained by dissipative processes such as gas cooling, supernovae explosions, star formation and other processes that decouple the dynamical evolution of the baryons from that of dark matter (Parry et al., 2012).

Springel et al. (2008a) and Navarro et al. (2010) performed a detailed numerical study of the distribution of mass and velocity dispersion of DM halos in the framework of the *Aquarius Project*. The Aquarius Project addresses the internal structure of halos by studying the highly non-linear structure of Galaxy-sized CDM halos in detail. The authors were particularly interested in the inner regions of these halos and of their substructures, where the density contrast exceeds 10^6, and the astrophysical consequences of the nature of dark matter may be most clearly apparent. In the highest resolution simulation the number of particles within a radius of 50 kpc was about 1.5 billion. The authors find that the mass profiles of halos are best represented by the Einasto profile (3.1). The shape parameter N varies slightly from halo to halo.

One possible way to solve the missing satellite problem is to assume that dark matter is not cold but warm. Warm dark matter particle mass is ~ 1 keV, whereas the particle mass of cold dark matter is ~ 1 GeV. In this case the power spectrum of density fluctuations is truncated. Warm dark matter (WDM) particles decouple from the other particles in the early Universe with relativistic velocities and become nonrelativistic when about a Galactic mass is within the horizon. This possibility was investigated by Polisensky & Ricotti (2011), Lovell et al. (2012) and a number of other authors.

Lovell et al. (2012) resimulated Aquarius N-body halos with the power spectrum suppressed at small scales, as expected in the WDM case, using a resolution much higher than in previous studies. The authors find that WDM halos form later and are less concentrated than CDM halos. They conclude that WDM is one possible explanation for the observed kinematics of the satellites, and the relatively small number of dwarf satellites.

6.2.7 Searches for dark matter particles

As discussed above, in the early 1980's it was clear that dark matter must be non-baryonic. The first natural candidate for DM particles was the massive neutrino. However, massive neutrinos (Hot Dark Matter) cannot form the dominating population of DM particles, since the large-scale-structure of the cosmic web would be in this case completely different from the observed structure. For this reason hypothetical weakly interacting massive particles were suggested, which form Cold Dark Matter. The CDM model satisfies most known astronomical restrictions for the dark matter.

Until recently it was thought that DM particles form a fully collisionless medium. However, it is natural to assume that in the most realistic cases, where the DM comprises some sort of elementary particle, those particles may have other, non-gravitational couplings to the rest of the matter. If this is the case, the phenomenology of DM could in principle be much richer. Indeed, there has been a lot of recent activity trying to detect DM particles in high precision nuclear recoil experiments.

Results from neutrino oscillation experiments require at least one of the neutrinos to have a mass not less than ~ 0.05 eV (e.g. Dolgov (2002)). This immediately implies that the corresponding density parameter $\Omega_\nu \gtrsim 0.001$, i.e. approaching the density parameter of the baryons visible in the form of stars! Although neutrinos cannot form the dominant component of DM, due to reasons discussed above, it shows that the general idea of the existence of dark matter in Nature is surely not fiction.

Springel et al. (2008b) studied the implications for the detectability of dark matter annihilation within the Milky Way's dark matter halo. Such detections could be indirect hints for the existence of the DM particles. Particularly interesting is the cosmic ray positron anomaly as revealed by the measurements of the PAMELA and HESS experiments. This anomaly could be an indirect indication for the existence of an annihilating or decaying DM particle with a mass at the TeV scale.

Our cosmology team has started collaboration with the high-energy physics team lead by Martti Raidal. I met Martti during my visit to Fermilab at 2000, and asked him whether he is interested in coming back to Estonia. He worked many years in leading high-energy physics centers, most notably at CERN. Now he is back and has created a very strong team of young physicists in Tallinn in the National Institute of Chemical Physics and Biophysics. Presently we have a joint Center of Excellence "Dark Matter, Astroparticle Physics and Cosmology" together with the Raidal team, who has access to CERN experiments, and represents the particle physics aspect of the problem. Our team adds knowledge in cosmology.

We have published a number of joint studies, among them by Hütsi et al. (2009, 2010, 2011), Tempel et al. (2012b), Tempel et al. (2012a), and Hektor et al. (2013).

The story of the birth of the last papers is interesting.

One of the recently analysed datasets of interest to investigate possible effects of dark matter comes from the Fermi Gamma-ray Space Telescope, launched on June 11, 2008. Its Large Area Telescope (LAT) can detect gamma rays in an energy interval from about 20 MeV to 300 GeV. Such gamma rays are emitted only in the most extreme conditions, by particles moving at a speed close to the speed of light. If the existing cosmological dark matter is a thermal relic consisting of weakly interacting massive particles, DM annihilations into standard model (SM) particles or gamma-rays should provide evidence of DM particle annilihilations. In this scenario the first emerging signal of DM annihilations is expected to appear either from Galactic centre or from other nearby DM- dominated objects.

Indeed, recently Weniger (2012) claimed that there is 4.5σ evidence of a monochromatic gamma-ray line from the Galaxy centre with an energy $E = 130\,\mathrm{GeV}$ present in the Fermi Large Area Telescope data. Christoph Weniger is a postdoc at the Max-Planck-Institut für Physik in Munich, a previously unknown young scientist. He published in arXiv his first papers on this subject on March 6 and April 12, 2012. An overview of Fermi LAT's search for signatures from dark matter annihilations was given by Bringmann et al. (2012), received by the publisher on March 26, 2012, and was sent immediately to arXiv. However, the physics community noticed the work only after Weniger gave a seminar talk at CERN on April 25. Martti Raidal and his collaborator Andi Hektor participated in the seminar. Next day they contacted Elmo Tempel via Skype and discussed if we can add something to the solution of the problem. Martti and Andi had no previous experience in cosmology, and Elmo in particle physics. Within about a week they found a way how to reduce Fermi data in a better way, and sent their analysis to arXiv on May 4 (Tempel et al., 2012b).

Initially the astroparticle community was rather sceptical about these results. Folks did not take the announcements seriously until Su & Finkbeiner (2012b) confirmed the detection; their paper was published on arXiv on June 7. One of the authors of the last paper is Douglas Finkbeiner, a recognised authority on the subject, working in the Harvard Center for Astrophysics. From now on the problem was taken seriously. Soon Su & Finkbeiner (2012a) detected that actually there is a double line spectrum with peaks at energies at 111 and 129 GeV in the Galactic center. This detected gamma ray signal can be explained by dark matter direct two-body annihilation into photons. Figure 6.4 shows the observed 110 and 130 GeV excess in Fermi LAT data.

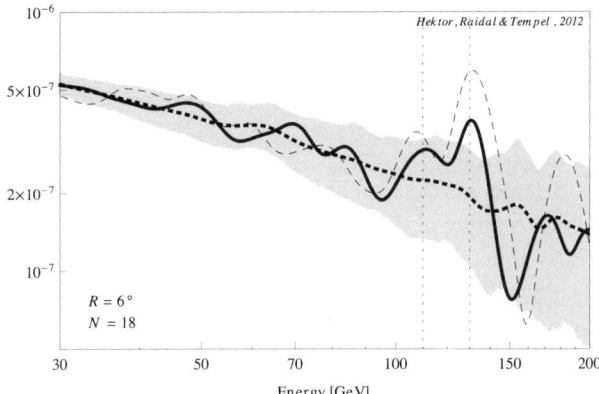

Fig. 6.4 Gamma-ray spectra for 6° regions around the 18 galaxy clusters as functions of photon energy (bold solid curve). The bold dashed line shows a fit to the background together with its 95% error band. The light dashed curve shows the reduced signal from the Galactic center for comparison (Hektor et al., 2013).

The 110 GeV and 130 GeV peak features are being searched for from other nearby dark matter dominated objects. Both peaks are visible from the stacked signal of 18 nearby galaxy clusters (Hektor et al., 2013). This paper was initially sent to "Nature", but was rejected since one referee suggested that before we can claim this result, confirmation from other experiments is needed. So the paper was sent to "ApJ Letters", where it was accepted immediately.

The double peak-like excess can be interpreted as a signal of DM annihilations into two channels with monochromatic final-state photons. The signal from galaxy clusters is boosted due to galaxy cluster subhalos. Since the signals from the Galaxy centre and from nearby galaxy clusters show exactly the same double peak structure, the signal must come from the same physics (Tempel et al., 2012a), showing very strong indication of a particle physics origin (Su & Finkbeiner, 2012b). Tempel et al. (2012a) concluded: *"The presence of double peak is a generic prediction of Dark Matter annihilation pattern in gauge theories, corresponding to $\gamma\gamma$ and γZ final states. Thus the two seemingly unrelated gamma-ray spectra, from the Galactic centre and from the galaxy clusters, favour the particle physics origin of the excess over any astrophysics origin"*.

This discovery has created an avalanche of studies. Almost all possible dark matter models were tested. Elmo had a talk on this topic in our astronomy seminar on January 16, 2013. Up to now 129 papers have been written on the subject, while the total number of authors and coauthors is 132. The papers by Tempel et al. (2012a,b) and Hektor et al. (2013) have over 100 citations combined. If these claims are true, this could be a strong evidence that DM is of particle physics origin, representing a breakthrough both in cosmology and in particle physics.

6.3 Alternatives to dark matter

The presence of large amounts of matter of unknown origin has given rise to speculations on the validity of the Newton's law of gravity at large distances. One such attempt is Modified Newtonian Dynamics (MOND), suggested by Milgrom & Bekenstein (1987). Indeed, MOND is able to explain a number of observational data without assuming the presence of some hidden matter. There exist also a large number of attempts to generalise MOND dynamics, one of them is the Tensor-vector-scalar gravity (TeVeS) model developed by Jacob Bekenstein (2004).

It is fair to say that in comparison to the dark matter paradigm the consequences of various modifications to Newtonian gravity have not been worked out in detail. Thus it still needs to be seen if any of those modified pictures could provide a viable alternative to the dark matter. However, one has to keep in mind that despite us having a good idea of what might make up dark matter, the dark matter paradigm is remarkably simple: one just needs an additional cold collisionless component that interacts only through gravity. Once this component is accepted, a host of apparent problems, starting from galaxy and galaxy cluster scales and extending to the largest scales as probed by the large scale structure and CMB, get solved. So in that respect one might say that there is certainly some degree of elegance in the dark matter picture. On the other hand, taking into account the simplicity of the dark matter paradigm, it is quite hard to believe that any alternatives described above could achieve a similar level of agreement with observational data over such a large range of spatial and temporal scales. Indeed, it seems that for different scales one might need "a different MOND".

However, there exist several arguments which make these models unrealistic. The strongest argument in favour of the presence of non-baryonic dark matter comes from the CMB data. In the absence of large amounts of non-baryonic matter during the radiation dominated era of the evolution of the Universe it would be impossible to get for the relative amplitude of density fluctuations a value of the order 10^{-3}, needed to form all observed structures.

The other strong argument in favour of the presence of some matter in addition to ordinary baryonic matter comes also from CMB data. The wavenumber of the first acoustic peak in the CMB spectrum is a very accurate indicator of the total matter/energy density of the Universe. Experiments show with great accuracy that the total density is equal to the critical cosmological density. On the other hand, both direct determinations as well as the nucleosynthesis constraints show that the density of baryonic matter is only about 4% of the critical density. In other words, there must exist some other forms of matter/energy than ordinary matter. The other forms are dark matter and dark energy. Dark energy causes the acceleration of the Universe, it was detected by comparison of nearby and distant supernovae by Riess et al. (1998) and Perlmutter et al. (1999).

There exist direct observations of the distribution of mass, visible galaxies and the hot X-ray gas, which cannot be explained in the MOND framework. One of such examples is the "bullet" cluster 1E 0657-558 (Clowe et al., 2006). This is a pair of galaxy clusters, where the smaller cluster (bullet) has passed the primary cluster almost tangentially to the line of sight. Weak gravitational lensing observations show that the distribution of matter is identical with the distribution of galaxies. The hot X-ray gas has been separated by ram pressure-stripping during the passage. This separation is only possible if the mass is in the collisionless component, i.e. in the non-baryonic dark matter halo, not in the baryonic X-ray gas.

The basic driving force behind the search for new laws of gravity is actually the understanding that something in our physics world view is missing. This is really the case. We still do now know what changes in our physics world view are needed. But when the nature of dark matter, and, more importantly, that of the dark energy is clarified, it is evident that this shall change our understanding of Nature.

6.4 Tartu Observatory in the late 1980's

6.4.1 *The singing revolution*

Our friend Mikk Sarv has lectured at the Viljandi Culture Academy of Tartu University in traditional music. As part of the teaching he organised every year in spring with his students walks on village roads singing ancient Estonian folk songs. This is an old tradition, meaning to bring a good harvest. People gathered in weekends and walked singing old songs from village to village, and from farm to farm. So Mikk has revitalised these ancient traditions in his teaching.

In the 1980's some other old traditions were revitalised: Hanseatic Days in Tartu, and Old City Days in Tallinn. Mikk with his friends thought about how they can contribute to Old City Days, and decided to walk singing ancient folk songs along Tallinn's city-walls, stopping at each of the eight main gates of the wall. Eight has an important meaning in Estonian folklore — in Estonian language there are eight cardinal points, each having its own meaning (so, between South and West we have a separate word for South-West, and so on, and each of these directions has its own cultural meaning — this is similar to other Ugro–Finnish languages). Mikk and his ensemble stopped at each gate and explained the meaning of the point as well as the story of the gate, accompanied with appropriate folk songs. Mikk wrote to me that it was very beautiful and sublime to walk during the night, to sing, and to think on our past and future.

On June 4, 1987 Mikk and his ensemble made the traditional walk through the Old City, and additional people joined them and also started to sing. They were Estonian musicians, singing popular songs. The joint singing continued through Town Hall Square until morning. The next evening they moved through the Old City to the Song Festival Arena, all the time new people accrued and joined the singing. At the Song Festival Arena they continued singing until the morning-light. The night was so emotional and joyful that musicians continued the singing night after night until the end of Old City Days, and decided to do it again in the next year.

Next year the leaders of various music ensembles prepared the concerts at the Song Festival Arena very carefully. They formed an informal headquarters to coordinate all actions. Soon they were informed that militia and the KGB will try to interfere. To avoid this they informed our radio and TV men, as well as Finnish TV. What Soviet authorities were most afraid of was the truth about the events within the USSR. Also a voluntary 'bodyguard' of several hundred young boys and girls was formed to keep order, and to tidy up garbage from the Song Festival place. The motto was 'everywhere it must be cleaner than before'.

On the first evening of the Old City Days, June 4, musicians walked through Old Tallinn to the Song Festival place, singing popular and folk songs. Tidings of the event had spread among the community, and much more people accompanied them than in the previous year. Estonian flags were not yet used, but many people had blue-black-white bands on their breast. Each evening there were more people gathering, and radio and TV reporters were there, most importantly from Finland and other countries. Some people had full Estonian flags, more and more every evening. One day the militia made an attempt to rend the flags, but immediately a dozen foreign reporters and cameramen were there, and the militia disappeared.

On the last evening Finnish companies had organised special tours, so that more than a thousand Finns participated. In total more than 70 thousand people gathered on Song Festival grounds. Estonian flags were openly carried, as well as flags of free Ukraine, Russia, Latvia and Lithuania. The spirit was high, people sang and cried from happiness. People linked their hands together and sang in chorus, a tradition had begun. This evening was broadcast live to Finnish TV, and suddenly all people started to chant: "Hyvää, Suomi! Hyvää, Suomi!" A bit later a telegram arrived with address "Tallinn Song Festival: Finnish people are with you, Estonian brothers". The message was read to the microphone, and people again chanted "Hyvää, Suomi!".

Fig. 6.5   Old Tartu Observatory on a 2011 night with the Estonia flag (author's photo).

The festival continued until 7 AM. Then the organisers got a message that the old leaders of the Estonian Communist Party were preparing to suppress the demonstrations. However, Gorbachev understood that this would be a catastrophe for the USSR, since everything that happened was known in the West. Thus he ordered that the previous General Secretary of the Estonian Communist Party be changed with a new one, Vaino Väljas, who understood the needs of the Estonian people. At this time Väljas was the Soviet Ambassador in Nicaragua. He was promptly called back, and on June 16 he arrived in Tallinn to get his new appointment.

The singing tradition at the Song Festival Arena was taken over by the Popular Front. On June 17, 1988, more than 100 thousand people gathered at the Song Festival Arena, to accompany our delegates to the Soviet Congress of People's Deputies. On September 11, 1988, a song festival, called "Song of Estonia", was held. This time nearly 300 thousand people came together, more than a quarter of all Estonians. Our political leaders participated, and were for the first time insisting on the restoration of independence.

A few words on our flag. Every nation has its symbols, and one of the most important symbols is the national flag. The Estonian national flag is blue-black-white: blue for the blue sky and the devotedness to our ideals; black for our black soul and national coat; white for our hope and aspiration for light and freedom. In the Soviet period the use of these colours in any way was forbidden; many brave boys were imprisoned for years for the use of these colours.

The very first official building which was decorated with our national flag was the old Tartu University Observatory. This was done on June 23, 1988, on Victory Day. The flag was raised by our astronomers Peeter Traat, Andres Kuperjanov and Kalle Jürgenson on their own initiative. On the same day the daughter of Peeter, Kristina, graduated from high school. This was chosen as an explanation in case the KGB asked questions.

Recently I met Andres Kuperjanov. He told me that the flag story was more complicated. A few days after the flag was raised it was removed and burned by KGB agents, who had climbed to the dome using the lightning rod. To avoid further removing of the flag, Peeter removed some fixing bolts of the rod and oiled it, so it was impossible to climb up outwards to the dome. Also astronomers formed a voluntary guard to watch the flag at night. The flag has stayed there until now; it is taken down only for replacements with a new flag. This happens 24 times a year. Now the replacement is made by Tartu municipal authorities; earlier astronomers used their own money to buy new flags.

In the Government Building in Tallinn the Estonian flag was raised on Independence Day on February 24, 1989.

6.4.2 *Academy of Sciences*

In 1981 I was elected to the Estonian Academy of Sciences. A few years later I was appointed to the post of the head of the division on Physics, Mathematics and Technical Sciences, which soon was divided into two divisions — one for Astronomy and Physics, and the other for all technical sciences. In the Soviet Union modern molecular biology and genetics were banned and replaced by Michurin–Lysenko type biology. In our academy also one special institute was formed to apply this form of biology. However, physicists working in nuclear research needed information on how radiation affects organisms, and thus inside physics institutes laboratories on molecular biology and genetics were quietly formed. This happened also in Estonia. Small laboratories of genetic studies were formed in the Institute of Physics and in the Institute of Chemical and Biological Physics. But soon they were too small and the time was ripe to organise a fully independent Institute — the Estonian Biocenter. The driving force for its formation came from Prof. Richard Villems, a good friend of Enn Saar. It was clear that our colleagues in the Academy from the Division of Chemistry, Biology and Geology were against the formation of this Center, so it was formed within our Division of Astronomy and Physics. Also the first new members of the Academy, specialists in modern biology, were elected to our Division. Presently the Estonian Biocenter is one of the most successful scientific institutions in Estonia. It is working in very close collaboration with Tartu University and with a number of the best centres in other countries.

In the late 1980's the time was ripe for another change in the life of the Academy. So far all Academies of Soviet republics were subordinated to the USSR central Academy of Sciences — we had to coordinate all our activities, including the elections of new members to our Academy. In 1990 we prepared a new bylaw of the Estonian Academy of Sciences where we eliminated all aspects of the subordination to the central Academy. We also wrote in the bylaw that the present Academy is the successor of the Estonian Academy of Sciences formed before the war, and closed by Soviet authorities in 1940. Soon Academies of other Soviet republics followed our example, but we were the first to do this.

I acted as chairman of the Division of Astronomy and Physics until 1995, after 12 years of service in this post. The work in the Academy was rather time-consuming, my Academy office was in Tallinn, and I had to drive to Tallinn every week for some days. After 1995 I withdraw from all administrative duties, both in the Academy and in Tartu Observatory. The new head of the Cosmology Department of the Observatory is Enn Saar.

6.4.3 *Towards an independent Estonia*

These years were rather tumultous in both our lives and in the Estonian Academy of Sciences. The scientists of the Academy were actively involved in public life. Here are some episodes from the life of the Academy, and of the Estonian independence movement.

In May 1987, students and intellectuals initiated a successful protest movement against Moscow's plans for large-scale, ecologically disastrous mining of phosphorites in north-eastern Estonia. Out of this effort grew the Estonian Greens Movement. Scientific arguments against the Soviet plan were collected by Endel Lippmaa, my colleague in the Estonian Academy of Sciences.

The next step was the Estonian IME program for economic autonomy, which won widespread acclaim in September 1987 as an attempt to solve national problems by making an Estonian contribution to economic reforms in the Soviet Union. Drawing on examples from China and Hungary, the proposal called for an end to central economic control over Estonia, a separate tax system, and the adoption of a convertible ruble. The idea was popular, not least of all because of the plan's name, "Isemajandav Eesti", whose acronym, IME, means "miracle" in Estonian. A committee was formed to prepare the reform plan. One of the leaders of the program, Professor of Tartu University Marju Lauristin, asked me to prepare the science program for IME. This program was later used to prepare the Science law in the Estonian Parliament.

In the late 1980's two important developments took place. In April 1988 the Popular Front of Estonia was formed, initially called "Popular Front for the Support of Perestroika". It was a major force in the Estonian independence movement. Popular Front participated in elections for the new Congresses of People's Deputies of USSR and the Estonian Supreme Soviet, and formed the last government in the Soviet period, which made actual preparations to achieve independence.

Radical nationalists formed in 1989 the Estonian Citizens Committee, a nonpartisan political movement, where Estonian emigrants in Western countries played a major role. The purpose was to register Estonian citizens, to carry out the elections of the Estonian Congress, and to convene the Congress as a legislative body. The Committee refused to participate in any Soviet organisation, including Estonian Supreme Soviet and government. The leaders of the movement did not understand that no country will recognise such movement as a legislative body since it had no real power over the country. The conflict between the Popular Front and Citizens Committee was dampened by the fact that a large number of people who wanted to restore Estonian independence participated in both movements.

In the autumn of 1988 a new constitution was being prepared by Soviet authorities who had the goal of significantly restricting the rights of the Union republics. This alarmed us all and made us rack our brains over what measures to take. One evening Valdur Tiit phoned me and the conversation soon turned to these problems. I cannot exactly recall who initiated the idea, but we decided to find Arnold Rüütel, the Chairman of the Estonian Supreme Soviet (President of the Republic in present terms). His home was close to Tiit's, and both of us had already known Rüütel for a long time, so there were no diplomatic obstacles to meeting him. We set to it straight away; I drove to Tartu and we knocked on Rüütels door. We told him at once that we had a serious issue to discuss and asked to talk without interference. He considered it for a while and took us to his sauna which he had built himself so he could be sure that no listening devices had been installed in it. We asked for his advice on how we could contribute in order to turn the political situation in our favour. Soon we came to the decision that the most effective way would be to make a declaration or appeal at the general meeting of the Estonian Academy of Sciences.

The next day I convened all members of the Estonian Academy from the Tartu region to the old University Observatory, and soon a project for the appeal was ready. This was the basis of the general assembly of the Academy taking place a few days later (November 2, 1988). Arnold Rüütel and Vaino Väljas were also present. They did not talk, but observed the ongoings intensely. The appeal we reached was the first step towards the declaration of sovereignty that was accepted at the Estonian Supreme Soviet (Parliament) some weeks later. We translated the appeal to Russian and on the same night I called Andrei Sakharov. I had met him a year before at the Friedmann conference of cosmology. Sakharov asked me to dictate the appeal so he could write it down. I had no doubt that our conversation was being recorded by "the competent authorities", but there was nothing to lose, since our appeal was going to come to general knowledge in any case.

The preparations for the declaration of sovereignty were known to high officials in Moscow, and Rüütel was ordered to Moscow by the prime minister of the USSR for questioning. Later Rüütel told me that he was threatened with ten years' imprisonment when he would continue preparations for the declaration. In the middle of the meeting the prime minister was called to Gorbachev where the top officials of the USSR discussed what to do. The danger of the dissolution of the USSR was so great that three options were discussed: immediate arrest of Rüütel, the organization of some 'accident' (car crash or something similar) to kill him, or let him go. Finally the foreign minister Shevardnadze convinced the rest of the officials that it is best to let him go, since in his opinion there is no hope that the resolution will pass in the Estonian Supreme Soviet. The other argument was that foreign press was already informed of our plans, and every other action would bring complications for the USSR. So after four hours of examination Rüütel was allowed to return home.

Rüütel did not tell anybody about the threats from Moscow and continued preparations for the Supreme Soviet meeting. He asked Vaino Väljas to preside the meeting — this helped to create a favorable atmosphere among russian-speaking delegates of the Supreme Soviet. Also a number of Estonian intellectual leaders were invited as speakers, in addition to members of the Soviet. They could not participate in voting but in their speeches gave support to the declaration. The declaration was very carefully worded to avoid attacks as we planned to separate from the USSR. The declaration asserted Estonia's sovereignty and the supremacy of Estonian laws over the laws of the Soviet Union. It also laid claim to the republic's natural resources: land, inland waters, forests, mineral deposits and to the means of industrial production, agriculture etc. in the territory of Estonias borders. The declaration was accepted by a large majority of voters on the meeting of the Supreme Soviet on November 16, 1988.

I remember this evening very well. I listened to the broadcast of the Supreme Soviet by radio. When the Estonian foreign minister Arnold Green recited the declaration with a solemn voice, I broke out into tears — the first time in my adult life. I understood that I had dreamed of a free Estonia all the time, but had suppressed this dream in everyday life. When the declaration was accepted, all astronomers living in the Observatory rushed out to congratulate each other.

But the story was not over. A problem was to publish the declaration immediately in the open press, since all declarations and laws come into force only after they are published. The declaration was published in a special issue of the newspaper "Rahva hääl" (Voice of the People).

Rüütel was immediately ordered to go to Moscow for questioning. He was again threatened and commanded to cancel the declaration. He refused to do so. Some time later the All-Union Congress of People's Deputies held its Meeting, and Rüütel was strongly advised to declare the abolishment of the Estonian Supreme Soviet declaration. In the Meeting of the Congress over 2500 delegates all over the USSR gathered. The Meeting was also attended by several hundred foreign journalists. But instead of nullifying the Estonian declaration, Rüütel justified it on the basis of economic as well as political arguments. When he finished, the large congress hall was dead silent. Finally Gorbachev clapped his hands a few times. Thereafter, a storm of applause emerged. Within a year or two almost all Soviet republics accepted their own declarations of sovereignty, most importantly the Russian Soviet Federative Republic. The dissolution of the Soviet Union had started.

This happened in 1988, a year before the velvet revolution in Prague and the fall of the Berlin wall.

At the same time (first days of November 1988) Remo Ruffini organised in Tallinn a small meeting to discuss the collaboration between Italian and Soviet cosmologists. With him was the director of the Vatican Observatory, George Coyne, and his friend Sergio Romano, the Italian Ambassador to the USSR. For diplomats visiting Soviet republics was a problem, but in this case Remo used his right to take with him some scientists, and Romano is a historian. After the meeting we walked in the beautiful nightly atmosphere of old Tallinn and noticed that in all churches there were people praying for the freedom of Estonia. In one church a folk-ensemble was singing songs of freedom. A year later I was visiting Italy, and Sergio Romano invited me to his home. We discussed the situation in Estonia, and Romano finally asked: *"What do you actually want?"* My response was: *"Our own money and passports, and economic independence."* Romano replied: *"But this means secession!"* I answered: *"Of course."* Romano is also a publicist, a bit later he wrote in an Italian newspaper a detailed review on the development in Soviet Union, citing our conversation.

On August 23, 1989 there was the 50th anniversary of the Nazi–Soviet Nonaggression or the Molotov–Ribbentrop Pact, which divided East Europe into spheres of influence between Nazi-Germany and the Soviet Union. This Pact allowed Germany to attack Poland, and Soviet Union to 'liberate' East-Poland, as well as to occupy of Estonia, Latvia and Lithuania. The Popular Front of Estonia jointly with similar Latvian and Lithuanian organisations prepared a peaceful mass demonstration to mark this event. From Tallinn over Riga to Vilnius a continuous chain of up to three million people was formed. The preparations to form the chain were very well made, the route was fixed and sections divided between various organisations who helped to transport people to the right place.

In 1989 academician Isaak Khalatnikov with his wife Valentina Glebovskaya had their summer vacation in our Observatory guesthouse. Valentina's roots come from a famous Polish family: one of her great-grandfathers is Hetman Khodkevich, the commander of the Polish–Lithuanian army during the Polish–Russian war in the early 17th century. Also we had a postgraduate student of Ruffini, Daniela Calzetti, visiting Tartu Observatory (now she is an astronomy professor at the University of Massachusetts, USA). So when I drove to our assigned place Valentina and Daniela were also with us. During the trip we stopped at the Helme cemetery and put flowers on the graves of my grandfather Jaan Lammas and grandmother Anna Lammas. One small section near the Latvian border was given to Tartu Observatory. We arrived a bit earlier. At the right moment all people linked hands and started to chant "freedom, freedom". This protest action against Soviet rule brought the national liberation movements into the spotlight of the world community. During the 13th Marcel Grossmann Meeting in summer 2012 in Stockholm I met Isaak and his wife Valentina. Isaak told me that his wife remembers with pride that she participated in this protest action.

In spring 1989 the new All-Union Congress of People's Deputies was elected. This time elections were almost free (at least in Baltic republics), and a number of intellectuals were elected to the Congress. Estonian Popular Front won the majority of seats reserved for Estonia. Among the elected delegates there were several academics from Tartu University: economist Mihhail Bronshtein, social scientist Marju Lauristin, chemist Viktor Palm and some others. Bronshtein in his speech justified the transition to a market economy. Palm was an initiator of the Interregional Group of Deputies (IRGD) and one of the co-leaders of the Group. The five co-leaders included human rights activist Andrei Sakharov and Boris Yeltsin. As its only non-Russian co-leader, Palm became an essential link between the Baltics group of deputies and the Russian reformers of the IRGD.

Delegates from the Baltic countries had a basic goal — to prepare our countries for separation from the rest of the Soviet Union. The first tactical goal was to make the agreement between the Soviet Union and Germany in 1939 public, and to void this agreement by the Congress. Under pressure from the Baltic countries a committee was formed to investigate the case of the Molotov–Ribbentrop Pact.

One member in the committee to investigate the Molotov–Ribbentrop Pact case was Endel Lippmaa, a prominent Estonian physicist. He got an invitation to a physics conference in USA, and used this chance to search in US archives for documents related to Soviet foreign policy in 1939. He found not only copies of the Molotov–Ribbentrop Pact and its secret Annex, but a number of other documents, showing preparations of the Soviet Union to get back territories the czarist Russian Empire had lost after WWI. He was clever enough to get for all important documents affirmation from the archives that the copies are correct. One argument of the Russian delegation was that the documents are falsifications by Germans. Lippmaa had foreseen this possibility. He organised an independent analysis of documents which showed that the secret Annex and open documents, written during the Molotov–Ribbentrop Pact signing ceremony, were typed with the same typewriter. In this way for every argument of the Russian delegation he had a well documented counterargument. So, on December 24, 1989 the All-Union Congress of People's Deputies recognised the presence of the secret Annex to the Molotov–Ribbentrop Pact. Also the Congress acknowledged the Pact and its Annex as null and void. This was a great shock for the Russian public since Soviet propaganda had always accused Nazi-Germany of unleashing World War II.

Chapter 7

The structure of the cosmic web

In the 1970's our main attention was devoted to the qualitative study of the distribution of galaxies and clusters of galaxies in space. This led us to the discovery of the cosmic web, and to an understanding of the meaning of the structure — the present structure is the remnant of the structure at the time of galaxy formation, and this structure itself depends on properties of dark matter, the dominant population of the Universe. As Zeldovich insisted during the Tallinn symposium, now we need more quantitative descriptions of the structure. This was our main goal in the 1980's.

7.1 Quantitative characteristics

7.1.1 *The search for quantitative characteristics*

We summarised our basic observational results on the presence of the cosmic web in a paper to "Nature" (Einasto et al., 1980b). Zeldovich suggested that we should write a similar paper on structure formation theories. The paper was ready in summer 1982, and was sent as a Letter to "Nature"; also we made a preprint of it. Here we showed the distribution of nearby Zwicky clusters and galaxies in the Coma and Hercules supercluster regions, and compared the distribution with basic structure formation scenarios.

Soon we got an answer from the "Nature" editor with a suggestion to write instead of a letter a review paper on voids, and to compare theoretical models with observations in more detail. So we started to think about how to improve the paper. The presentation of the distribution of galaxies and clusters in our previous papers was basically graphical. But in order to make comparisons with various models of structure formation quantitative tests were needed. This was emphasised already by Zeldovich (1978) in his Tallinn Symposium talk. Thus we started the search for quantitative methods to analyse the structure, which were sensitive to the features we had seen in the observed as well as in the model distributions.

To make room for quantitative analysis we excluded from the observational part of the paper the distribution of Zwicky clusters. This was used in our later paper on the Coma supercluster analysis (Tago et al., 1984). Only the distribution of galaxies in two thin sheets of thickness 10 Mpc, crossing the Virgo and Coma superclusters, remained. The new analysis confirmed our earlier finding (see Figs. 5.5, 5.4) that galaxy chains, joining the Virgo and Coma superclusters, are very thin.

In the search for methods to analyse the galaxy distribution Zeldovich suggested using the percolation method. This method makes use of the Friends-of-Friends algorithm to collect particles to a system. Let us draw a sphere of radius r around each sample particle (in our case a galaxy or a simulation particle). If within this sphere there are other particles they are considered as belonging to the same system, i.e. they are considered as "friends". Now draw spheres around all new neighbours and continue the procedure using the rule "any friend of my friend is my friend". The procedure stops when for a given neighbourhood radius no more new members can be added — a system is identified.

This procedure is repeated using increasing neighbourhood radii. If the radius r is small enough, all particles are isolated, i.e. there are no systems of particles, and the length of the longest system is zero. With increasing radius more particles join to form systems. First high-density cores of clusters are formed, thereafter regions of lower spatial density join to systems found with smaller radius, see Fig. 7.1. At each radius the longest system is found. At a certain radius, called the percolation radius, the longest system spans the whole volume of the sample studied.

Next we tried the correlation function. We already knew that the correlation analysis, used by Peebles and many other investigators for the apparent two-dimensional distribution of galaxies, was not very sensitive to the presence of filaments, which we had seen in the spatial distribution, because the correlation function contains no phase information, see below. Now we applied it again, but using three-dimensional data we had for galaxies as well as for models. The left panel of Fig. 7.2 shows the spatial correlation functions for three samples. The absolute magnitude limited observed sample around the Virgo supercluster (using CfA redshift survey, complete up to $m = 14.5$) is designated as O. The sample A was calculated using the simulation of the adiabatic model by Klypin & Shandarin (1983). H is for the sample generated using the method to find a sample of particles in the hierarchical clustering scenario (it is not an actual dynamical simulation).

To our surprise we found that the correlation function can also be used to find differences in the spatial structure of our samples. The most important feature of the O and A samples is the presence of a knee in the correlation function, absent in the

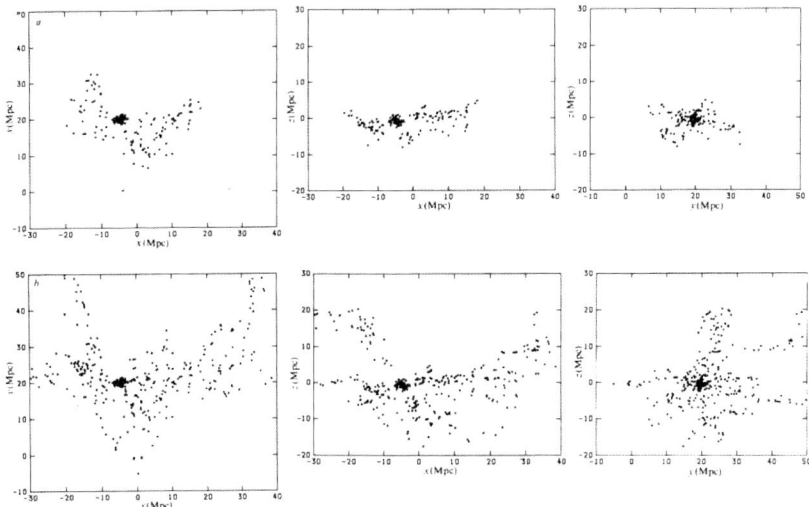

Fig. 7.1 Three views of a connected system of galaxies around the Virgo cluster. Upper panels are found with a neighbourhood radius $r = 4$ Mpc, lower panels with radius $r = 5$ Mpc. Using the smaller radius the central dense part of the Virgo supercluster is collected to the system, it consists of two sheets of galaxies on both sides of the Virgo cluster. The larger neighbourhood radius collects to the supercluster also the network of galaxy strings or filaments, which form a spider-like configuration with several "legs". All distances are expressed using the Hubble constant $h = 0.5$ (Zeldovich et al., 1982).

H model. At small distances the correlation function is sensitive to the distribution of galaxies/particles at small mutual distances. At such small distances the majority of galaxies belong to clusters and groups, which have an almost spherical shape. At larger distances the correlation function feels the presence of strings/filaments of galaxies, which are essentially one-dimensional. Thus the geometry of the structure changes when we move from small to large mutual distances of galaxies/particles. The hierarchical model has no strings/filaments, thus the correlation function is featureless. As we see later, this behaviour is one of the characteristics of the structure's topology.

The FoF method allows us to use the length of the largest system as a test. The right panel of Fig. 7.2 shows the maximal lengths of systems of galaxies/particles as a function of the neighbourhood radius r (in this Figure distances are expressed using the Hubble constant $h = 0.5$). Here we compare the behaviour of four samples: the observed sample O, the models A and H, and the Poisson sample P.

Galaxies as well as particles in simulations are clustered, this means that at small radii r the length of the longest system grows with increasing r faster than

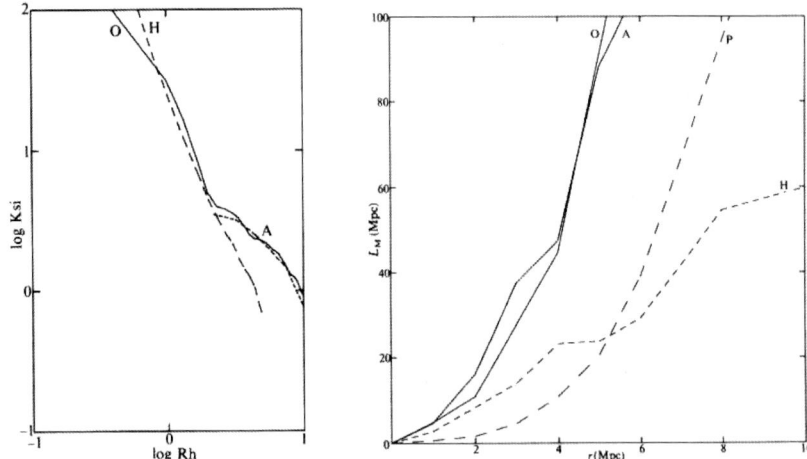

Fig. 7.2 Left panel: the correlation function of the observed sample O around the Virgo cluster (cube of side 80 Mpc), of the sample generated by the hierarchical clustering model H, and of the adiabatic model A. Right panel: the maximal length L_M of connected regions as a function of neighbourhood radius r for four catalogues: O, A, H, and P (Poisson model). All distances are expressed for Hubble constant $h = 0.5$ (Zeldovich et al., 1982).

in the Poisson sample. But at larger radii the behaviour of samples is different. In the observed sample O as well as in the model sample A there are filaments joining clusters to a network. These filaments make the formation of longer systems easy, and the length of the longest system grows more rapidly than in the Poisson case. Fig. 7.2 shows that samples O and A have almost identical growth. In contrast, in the sample H at larger r the growth of the length L with radius r is *slower* than in the Poisson sample. The reason is simple: the density of the field particles of the sample H is lower than in the Poisson sample, since a large fraction of particles are used in clusters. Thus we see that this test is sensitive to the presence of filaments which join clusters to a connected network.

As a further test I suggested using the multiplicity function of systems of galaxies/particles, found for various neighbourhood radii r. At each radius r we counted the number of systems of certain multiplicity (number of galaxies/particles). The frequency of systems of various multiplicity for the neighbourhood radius $r = 5$ Mpc (for Hubble constant $h = 0.5$) is shown in Fig. 7.3. The multiplicity is expressed in a logarithmic scale, in powers of 2: 0 corresponds to $2^0 = 1$, i.e. isolated galaxies, 5 to $2^5 = 32$, i.e. medium rich systems, etc. Actually a histogram of multiplicities is shown; between indices i and $i + 1$ all systems with multiplicities between 2^i and 2^{i+1} are counted.

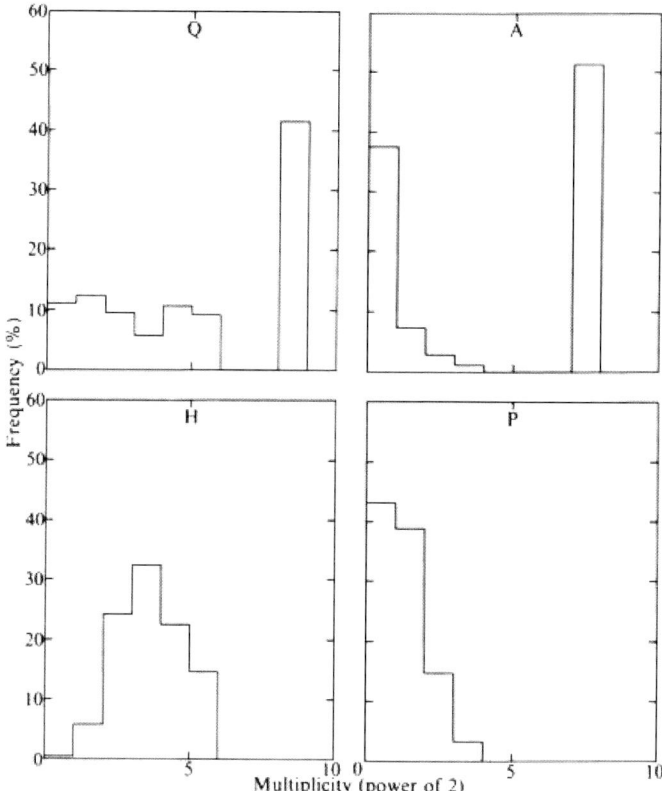

Fig. 7.3 Distribution of galaxies according to the multiplicity of the system. Neighbourhood radius $r = 5$ Mpc (for Hubble constant $h = 0.5$). Multiplicity is expressed in powers of 2 (5 corresponds to $2^5 = 32$). Samples are designated as in Fig. 7.2: O — observed, A — adiabatic model, H — hierarchical clustering model, P — Poisson model (Zeldovich et al., 1982).

This Figure shows that all samples studied have different distributions of multiplicities. The observed sample has approximately an equal fraction of systems of various richness, i.e. there exists a fine structure of systems of galaxies of different richness. The largest fraction of galaxies belongs to one single large system — the Virgo supercluster. The A sample has also one large system, but the distribution of smaller systems is more similar to the Poisson sample distribution, i.e. there are almost no systems of intermediate richness — small-scale filaments.

One more important difference: in the A sample there exists a large fraction of isolated particles — these are void particles, forming a smooth low-density population in voids, absent in the O sample. The H model has no superclusters and no population of smoothly distributed isolated void particles; multiplicities have a peak at medium multiplicity level.

Our results were summarised in the revised version of the article by Zeldovich et al. (1982); a more detailed discussion of the analysis was published by Einasto et al. (1984). My original suggestion was to title the paper "Giant voids in the Universe: an eyewitness story of galaxy formation"; this title was actually used in the preprint. The last part of the title emphasised the similarity of the present structure to the structure at its formation. However Zeldovich wanted to have a simpler title, so the last part was omitted. The theoretical part of the paper was written by Zeldovich, while the numerical simulations were made by Shandarin and Klypin. My contribution was the observational part, the quantitative comparison of observations and models, and the final polishing of the paper.

At the last stage I had a problem: initially Zeldovich did not realise that our results show that his pancake model in its original form also has serious difficulties — the absence of fine structure in superclusters, seen both in pictures and in the multiplicity test. To avoid a conflict we formulated this result as follows: *"Astronomers usually question which formation scenario is correct. But both basic scenarios have weak points — perhaps it would be more correct to ask, which physical processes have led to the formation of structure in the Universe"*. In the original adiabatic pancake scenario small-scale waves are cut. In this aspect the Zeldovich scenario is practically identical to the neutrino-dominated dark matter scenario. This possibility was discussed in the paper.

In late 1981 I was visiting ESO to analyse the large scale structure. In collaboration with Dick Miller we prepared several movies where the third dimension was visualised by rotation of the galaxy sample (Einasto & Miller, 1983). I used the Huchra ZCAT data to compile several absolute magnitude (volume) limited catalogues of galaxies around the Virgo cluster, in the region between the Virgo and the Coma superclusters, and in the large void between these superclusters, the Northern Local Void. The program to rotate the galaxy sample was written by Dick Miller; he also prepared the original film by photographing individual plots of each time step after rotating the file on a computer monitor. The final copy of the film was made a few months later in the Tartu University photo-labor.

The first demonstration of the original movie was in the ESO conference room on December 22, 1981. All participants of the project looked with excitement and curiosity at how the galaxies are *actually* distributed in space — the title of the movie was *"The Universe as it is"*.

The first parts of the movie showed the structure of the Virgo supercluster; some views of this supercluster are shown in Fig. 7.1. The most important impression from the movie was: there exists no isolated field galaxies. All galaxies seen in the movie belong either to clusters/groups forming the cores of superclusters, or to filaments/chains of galaxies. The general impression of the structure of the Virgo supercluster showed its spider-like distribution of galaxy strings, which started from the Virgo cluster in various directions. A similar movie was made by Brent Tully using his HII radio observations of distances of galaxies in the Virgo supercluster; he showed the movie in the Tallinn symposium on 1977. Our movie contained more galaxies, but the general appearance was very similar.

Other parts of the movie showed large regions behind the Virgo cluster in the direction of the Coma and the Hercules superclusters, and the very large void between these superclusters. Here a thin network of galaxy filaments joining the Virgo supercluster with the Coma and Hercules superclusters was clearly visible. No sheets of galaxies which could separate voids were found. Our movie was screened at the IAU Symposium on Large-scale Structure in Greece in summer 1982, see below. The movie is included in the website accompanying this book. Using a bit more complete sample of nearby galaxies in ZCAT I prepared jointly with Dick Miller and his students a new version of the movie. The new version is called *"The real Universe"*. The final version was prepared at the NASA Ames Research Center, and both versions were digitised at the Tartu University Media Center.

To check the possible presence of galaxy walls which could isolate neighbouring voids I developed a simple algorithm to study the connectedness of empty or filled regions in the density field. My results showed that there exists no isolating surfaces between large voids defined by superclusters. More on this in the next section.

At this time Jan Henrik Oort was preparing a review paper on superclusters (Oort, 1983). To discuss the structure of superclusters, Oort invited me to the Netherlands. However, under Soviet rules I had no permission to visit the Netherlands during this visit, so we agreed with Oort to meet in Bonn, the German observatory closest to the Netherlands. I gave a seminar talk at Bonn University Observatory on December 28, 1981, and screened our recent movie.

Later we continued the discussion with Jan Oort privately. We agreed that there exists strong evidence for the formation of galaxies in chains: the velocity dispersion of groups and galaxies perpendicular to the chain axis is practically zero; main galaxies of clusters are elongated along the axis of the chain, as seen in the Perseus chain by Jõeveer et al. (1978) and Einasto et al. (1980a). Further, there are practically no galaxies in voids, but low-density primordial matter should exist there according to numerical simulations.

These results were important in several aspects. First, this was the earliest demonstration that no galaxies form in voids, i.e. galaxy formation is a threshold phenomenon: in real samples voids are empty; in simulations voids contain a rarefied sample of smoothly distributed isolated particles. This is the basic property of biased galaxy formation, see below. Secondly, galaxy formation occurs *in situ* in chains and clusters. Third, all previous structure formation scenarios had weak points, and a new better scenario had to be suggested. When I read the published version of the Oort (1983) review, I was happy to notice that Jan used in the review our pictures on the distribution of galaxies and clusters, and also Figures (7.2, 7.3) to show quantitative differences between observations and main theoretical scenarios. Also the general style of the review was in the same spirit as our discussion in Bonn.

We had our discussion in a cafe near the Bonn railway station. It was a beautiful winter day, and when scientific discussions were finished, Jan noted that in cold winters there is a skating tour in Holland. The tour is called Eleven Cities Tour, and it is almost 200 kilometres long on frozen canals, rivers and lakes between the eleven historic cities in the North of the Netherlands. The tour starts very early in the morning, and Jan said that it is very nice to skate through the winter landscape. To my question of whether Jan covered the full distance, he modestly answered: *"in the last time only two thirds"*. The length of our Tartu ski marathon is 63 km, so he skated a distance twice the Tartu marathon!!! And Jan Oort was 82 years old when we discussed the structure of superclusters in Bonn.

I presented our results at the IAU Symposium on the Early Evolution of the Universe and its Present Structure, which took place in Crete after the IAU General Assembly in 1982 (Einasto et al., 1983b). Our main conclusions were:

(1) the basic structural elements in the Universe, larger than clusters of galaxies, are strings/filaments of galaxies, groups and clusters;
(2) no large sheets (flat pancakes) of galaxies have been found, small sheets surround some rich clusters near centres of superclusters;
(3) galaxy strings/filaments connect superclusters to a single intertwined lattice;
(4) no theoretical scenario proposed so far explains all observed clustering properties.

For further discussion of our results reported in 1982 at the IAU General Assembly and Crete Symposium see the sections on biasing and walls.

7.1.2 *Topology of the cosmic web*

According to Zeldovich's ideas the evolution of density perturbations starts from pancaking to form flat sheets (walls) (Zeldovich, 1970, 1978). Filaments form by the flow of pre-galactic matter towards the crossing of sheets, and clusters by the flow of matter towards the corners of the cellular network. The first numerical simulations were two-dimensional (Zeldovich, 1978). If extrapolated to three dimensions, the structure resembles cells, with clusters of galaxies forming cell corners, galaxy filaments the cell edges, sheets the cell walls, and voids the cell interiors.

There are two possibilities to interpret the 3-dimensional distribution of particles in simulations and real galaxies. One possibility is to understand the cellular structure as a honeycomb, where cell walls isolate neighbouring voids. This corresponds to the classical pancake scenario, as described above. A different structure is realised if cell walls are so thin that galaxies do not form there. In this case cell walls do not form continuous and isolating surfaces: both galaxy systems and voids form connected intertwined regions.

When we started our study of the distribution of galaxies we did not have any prejudice to any theoretical scenario. We wanted to understand how the structure actually looks like. Already our first results indicated that the dominant structural elements are chains or filaments of galaxies, groups and clusters, arranged in superclusters, and that the space between filaments and superclusters is almost void of galaxies (Jõeveer et al., 1977; Jõeveer & Einasto, 1978; Jõeveer et al., 1978). Filaments form a more or less continuous network. The similarity of the observed structure to the simulated structure was obvious, thus we used the term "cellular" for the observed structure. We interpreted the cellular structure as a 3-dimensional lattice or supercluster–void network, filaments forming rods of the lattice, clusters and superclusters as knots, and voids as the interiors of cells.

In the early 1980's new redshift data available — the first releases of the Harvard Center of Astrophysics Redshift Survey, complete up to 14.5 magnitude, and additional data from the redshift compilation ZCAT, prepared in Harvard by John Huchra and his students. New data allowed us to investigate in more detail three superclusters and their vicinity — the Virgo, the Coma, and the Perseus–Pisces superclusters, and galaxy chains/filaments joining these superclusters to a connected network (Einasto et al., 1980a, 1984; Tago et al., 1984).

We focused our attention on the question: Are principal structural elements filaments or rarefied sheets, forming surfaces of cell walls? As described above, I made in collaboration with Dick Miller and John Huchra for this purpose movies of the distribution of galaxies in the Virgo supercluster, and in the large void behind the Virgo supercluster in the direction of the Coma and Hercules superclusters (Einasto & Miller, 1983). Also I developed a program to find connected regions in the density field of galaxies, used in the study of the Coma supercluster by Tago et al. (1984).

We did not find evidence for the presence of wall-like sheets of galaxies between voids — the dominating structural elements were chains (filaments) of galaxies and clusters. Sheets of galaxies exist inside superclusters, an example is the Virgo supercluster, see Fig 7.1. However, sheets do not separate neighbouring voids. These results were discussed by Zeldovich et al. (1982), Tago et al. (1984), Einasto & Miller (1983), and by Einasto et al. (1983b) at the IAU Symposium in Crete in summer 1982. Our main conclusions were: *the topology of the cosmic web differs from a simple honeycomb-like topology with a connected network of filaments, and disconnected voids. The connectivity test also indicated that voids form one large connected region.*

In the early 1980's several groups elaborated programs to simulate the evolution of structure using the fast Fourier transform to solve the Poisson equation. First three-dimensional numerical simulations of the evolution of the cosmic web by Klypin & Shandarin (1983), Centrella & Melott (1983), and White et al. (1983) show the formation of a cellular structure, as expected from earlier simpler two-dimensional models by Doroshkevich et al. (1980). These simulations were made with the power spectrum cut on small scales. This spectrum corresponds to the neutrino-dominated dark matter scenario, as explicitly stated by all three teams. An essential feature of the evolution in this scenario is the formation of flat objects in the non-linear stage of the evolution, called pancakes by Zeldovich. The subsequent evolution leads to the intersection of the pancakes along filaments, and to the formation of a cellular structure, as predicted by the Zeldovich (1970) analytical arguments.

As discussed above, the models calculated for a neutrino-dominated dark matter Universe had serious difficulties. Adrian Melott simulated the formation of the cosmic web using both the neutrino-dominated (or Hot Dark Matter — HDM) and the axion-dominated (or Cold Dark Matter — CDM) types of dark matter. In 1983 he visited Moscow and Tallinn to discuss his models. A comparison of both simulations with observations showed that the axion-dominated CDM model agrees with observations quantitatively, and thus must be taken seriously (Melott et al., 1983).

Now we had a good model for comparison, and continued the study of the connectivity of both observational and model samples. But here we had a difficulty — so far we had a program by Enn to find the connected systems of *particles*, but we were also interested in the connectivity of empty regions. So, after some thinking Enn invented another program, which determined the connectivity of regions of the density field. It was much faster than my own similar code used in my preliminary determination of the connectivity. Connected systems could be defined either as over- or under-density regions, using a fixed threshold level. The Enn program is extremely fast and has been in use until now.

Equipped with these new possibilities we continued the connectivity study of observed and model samples. We used various density thresholds to define over- and under-dense regions. New calculations confirmed our previous visual impression and preliminary connectivity calculations, that there are no walls isolating voids, and that superclusters are joined by galaxy filaments to a connected network — the cosmic web. Both in the observed and the model samples voids form just one connected region. Similarly, at a certain threshold density the largest over-density region spans the whole sample volume under study.

Similar results were obtained by Shandarin & Zeldovich (1983) using computer simulations of adiabatic structure-formation models and their percolating properties.

We described our new results at the Crete Symposium (Einasto & Miller, 1983; Einasto et al., 1983b,a) and elsewhere (Einasto et al., 1984; Tago et al., 1984), but did not make any attempt to write a special paper to describe the topology of the cosmic web. But this was an error. During a visit to ESO in 1985 I discovered on the preprint shelf the preprint of the paper by Gott et al. (1986). In this study the same observational database was used, as well as the same simulation by Melott. The connectivity of high- and low-density regions was defined as the topology of the large-scale structure. So far "topology" meant the general character of the universe, an open or closed one etc. Now this term was used to describe properties of the cosmic web, not the universe as a whole.

Gott et al. compared our first results of the cellular model (Jõeveer & Einasto, 1978) with the hierarchical clustering model by Soneira & Peebles (1978). They described our model as a "Swiss cheese" model, in which isolated voids are surrounded by high-density medium. Next authors described in detail our movie shown at the IAU symposium in Crete (Einasto & Miller, 1983) which shows that real data support neither the honeycomb-like cellular model nor the hierarchical clustering model. The authors used the name "sponge" topology to describe the actual structure, where both the high- and low-density regions are multiply connected. To describe quantitatively this property the authors used the genus of the density field.

Reading this preprint I understood that it was high time to describe our own results studying this problem. I discussed the issue with Enn Saar and my other colleagues, and we decided to make the analysis a bit differently. Soon our paper was ready and was submitted to "Monthly Notices". Also we prepared a preprint (Einasto et al., 1986a) which was sent to all observatories (in the published version Gott et al. (1986) refers to this preprint).

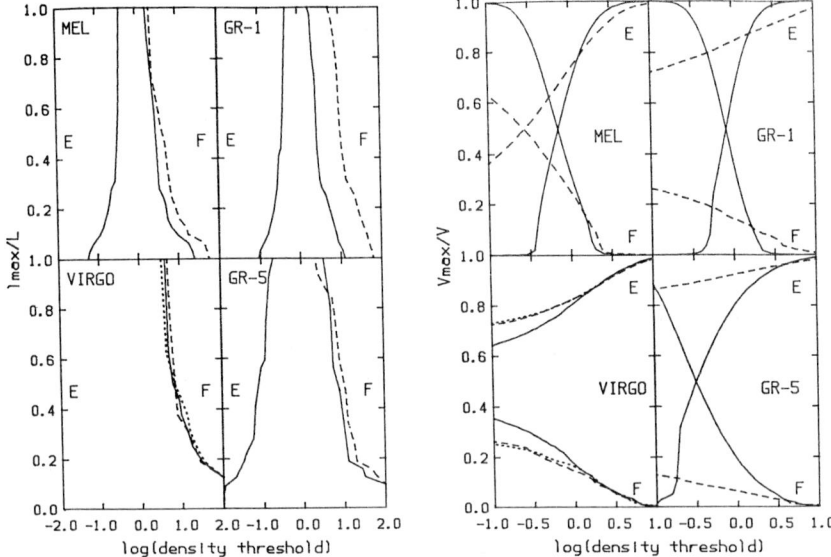

Fig. 7.4 Left panel shows the length of the largest system (in units of the box size) versus the density threshold (in units of the mean density of the sample). E (empty) denotes low density regions, F (filled) high density regions; MEL, GR-1, GR-5, and VIRGO denote Melott simulation, Gramann simulation at expansion factors 1 and 5.2, and the observed sample around the Virgo supercluster, respectively. In models solid lines indicate unbiased samples with all test particles included, dashed lines biased samples, where particles in low-density regions have been removed; for the meaning of lines of observed sample see text. Right panel shows the filling factor (in units of the total volume of the sample) versus the density threshold. Designations as in the left panel (Einasto et al., 1986a).

Our basic observational samples were centered on the Virgo cluster, had a length $20\,h^{-1}$ Mpc, and were volume limited. The sample A had the apparent magnitude limit $m = 14.5$ of the CfA survey; the sample B contained galaxies in magnitude interval $14.5\ldots 15.0$ and was composed on the basis of a dwarf galaxy radio survey; the sample C was the sum of samples A and B. The number of galaxies in these samples were 524, 486, and 1010, respectively. In Fig. 7.4 these samples are plotted by dashed, dotted, and solid lines.

The Melott simulation sample was calculated for the standard CDM model with density parameter $\Omega_m = 1$, it had cube length $32\,h^{-1}$ Mpc, and had 64^3 particles in a 64^3 mesh. The other model was calculated by Mirt Gramann (1987), it was a ΛCDM model with density parameters $\Omega_m = 0.2$, $\Omega_\Lambda = 0.8$, had a cube length $40\,h^{-1}$ Mpc, and contained 64^3 particles in a 64^3 mesh. For this model two epochs were used, the early epoch at the end of the linear regime of structure growth (expansion parameter $a = 1$), and the present epoch at the expansion parameter $a = 5.2$, designated as GR-1 and GR-5 respectively. In models we used all test particles, which corresponds to unbiased samples, and biased samples, where particles in low-density regions were removed.

Observed density fields were calculated with a top hat smoothing using a kernel of size $1.25\,h^{-1}$ Mpc; for model samples the kernel size was $1\,h^{-1}$ Mpc. These smoothing lengths applied correspond to the effective size of galaxy groups and clusters; they are approximately equal to the sizes of dark matter halos, the dominant population in the Universe. The biasing threshold level was chosen ≈ 1.5 in mean density units, in this case the biased samples contain about $1/8$ of all test particles. By definition these particles are located in high-density regions. The shape of biased model high-density regions is very similar to the shape of observed galaxy samples in the smoothed density field. It should be mentioned that this topology study was the first application of the Mirt Gramann (1987) model with the cosmological constant.

To study the topology of model and real samples we used high-resolution density fields, and divided it at varying threshold density level to cells of 'empty' or 'filled' regions, depending on whether the density was lower or higher than the threshold density, respectively. Next we found systems, consisting of empty cells (voids) and of filled cells (galaxy systems). Neighbouring empty (or filled) cells belong to one system when they have at least one common sidewall. In such a way we find all individual voids and galaxy systems in the sample. Finally we find the largest system, and calculate its length (the maximal length in x, y, or z, in units of the cube length of the sample), and its filling factor (the volume in units of the total volume of the sample). The largest systems were found for both voids (E) and clusters (F) for a large range of threshold densities.

Our principal results are shown in Fig. 7.4. The Figure shows that the behaviour of voids and clusters of unbiased models for various threshold densities is symmetrical with respect to the mean density. The length and the volume of voids (clusters) is small at small (high) threshold density. When the threshold density approaches the mean density, the largest voids (clusters) rapidly increase in length and volume.

The biased model samples and real samples have completely different bahaviour. For all threshold densities considered the largest void crosses the whole sample volume. The dependence of the length and the volume of the largest cluster on the density threshold is approximately the same (at the present epoch) for all model and real samples. However, note the differences in the volume of voids and clusters in the biased standard CDM and the ΛCDM models. The last model behaves in this respect very closely to the observed sample, while the standard CDM model does not. This was the first quantitative evidence which showed the superiority of the ΛCDM model over the standard CDM model.

The main conclusion of our topology study was that the topology of both voids and clusters depends strongly on the biasing and on the density threshold, applied in the definition of voids and clusters. A honeycomb-like cellular topology with isolated voids is seen only in unbiased model samples (all particles are present) using a very low density threshold. At medium density thresholds all model and real samples have a sponge topology, i.e. both voids and clusters are multiply connected, and form percolating systems which span the whole volume under study. At high threshold densities the topology of clusters is of the type "islands in the ocean".

We sent our topology paper to "Monthly Notices", but received a very negative referee report. We spent about half a year to make a new analysis. To our surprise now the referee again rejected the paper asking us to make changes almost opposite to his previous suggestions. So we did not know what to do. We had a number of other projects running, so the revision of the paper stopped. As a result we did not revise the paper again — we lost the "battle" with the referee. Here again the words of Öpik (1977) are relevant — it is difficult to publish results which do not fit into the conventional world picture or paradigm. Now the preprint of the paper is available on our website; the actual preprint was sent to all observatories immediately after it was printed.

Rich Gott and his collaborators continued the study of the topology of the cosmic web (Park & Gott, 1991; Park et al., 1992b; Vogeley et al., 1994b), and now this is a respectable topic of cosmological studies.

7.1.3   *Fractal properties of the cosmic web*

The correlation function is probably the most commonly used statistic in cosmology. In the mid 1980's it was clear that filaments and voids are important ingredients of the cosmic web, and we started to look at how they influence the correlation function. We used the CfA redshift survey with magnitude limit $m_B = 14.5$, and ZCAT compilation by John Huchra. The observed sample was divided into a number of cubic and conic volume limited samples of various depth and absolute magnitude limits. Velocities had been corrected for solar motion, Virgo-centric flow, and peculiar velocities in groups and clusters. Also a sample of Abell clusters was used. For comparison we used unbiased and biased model samples based on the HDM, CDM, and hierarchical clustering models, calculated by Adrian Melott.

Our analysis confirmed earlier results by Zeldovich et al. (1982), Melott et al. (1983) and Einasto et al. (1984), that all galaxy correlation functions have a shoulder as seen in Fig. 7.2. This phenomenon is due to the presence of galaxy systems of different shape: spherical or slightly elongated in densely populated clusters, and extremely elongated and less dense in filaments.

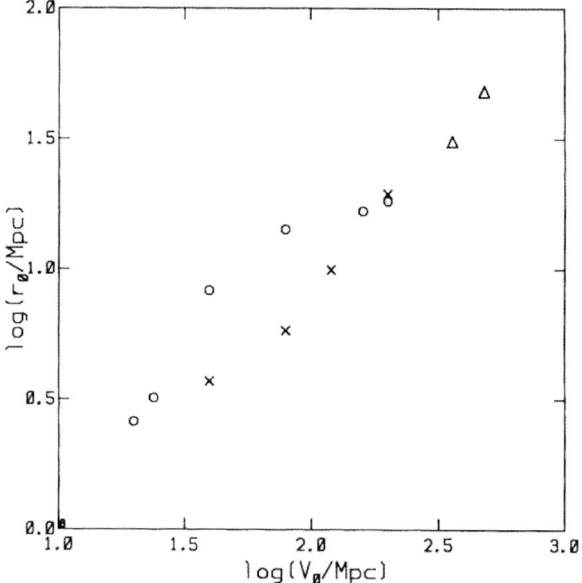

Fig. 7.5 The correlation length, r_0 versus the cube length L or limiting redshift V_0. Open circles are samples in the direction of the Coma supercluster; crosses, samples in the direction of the Perseus supercluster; triangles, samples of Abell clusters of galaxies (Einasto et al., 1986b).

Next we found that the correlation length increases with the sample size, see Fig. 7.5 (Einasto et al., 1986b). To check the dependence of the correlation function on the limiting magnitude, we calculated for the sample of limiting redshift $V_0 = 2000$ km/s correlation functions for different limiting absolute magnitudes (-17.5, -18.5, -20.0), and found no difference. Thus our interpretation was that the increase of the correlation length with the sample size is due to the increasing role of voids in samples of larger sizes. Our data suggested that the correlation length of a fair sample of the universe is $r_0 \approx 10\,h^{-1}$ Mpc, about twice the generally accepted value. Later we found, using more complete samples of galaxies, that the traditional value, $r_0 \approx 5\,h^{-1}$ Mpc, is correct.

The paper by Einasto et al. (1986b) triggered a number of subsequent studies of the correlation function. Initially we did not understand that our results actually hint at the presence of the fractal nature of galaxy distribution. The fractal description of galaxy distribution was suggested by Mandelbrot (1982). Remo Ruffini followed our studies carefully; once he invited Enn and me to participate in the Marcel Grossmann Meeting. We could not attend, but our talk was included in the proceedings (Einasto & Saar, 1982). When the preprint of our correlation function paper (Einasto et al., 1986b) was distributed, he immediately contacted us. He said that our results confirm the fractal nature of the galaxy distribution, and we started a further joint study of the fractal nature of the galaxy distribution (Calzetti et al., 1987).

Pietronero (1987) used our results to suggest that there exists a single fractal (self-similar) structure extending from the galaxy scale up to the present limits of observation. Pietronero emphasised that the present observational data do not support the existence of a length-scale above which the distribution of matter becomes homogeneous. If I understand his opinion correctly, the fractal nature poses serious doubts about the current determinations of "a fair sample" from which one deduces the average mass density and the cosmological density parameter Ω. Pietronero and a number of other astronomers suggest a *fractal cosmology* in which the structure of the whole Universe is characterised by a single fractal. According to this cosmology concept the mean density of matter is almost zero.

In spring 1987 Enn Saar and I had an invitation from Bernard Jones to visit Nordita in Copenhagen. The possible fractal nature of the galaxy distribution was at this time very topical, and we started to study this phenomenon in more detail. Soon Vicent Martinez, a postdoc from Valencia, joined our team. My task was the collection and analysis of observational data; Enn, Bernard and Vicent studied the problem from the theoretical point of view. Soon it became clear that the distribution of galaxies in space cannot be described by a simple fractal (as already suggested by the shape of the correlation function); instead a multifractal description is needed. In the fractal description an effective fractal dimension of galaxy systems is defined. As mentioned above, on small scales galaxies are well clustered: they form groups and clusters, which have an approximately spherical shape. At larger distances the role of filaments dominates. Filaments are essentially one-dimensional, thus the effective fractal dimension changes. This is clearly seen in the correlation function, see Fig. 7.2. The results of our common study were published by Jones et al. (1988).

In summer 1987 near Balaton Lake an IAU Symposium devoted to the Large Scale Structure was held. We had a talk on the fractal nature of the distribution of galaxies (Einasto et al., 1988). According to our analysis the structure of the Universe can be described as a multifractal, i.e. the fractal dimension is different on different scales. The talk was given by Bernard.

At the Symposium one of the main specialists of the fractal theory, Benoit Mandelbrot, participated. After the Symposion a smaller group of astronomers had in Budapest a workshop. Here the main speakers were Yakov Zeldovich and Benoit Mandelbrot. Zeldovich disliked the fractal description for two reasons. First, in his opinion the fractal description contains no hints to the physical nature of the distribution of galaxies and the evolution of the structure of cosmic web. Second, if the structure is described by a simple fractal, then the mean density of the Universe is zero, which contradicts all other independent data. Zeldovich stressed that the fractal description can be valid only in a limited range of scales, because on very large scales the distribution of galaxies is rather uniform — there are no extremely large systems of galaxies. The cosmic web itself is the largest system, and on large scales it is statistically homogeneous.

A few years later Benoit Mandelbrot visited the Soviet Union and invited me to Leningrad to discuss the fractal nature of galaxy distribution. We agreed that the fractal nature of the distribution is clearly evident. I had the same opinion as Zeldovich that there exists rather well-defined lower and upper limits of scales where it can be used. The lower limit is probably the typical scale of galaxies with their dark halos, $\sim 0.1\,h^{-1}$ Mpc. The upper limit is given by the characteristic scale of the supercluster–void network — $\sim 100\,h^{-1}$ Mpc.

Together with Anatoli Klypin we continued the study of fractal properties of galaxy systems. This time our attention was devoted to the Local and Coma superclusters and their environment (Klypin et al., 1989). We noticed that with increasing depth of the sample the ratio of the volume occupied by systems of galaxies to the full volume of the sample — the filling factor — decreases. This phenomenon is the basic property of the fractal distribution. This fact hints also at the self-similarity of systems of galaxies of various scale. Einasto et al. (1989); Einasto & Einasto (1989) studied the self-similarity of voids in galaxy distribution. This is another manifestation of the hierarchical nature of the distribution of galaxies and voids.

To investigate the scaling law of the distribution of galaxies Martinez & Jones (1990) reanalysed data by Einasto et al. (1986b) and confirmed the increase of the correlation length with the size of the sample. However, they suggested that the increase of the correlation length may be due to the increase of the luminosity limit of subsamples, because all samples are volume (absolute magnitude) limited.

Maret Einasto (1991) analysed the behaviour of the function $g(r) = 1 + \xi(r)$ for volume limited samples of various size and absolute magnitude limit. The apparent magnitude limit used was $m = 14.5$. The slope of the function $g(r)$ in log–log representation, γ, is related to the fractal (correlation) dimension D of the sample as follows: $D = 3 - \gamma$ (Pietronero, 1987; Martinez & Jones, 1990). This time a high-luminosity galaxy subsample was also used; this sample had a larger correlation length. The slope of the function $g(r)$ changes at the scale $\sim 3\,h^{-1}$ Mpc. The change of the slope of the correlation function at this scale is visible also in Fig. 7.2 by Zeldovich et al. (1982). The increase of the amplitude of the correlation function with luminosity is due to the biasing problem, see below.

Similar analyses were made by Martinez et al. (1990); Martinez & Jones (1990) and Guzzo et al. (1991) with rather similar results. Guzzo et al. (1991) find for small and large separations the correlation dimensions $D \approx 1.2$ and $D \approx 2.2$, respectively.

Einasto et al. (1997b) estimated the fractal dimension of the sample of clusters of galaxies, using the correlation function. The fractal dimension of the sample of all clusters is $D \approx 2$, and for clusters which belong to very rich superclusters, $D_{\mathrm{SCL8}} \approx 1.4$. Thus structures delineated by very rich superclusters are more one-dimensional than two-dimensional, as in the case of structures defined by all clusters.

These results raise the question: What is the size of a representative sample of the Universe? The galaxy correlation function has the correlation length $r_0 \approx 5\,h^{-1}$ Mpc, on scales $r \geq 5\,r_0$ the correlation function is close to zero. Thus it was thought that the size of a representative sample is of the order of $5r_0 \approx 25\,h^{-1}$ Mpc. The presence of the cosmic web with filaments, superclusters and voids of size up to $\approx 100\,h^{-1}$ Mpc suggests that samples smaller than this size are probably not representative enough. Einasto & Gramann (1993) found that the transition scale is at least $175\,h^{-1}$ Mpc, see below. If we expect that in a representative sample also superclusters of galaxies of various richness should be present, then the transition scale to homogeneity is evidently still higher, since extremely rich superclusters are very rare (Einasto et al., 2006).

7.1.4 *Physical biasing*

One problem to solve was to find some explanation for the absence of galaxies in voids. Observational data show that there are no galaxies in voids except galaxy filaments joining superclusters to the web, as seen in Figs. 5.5 and 5.4. In contrast, numerical simulations show the presence of a rarified population of test particles in voids, see Fig. 5.3. In the first approximation the absence of galaxies in voids was explained by Einasto et al. (1980a). Enn Saar developed an approximate analytical model of the evolution of density perturbations in under- and over-dense regions based on Zeldovich (1970) ideas. He found that the matter flows out of under-dense regions and collects in over-dense regions until it collapses (pancake forming) and forms galaxies and clusters. In under-dense regions the density decreases continuously, but never reaches zero. In other words: there must be primordial matter in voids, see Fig. 7.6. Galaxy formation occurs not everywhere but only in regions where the matter has collapsed.

The referee of the paper Einasto et al. (1980a), Michael Fall, asked to exclude most theoretical considerations from the paper, thus the density evolution formula was published many years later when we returned to the void evacuation

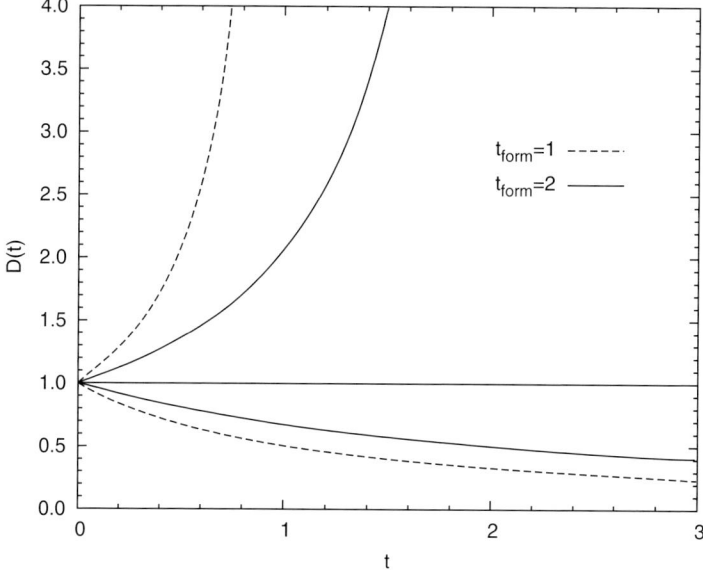

Fig. 7.6 The evolution of over- and under-densities according to an analytical approximation found by Enn Saar. The time t is shown in arbitrary units, the density D is expressed in mean density units. Higher over-density regions collapse earlier, compare density changes corresponding to $t_{\rm form} = 1$ and $t_{\rm form} = 2$. At structure formation time the density in under-dense region is $D(t_{\rm form}) = 0.5$. The evolution was calculated using formulae given by Einasto et al. (1994a).

problem (Einasto et al., 1994a). In autumn 1980 I was invited to visit the Institute of Astronomy, Cambridge, and gave a seminar talk on the large-scale distribution of galaxies. After the talk Michael Fall asked: *"all this looks nice, but how can you explain filamentary superclusters and voids?"* My answer was: *"my task is to find how the structure of the Universe actually is, the explanation of the observed structure is the task of theorists"*. At this time we had just started to find quantitative tests to compare observations with theory, so it was too early to make more conclusive statements.

Soon results of such tests were available (Zeldovich et al., 1982). The multiplicity test shows clearly the difference between observational and model samples. As already mentioned, this test suggests that galaxy formation is a threshold phenomenon: *in low-density regions no galaxy formation occurs at all, here the matter is still in pre-galactic unclustered form*. In the following years, when comparing observations with simulations, we always excluded non-clustered particles in low-density regions to get a simulated sample, which could be compared with observed galaxy samples (Melott et al., 1983; Einasto et al., 1986a; Einasto & Saar, 1987; Einasto et al., 1989).

The term "biasing" was introduced by Kaiser (1984) to denote the difference between the correlation functions of galaxies and clusters of galaxies. The correlation function and the power spectrum are connected by the Fourier transform, thus a similar difference exists between the power spectra of galaxies and clusters. In the following years the term "biasing" was used in a broader context to quantify the difference between the distributions of various populations: dark matter, galaxies of various luminosity, clusters of galaxies etc.

In our simple analysis of the evolution of over- and under-dense regions we noticed that in over-dense regions matter contracts to form halos which can evolve into galaxies or clusters, while in under-dense regions the density decreases, and there are no conditions to form galaxies. This picture seems to be evident, as it is well-known that only density enhancements in excess of a factor 1.68 relative to the mean density can collapse during Hubble time (Bardeen et al., 1986). This biasing scheme was used by White et al. (1987) to find simulated galaxies in N-body experiments.

Another possibility to define the biasing parameter b is to use the ratio of the density contrast of galaxies and matter at location $\mathbf{x}$: $\delta_{gal}(\mathbf{x}) = b\delta_m(\mathbf{x})$. This definition is based on the tacit assumption that galaxies are randomly placed. In other words, voids are just regions of lower density of galaxies. But there is a problem with this interpretation. Observations show that there are no galaxies in voids, except faint galaxy filaments crossing voids, determined by clusters or superclusters. Thus we expect $b = 0$ in void regions outside filaments. If the distribution of galaxies follows the distribution of matter in high-density regions, then in these regions $b = 1$. In order to apply the above formula, and to find the mean value of the biasing parameter, most authors apply smoothing of the density field using a rather large smoothing length (up to $8\,h^{-1}$ Mpc), so that the density of galaxies is everywhere non-zero. But this procedure smoothes galaxies into regions which are actually empty, thus it does not take into account the actual distribution of galaxies.

To avoid this difficulty it is better to define the biasing factor b using power spectra of matter and galaxies, which can be determined also for cases where regions of zero density of objects of interest exist. The biasing factor can be determined for different populations using varying density thresholds to define populations of different type (galaxies of different luminosity, galaxy systems of different richness etc.).

Einasto et al. (1999a) found the following relation between the power spectra of the matter, $P_m(k)$, and that of the clustered population (galaxies or clusters of galaxies), $P_c(k)$:

$$P_m(k) = \langle |\delta_m(k)|^2 \rangle = F_c^2 P_c(k), \tag{7.1}$$

where k is the wavenumber, and F_c is the fraction of matter in the clustered population. The last equation gives for the bias factor of the clustered population $b_c = 1/F_c$. In other words, the biasing factor depends on the fraction of matter in the clustered population, at least in the first approximation.

These equations show that the subtraction of a homogeneous population from the whole matter population increases the amplitude of the spectrum of the remaining clustered population. In this approximation biasing is linear and does not depend on scale. These equations have a simple interpretation. The power spectrum describes the square of the amplitude of the density contrast, i.e. the amplitude of density perturbations with respect to the mean density. If we subtract from the density field a constant density background, but otherwise preserve density fluctuations, then amplitudes of *absolute* density fluctuations remain the same, but amplitudes of *relative* fluctuations with respect to the mean density increase by a factor which is determined by the ratio of mean densities, i.e. by the fraction of matter in the new density field with respect to the previous one (Einasto et al., 1999a).

To check the above relations Einasto et al. (1999a) performed numerical simulations. The relations are identical in the 2-D and 3-D cases, thus a 2-D simulation with 512^2 particles and cells was made to obtain a better resolution. Simulation particles were divided into populations according to values of the local density. Particles with low density values, $\varrho < \varrho_0$, were called *void particles*; all others are called *clustered particles* of various density. ϱ_0 is the threshold density to separate clustered and unclustered particles. If we consider all galaxies, including the faintest dwarf galaxies, then it is natural to accept $\varrho_0 = 1$. For galaxies of higher luminosity and for groups/clusters $\varrho_0 > 1$. Distributions of particles of various threshold density are shown in Fig. 7.7. For comparison also a sheet of galaxies of various absolute magnitude is shown, which crosses the Local, the Coma and the southern corner of the Hercules supercluster.

Power spectra and respective bias factors (as a function of the wavenumber k) for various samples of simulation particles are shown in Fig. 7.8. Galaxy and cluster samples are the same as used in Fig. 7.7. Calculations show that the biasing factor can be indeed calculated from the fraction of particles in the respective sample, for details see Einasto et al. (1999a).

Figure 7.8 shows that all power spectra of samples of clustered particles are similar to the power spectrum of the matter, but have higher amplitudes. This similarity of the shape of the power spectra shows that biased galaxy samples

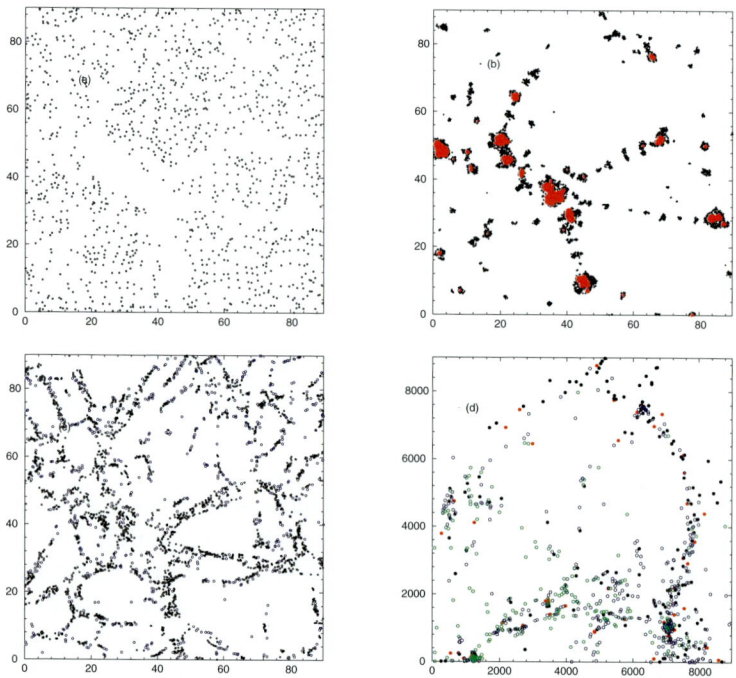

Fig. 7.7 The distribution of simulated and real galaxies in a box of side-length $90\,h^{-1}$ Mpc. Panel (a) gives particles in voids ($\varrho < 1$); panel (b) shows the distribution of simulated galaxies in high-density regions: galaxies in the density interval $5 \leq \varrho < 20$ are plotted as black dots, galaxies with $\varrho \geq 20$ as filled (red) regions; panel (c) shows field galaxies in the density interval $1 \leq \varrho < 1.5$ (open blue circles), and $1.5 \leq \varrho < 5$ (dots). Densities are expressed in units of the mean density of the Universe. Panel (d) shows the distribution of galaxies in supergalactic coordinates in a sheet $0 \leq X < 10\,h^{-1}$ Mpc, horizontal and vertical axes are supergalactic Y and Z, respectively; bright galaxies ($M_B \leq -20.3$) are plotted as red dots, galaxies $-20.3 < M_B \leq -19.7$ as black dots, galaxies $-19.7 < M_B \leq -18.8$ as open blue circles, galaxies $-18.8 < M_B \leq -18.0$ as green circles (absolute magnitudes correspond to Hubble parameter $h = 1$) (Einasto et al., 1999a).

and even samples of particles in clusters contain all essential information on the amplitudes of density waves of different scales (excluding the shortest waves, which can be followed only by dwarf galaxies). From the difference in the amplitude of the power spectra of these populations with respect to the power spectrum of matter we derived the biasing parameter as a function of the wavenumber. The results are plotted in the right panel of Fig. 7.8. For most samples the biasing parameter is almost constant. As expected, the deviations are larger on small scales where the cluster sample contains almost no test objects.

The overall amplitude shift is determined by the fraction of matter in the clustered population associated with galaxies or systems of galaxies. In other words,

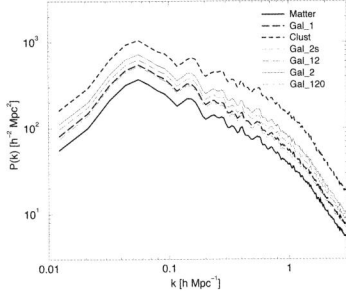 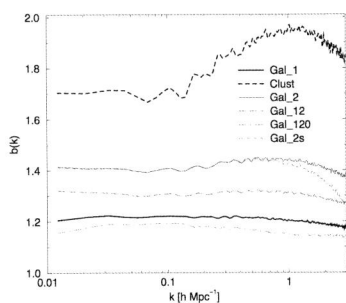

Fig. 7.8 Left: Power spectra of simulated galaxies. The solid bold line shows the spectrum derived for all test particles (the matter power spectrum); dashed and dotted bold lines give the power spectrum of all clustered particles (sample Gal-1) and clustered galaxies in high-density regions (sample Clust). Thin solid and dashed lines show the power spectra of samples of particles with various threshold densities and sampling rules. Right: the biasing parameter as a function of wavenumber, calculated from definition given in the text. Samples and designations are the same as in the left panel (Einasto et al., 1999a).

populations with higher values of power spectra (and correlation functions) are not more clustered, as traditionally interpreted for the cluster correlation function (Bahcall & Soneira, 1983; Klypin & Kopylov, 1983), but are more strongly normalised to take into account the smaller fraction of objects in their samples. The definition of the biasing factor using power spectra of populations is in harmony with the original definition by Kaiser, since the correlation function and the power spectrum express the same property of the distribution, one in real space, the other in Fourier space.

This explanation of the correlation function of clusters of galaxies indicates that there is no need to introduce two different power spectra for galaxies and clusters with different cut-off scales, as suggested by Dekel (1984).

Our simple biasing algorithm does not take into account physical processes of galaxy formation which influence masses and luminosities of galaxies. One of these processes is the supernova-driven wind which blows out interstellar gas from galaxies and is very important in the formation of dwarf galaxies (Dekel & Silk, 1986). A detailed discussion of related problems is, however, outside the scope of this review.

In connection with the biasing problem it is appropriate to ask the question: What is the fraction of matter in voids? This problem was studied by Einasto et al. (1994a) using numerical simulations of the ΛCDM Universe with density parameter $\Omega_m = 0.2$, for comparison also models with $\Omega_m = 1.0$ were calculated.

The void matter was defined by the high-resolution density field, with local density $\varrho \leq 1$, where ϱ is the density in units of the mean density of the sample. Hydrodynamical simulations confirm that in low-density regions galaxies do not form (Cen & Ostriker, 1992, 2000). By definition, half of the particles are initially located in regions with densities less than the mean density. During the evolution matter flows out from voids to systems of galaxies. How much matter is presently located in voids and in the clustered population depends on the parameters of the model. For the $\Omega_m = 0.2$ ΛCDM model Einasto et al. (1994a) found for the fraction of matter in voids $F_v = 0.15 \pm 0.05$.

The above formula takes into account the major factor — the absence of galaxies in voids. Differences in the distribution of dark matter and matter associated with galaxies in high-density regions influence the biasing parameter much less, as our calculations show. What is important here is the use of a high-resolution density field with smoothing scale of the order of the scale of actual structures — galaxies with dark halos and groups/clusters of galaxies. Using such a small smoothing scale we get for voids a zero density of matter associated with galaxies.

Once I discussed with Enn Saar this problem. He argued that in mathematical statistics a zero density has no meaning, since in this case it is impossible to estimate errors. Thus large smoothing is needed to have everywhere non-zero density of the clustered matter. But it seems to me that biasing is not a statistical, but a hydrodynamical problem. During the early stages of the evolution of the Universe there were no galaxies, only the primordial gas, a mixture of dark matter and hydrogen–helium baryonic gas. In this gas the density of heavier chemical elements was zero. Also, remember the approach of Zeldovich to the evolution of the early Universe — he considered the problem as a hydrodynamical one, particles were used only as markers of positions, but the formula for the evolution were hydrodynamical.

In this approach the numerical value of the biasing factor depends on the density threshold ϱ_0, used to separate clustered and unclustered matter. Our numerical simulations show that the distribution of local densities (smoothed with a small kernel) is a smooth function of the density, i.e. there exists no natural value for the threshold density, which could be found from the density distribution itself. Thus the value of the threshold density must be calculated separately, taking into account processes which lead to the formation of galaxies. One such possibility is the Press & Schechter (1974) algorithm of galaxy formation (see also Bardeen et al. (1986)). According to this approximation galaxies have time to collapse in the Hubble time, if the local density of the contracting cloud has a density exceeding 1.68 times the overall mean density. I used various density thresholds in the interval from

1 to 1.68 to find the biasing parameter. Results show that the biasing parameter changes only a little when the threshold is varied within this small interval.

To conclude the discussion of the biasing problem I can say that our results have two consequences:

(1) the biasing parameter b is not a free parameter which can be chosen to bring models into agreement with observations, as done in early simulations of CDM models of critical cosmological density (Davis et al., 1985; White et al., 1987); actually it depends on the fraction of matter in the clustered population;
(2) there is no possibility of hiding large amounts of dark unclustered matter in voids; most of the low-density matter has been 'eaten' by galaxies and systems of galaxies, increasing the fraction of the clustered matter. This excludes CDM models with $\Omega = 1$, since the observed value of the density of matter associated with galaxies is $\Omega \approx 0.2$ (Einasto et al., 1974d; Ostriker et al., 1974).

7.1.5 *Power spectra of galaxies*

In numerical simulations of the evolution of the structure in the universe for every time step both the mass density and the gravitational potential fields are calculated; the potential field is needed to find the velocities of particles. Using these data it is easy to calculate also the power spectrum of the density field for every evolution step. During our visits to the Cambridge Institute of Astronomy with Mirt Gramann we compared results of simulations with real observational data. Soon we understood that the same technique can be used to calculate the power spectra for galaxies and clusters of galaxies. We adapted the simulation program to apply it to volume limited samples of galaxies and clusters of galaxies. We presented the principal results of our analysis in October 1990 at the Ringberg workshop and at a conference in Rome, and in March 1991 at the 2nd DAEC workshop in Paris. More detailed versions were published by Gramann & Einasto (1992) and Einasto et al. (1993).

Our analysis had two goals. The first goal was to compare the power spectra for galaxies of various luminosity and for clusters of galaxies to detect the possible variations of the power spectra with sample location, luminosity limit, richness etc. The second goal was to compare the real power spectra with spectra for various cosmological models. For an ideal sample of infinite volume the power spectrum is directly related to the correlation function, as they form a Fourier transform pair. But in reality the situation is more complicated. The correlation function characterises the structure of galaxy systems well on small scales, whereas the structure on large scales can be better described in terms of the power spectrum.

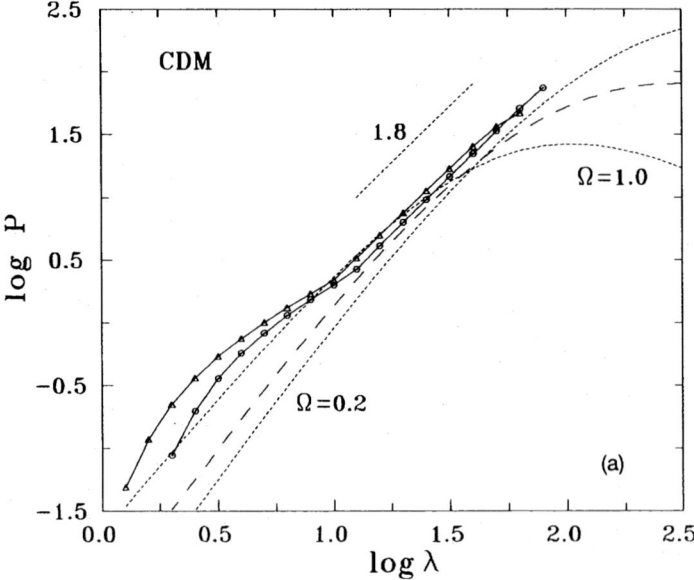

Fig. 7.9 Comparison of the observed power spectra with the predictions in different CDM models. Triangles and open circles show the observed spectra in the Perseus and Virgo-Coma regions, respectively. The linear predictions of models are plotted as follow: CDM model with $\Omega = 1$, $h = 0.5$ — dotted line; CDM $\Omega = 0.2$ model for $h = 0.5$ and $h = 1$ with a dotted and dashed line, respectively. Deviations of models from observations on small scales almost disappear when non-linear evolution of models is taken into account. On large scales the evolution is linear, the absence of large-scale power in the $\Omega = 1$ model is clearly seen (Gramann & Einasto, 1992).

Gramann & Einasto (1992) and Einasto et al. (1993) calculated power spectra for various volume limited galaxy samples in the Virgo-Coma and the Perseus supercluster regions, and for Abell clusters of galaxies. The analysis showed that the observational power spectra for galaxies of various luminosity limits, and for clusters of galaxies have a similar form in the log–log presentation, but have different amplitudes. This effect is easily explained by different fractions of matter in the clustered populations, i.e. by different biasing levels as discussed in the previous section.

We compared the observational power spectra with the spectra of the standard CDM model ($\Omega = 1$), and of the open CDM model with $\Omega = 0.2$ and for various Hubble parameter values. The results of the comparison are shown in Fig. 7.9. In this Figure we showed linearly evolved models, but in a more detailed comparison we used ΛCDM models with $\Omega_m = 0.2$ instead of open CDM models with a similar density parameter. The Figure shows that the standard CDM model with

$\Omega = 1$ has much less power on large scales than the real data and open CDM (or ΛCDM) models with $\Omega_m = 0.2$. Our finding of the absence of large-scale power in the standard CDM model confirmed independent analyses by Efstathiou et al. (1990) and Peacock (1991). Efstathiou et al. calculated angular correlations for the deep APM galaxy survey and for the large-scale clustering of IRAS galaxies. Peacock derived power spectra for the CfA galaxy survey and for the IRAS QDOT infrared galaxy survey. In all these studies the deficit of the large-scale power in the standard CDM model was clearly seen.

Gramann & Einasto (1992) analysed also correlation functions of various galaxy samples. All samples had a characteristic shoulder on a scale $r \approx 3\,h^{-1}$ Mpc, visible already in our previous data, see Fig. 7.2. This shoulder corresponds to the transition from almost round clusters/groups of galaxies to elongated filaments of galaxies. A similar change is observed in the power spectrum at a scale $\lambda \approx 10\,h^{-1}$ Mpc; this scale is usually interpreted as the scale of transition from the linear evolution on larger scales to the non-linear evolution on smaller scales.

An independent analysis of power spectra was made by Vogeley et al. (1992) for galaxies in the second CfA redshift survey, and by Park et al. (1992a) for galaxies in the Southern Sky Redshift Survey. Changbom Park developed a very accurate method to calculate power spectra for observed galaxy samples. Both galaxy samples show that at small scales ($\lambda \leq 30\,h^{-1}$ Mpc) the observed power spectrum is consistent with the standard CDM model, but on larger scales data indicate an excess of power over the standard model. The open CDM model with $\Omega h = 0.2$ is consistent with the observed spectrum over all scales.

The characteristic diameter of supervoids between rich superclusters is about $100\,h^{-1}$ Mpc (Jõeveer et al., 1978), the same scale was found by Broadhurst et al. (1990) between peaks in the galaxy distribution towards Galactic poles. This rises the question: What is the scale of a homogeneous Universe? This problem was investigated by Einasto & Gramann (1993) by studying the behaviour of the power spectrum for galaxies on large scales. The transition scale to homogeneity was defined as the scale where the spectral index of the matter power spectrum reaches the value $n = 1$, predicted by the Harrison-Zeldovich model. We used various observational samples to calculate the power spectra (Abell clusters, CfAII galaxies, QDOT galaxies), and compared the observed power spectra with models. Our result was that at scale $\lambda \approx 175\,h^{-1}$ Mpc the power spectrum reaches the Harrison-Zeldovich form.

To see the influence of large-scale waves to the density fields of matter and of gravitational potential Einasto & Gramann (1993) calculated the evolution of two-dimensional CDM models of size $L = 512\,h^{-1}$ Mpc for a 512^2 mesh with 512^2 particles. Two cases were studied, one with the full CDM-type power spectrum, and

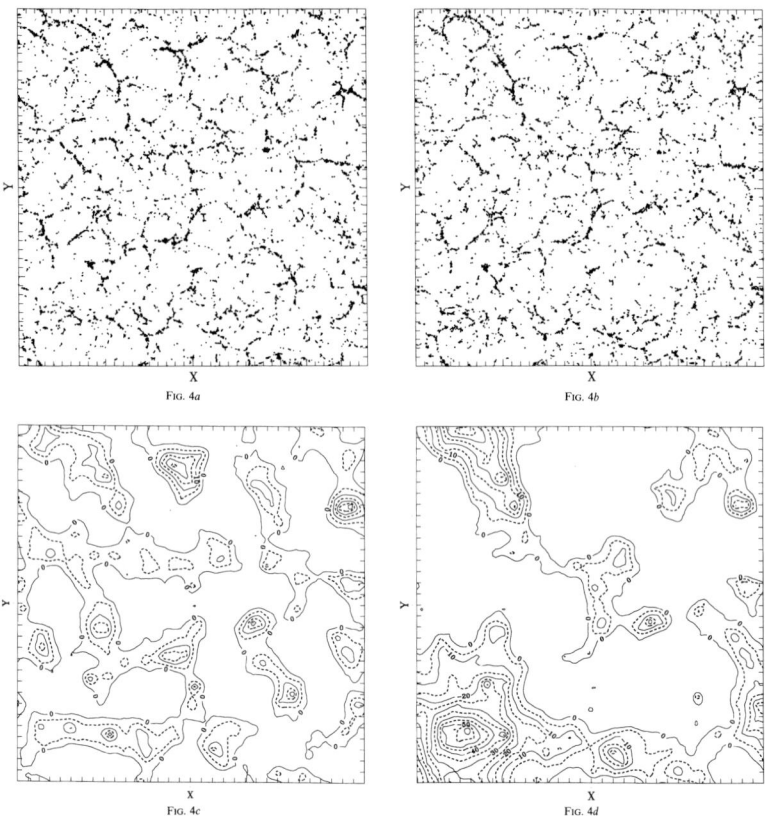

Fig. 7.10 The upper panels show the distribution of simulated galaxies in two-dimensional models (particles in voids are removed), the lower panels show the respective potential wells — regions of negative gravitational potential. In the left panels we plot data for the model with a sharp truncation of the spectrum, in the right panels for the CDM model with a smooth transition to the Harrison-Zeldovich spectrum on large scales. Equipotential lines have been drawn in both panels at the same levels (in dimensionless units) (Einasto & Gramann, 1993).

the other with the power spectrum cut on a scale $\lambda = L/4$, so that the amplitude of all waves of scales larger than the cutoff scale was put to zero. Both models had the same realisation of initial density fluctuations. The final distribution of particles in clustered regions, and the shape of potential wells corresponding to high-density regions is shown in Fig. 7.10. The Figure shows that the distributions of clustered particles in both simulations are very similar, however the shape of potential wells is very different. In the model with the full power spectrum potential wells are deeper, and their mutual distance from each other is larger than in the truncated model. This exercise shows that large waves play an important role in the formation of the structure of the Universe.

In late 1990's I turned to the study of the power spectrum again, motivated by the BEKS effect (Broadhurst et al., 1990; Einasto et al., 1997a). Together with my collaborators I tried to find the present power spectrum for galaxies (Einasto et al., 1999c), to calculate more accurately the biasing correction (Einasto et al., 1999a), and to find the primordial power spectrum of matter (Einasto et al., 1999b). However, the data available at this time were insufficient to find a much better power spectrum than had been found earlier. The modern SDSS data yield much better power spectra, but the detailed description of these new results is beyond the scope of this book.

7.2 Redshift surveys and catalogues

7.2.1 *Redshift surveys*

Our first 3-dimensional pictures of the distribution of galaxies were based on the Second Reference Catalogue of Galaxies by de Vaucouleurs et al. (1976), as well as on the published redshifts of nearby clusters and Markarian and other active (radio) galaxies. As new redshifts became available (Sandage & Tammann (1981), the first Harvard Center for Astrophysics redshift survey, and ZCAT by John Huchra 1981), we made more detailed analyses of the Local, Coma, Perseus and Hercules superclusters, and the huge void between these superclusters.

In 1979 at Princeton a conference was held to discuss observing programs for the Hubble Space Telescope. I had the luck to be one of the participants. After the conference our small delegation from the Soviet Union made a trip to Center for Astrophysics of Harvard University. We had a chance to meet Riccardo Giacconi, who was just discussing first results of the Einstein X-ray orbiting observatory. Also I had long discussions with John Huchra, one of the initiators of the First Harvard Redshift Survey. He showed me a model of the distribution of nearby galaxies, made of small plastic balls, similar to our model made a few years earlier. Huchra's model was better made — the balls were smaller and better fixed, so it was possible to put the model in the lobby of the Center. The filamentary distribution of galaxies in the nearby Universe was very well seen.

In autumn 1982 I got an invitation to attend the Texas Symposium which was held in Texas. The organisers were ready to cover all my travel expenses to the Symposium and to other places in USA of interest. Just in summer 1982 I had a chance to discuss during the IAU Symposium in Crete the large scale distribution of clusters with George Abell, who was preparing the Southern extension of his cluster catalogue. I phoned George and asked, would it be possible to have a visit to the University of California in Los Angeles, where he worked? His reply was very positive. Thereafter I phoned Harvard and asked whether a visit to Center for Astrophysics would be possible to discuss the distribution of galaxies. Here I got also a positive response. In short, I had the chance to visit several major astronomical centres in the USA in a three month visit.

In Austin I had long discussions with Gerard de Vaucouleurs. Thereafter, in the trip to Los Angeles I had a stop at Tucson, Arizona, where the headquarters of the Kitt Peak Observatory is located. Here I visited Kitt Peak Observatory headquarters. I met Bart Bok, and discussed with him the nature of dark matter. He agreed with my arguments that dark matter is a new population, different from the conventional halo consisting of old metal-poor stars, thus it is better to use the term "corona" instead of "halo" to avoid confusion.

Also I had an agreement with John Huchra to visit the Mount Hopkins Observatory, where the 60 inch telescope used by John and his team to collect redshifts of galaxies for the Harvard Redshift Survey is located. John arrived, rented a huge SUV, and we drove up to the mountain. Margaret Geller was already there. Here I was witness how well survey observations were prepared and made. Finding charts of galaxies were made from Palomar Sky Survey prints with exact coordinates. So when the recording of the spectrum of a galaxy was finished, it was possible to turn the telescope very fast to the next object and to continue the observation.

It was autumn and galaxies in the Perseus region were in the program. I saw in their list a small group of galaxies close to the Milky Way zone of avoidance. In our study of the Perseus–Pisces supercluster we noticed that a filament of poor nearby clusters extends in the Northern direction away from the Perseus main cluster. This group was just one of these groups we noticed in our study (Zwicky cluster 22 in Fig. 5.7). So I said that the expected redshift of these galaxies is about 5,000 km/s. John was surprised and asked how I know this. Then I told him the results of our study of the Perseus–Pisces supercluster, and the presence of several filaments, in addition to the main cluster chain seen in Figs. 5.5 and 5.4. Observations confirmed that this chain of groups, seen in Fig. 5.7, is really part of the Perseus–Pisces supercluster.

In Los Angeles I had long discussions with George Abell about the Southern Survey of rich clusters of galaxies. Our experience has shown that all clusters are markers of the cosmic web, thus it makes sense to include into the catalogue also clusters they notice during the search, but which have less galaxies than their criterion for inclusion. Of course, these clusters do not form a statistically homogeneous sample; their role is to be the markers of the web.

A month later I was in Harvard and discussed our quantitative analysis of the cosmic web. The Harvard team was preparing for the Second Redshift Survey. We discovered that our approaches to the study of the large scale distribution of galaxies were very similar. So far most observers studying the distribution of galaxies in the mid 1970's concentrated their efforts on studying the *environment* of rich clusters. These studies confirmed the existence of superclusters — i.e. systems of galaxies around rich clusters. These studies found also the existence of voids in the distribution of galaxies. One of the first studies to apply this approach was made by Gregory & Thompson (1978). But these studies did not yield information on the presence of the cosmic web.

In contrast, both the Tartu and the Harvard teams understood the need for *wide field* surveys, covering a contiguous and substantial area on the sky, and the full range of galaxy densities from the richest clusters to the emptiest voids. These observations could provide the key to a better theoretical understanding of the history of the formation of the large-scale structure of the Universe. Our Tartu team started the study in the mid 1970's by collecting data from all possible sources in the whole Northern hemisphere. Almost at the same time the Harvard team started the First Redshift Survey, covering the whole Northern hemisphere, up to apparent magnitude 14.5. In the Second Survey the Harvard team planned to observe the same area again but with the magnitude limit 15.5.

These two approaches, the detailed study of smaller deeper areas and the study of less deep but larger contiguous areas, complement each other. The second approach allowed us to see the *global* features of the cosmic web earlier. The price was that wide surveys did not initially give an answer to the question, how *empty* voids actually are, because in wide field surveys it was not possible to have as faint a magnitude limit as in small-area studies for a given amount of observing time.

One item we discussed in Harvard was: how to plan observations in such a way that already the first slice observed can give significant results to understand major properties of the cosmic web. One favorable region is a slice which crosses the Coma, Hercules, Perseus–Pisces and Ursa-Major superclusters, and galaxy filaments joining these superclusters. This slice also crosses several voids: voids between the Local supercluster and the Coma and the Perseus–Pisces superclusters, as well voids behind the Perseus–Pisces and Hercules superclusters. The last one is called the Bootes Void.

Results of the first slice of the Second Harvard Redshift Survey were very interesting (de Lapparent et al., 1986). The slice covered a strip of the Northern galactic hemisphere between declinations $26°.5$ and $32°.5$ of length $117°$. Basic structural elements of the cosmic web were clearly visible. I was happy to read that authors thanked me for discussion. What surprised me a bit was the absence of references to our early papers (Jõeveer & Einasto, 1978; Jõeveer et al., 1978; Einasto et al., 1980a), only our main review paper (Zeldovich et al., 1982) to describe the cellular character of the structure was discussed.

de Lapparent et al. (1986) emphasised the sharpness of structures, and suggested that this hints to the presence of hydrodynamical processes during the galaxy formation. The authors argued that the sharpness can be explained in the framework of the Ostriker & Cowie (1981) explosive model of structure formation.

At the same time our team compared the observed structure with the ΛCDM model by Gramann (1987). Our quantitative comparison suggests that the ΛCDM model explains the observed structure very well (Einasto et al., 1986a), confirming our earlier comparison of the CDM model with observations (Melott et al., 1983). The sharpness of structures was not addressed in our model. Modern high-resolution ΛCDM simulations explain this feature very well.

At the IAU symposium on large-scale structure in Hungary in 1987 I met Margaret Geller and congratulated her for the very important results of the CfA 2nd Redshift Survey. A popular discussion of these results by Geller & Huchra (1989) received great attention. The astronomical community as well as the general public recognised the presence of voids, superclusters and walls in the galaxy distribution after these publications, about ten years later than the actual discovery of the comic web, discussed in the Tallinn IAU Symposium.

In the following years a number of other surveys covering large areas of the sky were made. John Huchra initiated a near-infrared survey of nearby galaxies, the Two Micron All-Sky Survey (2MASS). The advantage of this survey is the coverage of low galactic latitudes up to 5 degrees from the Galactic equator. The filamentary character of the distribution of galaxies is very well seen. During the Aspen workshop on voids in summer 2006 John told the story of this Survey. His initial suggestion was to make the Survey from space. But one referee, who gave a very positive opinion to the project, suggested that the Survey can be done using ground based observations. Thus the space project was rejected. The same referee later participated very actively in the Survey. However, from ground the Survey took much more time, and the total costs were approximately the same as the expected costs from the space version.

The largest so far wide area surveys are the Two-degree Field Galaxy Redshift Survey made with the Anglo–Australian 4-m telescope, and the Sloan Digital Sky Survey. When visiting Australian observatories in 1989 I had a chance to see the giant fiber robot later used for the 2dF Survey. Our Tartu team has made extensive use of both these surveys, which allowed us to investigate the structure of the cosmic web in great detail.

In recent years a number of new deep wide-field redshift surveys have been initiated. One survey is directed to the study of baryonic acoustic oscillations (BOSS), first results of this survey were discussed by White et al. (2011). BOSS is part of the SDSS-III survey; it makes use of luminous galaxies selected from the SDSS images to probe large-scale structure at intermediate redshift, $z < 0.7$. It uses the same telescope as SDSS-I and II, but a better spectrograph with 1000 fibers instead of 640 in previous surveys. Another new surveys is directed to find very large structures of the Sloan Great Wall type. This HectoMAP project was discussed by Geller et al. (2011), it has the goal to measure redshifts of red galaxies up to magnitude $r = 21$ in an area covering 50 square degrees in a 1.5 degree wide strip. Observations are made with the 6.5-m Multi Mirror Telescope on Mount Hopkins, Arizona, using a 300 fiber robotic spectrograph.

7.2.2 Catalogues of groups and clusters of galaxies

Most galaxies are located in systems of various richness — groups, clusters and superclusters. In our early studies we noticed that giant galaxies are surrounded by dwarf galaxies, and that companion galaxies are segregated by morphology: elliptical companions lie closer to the main galaxy than spiral and dwarf irregular galaxies (Einasto et al., 1974c, 1975b). Thus groups of galaxies have a certain structure; additionally they contain dark matter which can be considered as the extended corona of the main galaxy. The dark corona is so large that practically all dwarf companion galaxies are located inside the corona. If we consider the corona as a population of the main galaxy, then this means that dwarf companions lie inside of the main galaxy. To avoid confusion, we called such systems "hypergalaxies" (Einasto et al., 1974a), and prepared lists of hypergalaxies (Einasto et al., 1977; Vennik, 1984). This term has not been accepted by the astronomical community, probably because hypergalaxies are actually small groups of galaxies with one concentration center. The difference between groups in general lies in the fact that in hypergalaxies there is only one centrally located bright galaxy, and that companion galaxies are clearly morphologically segregated. Large groups contain several subgroups, and the segregation of satellite galaxies by morphology is not so clear.

Our main goal in the late 1990's and early 2000's was to investigate the general properties of the cosmic web. At this time fairly deep wide-field galaxy redshift surveys were already available, which allowed us to use galaxies as indicators of the web. The first such survey was the Two-degree Field (2dF) Galaxy Redshift Survey. This Survey allowed us to prepare group and supercluster catalogues, and to study properties of superclusters and the web in general. The first step in such studies is the preparation of a group catalogue to suppress the Finger of God effect (random motions of galaxies in groups) which distorts the estimation of distances using redshifts as distance indicators.

In the preparation of the group catalogue the method used to combine galaxies into groups plays a central role. Our earlier experience for nearby groups has shown that in rich groups bright galaxies are concentrated toward the group center, thus to find the group we can use the FoF method with constant neighbourhood linking length (radius). We applied the same scheme to the 2dF group catalogue and sent the paper to Monthly Notices. However we got a negative referee report. The referee was Vincent Eke, who has just compiled a group catalogue for the same 2dF Redshift Survey (Eke et al., 2004). His major criticism was the use of constant linking length in group finding. Thus we started to investigate the role of variable linking length in more detail.

Already in the early 1980's several group catalogues had been constructed. One of the most widely known was the catalogue by Huchra & Geller (1982). Huchra & Geller applied a variable search radius depending on the mean volume density of galaxies at the distance of the group. The linking length was taken as $l \sim f^{-1/3}$, where f is the selection function of galaxies at the particular distance from the observer. This scaling corresponds to the hypothesis that with increasing distance both the galaxy field in general, and that of the groups, are diluted in the same way by the absence of fainter galaxies at larger distance.

To see the difference between various group selection methods we compiled two versions of the group catalogue, one with a constant search radius, and the second one with a variable search radius, which increases with the distance according to the decrease of the mean number-density of galaxies, as done by the Eke team. Results of this test are shown in the Fig. 7.11 for the rich cluster Abell 933. We see that using a constant linking length the cluster A933 is divided into several concentrations. In contrast, if a variable linking length is used, depending on the number density of the whole sample, then a large filament around the cluster is also included into the system. The reason for this increase is the location of the cluster A933 in a rich supercluster.

In other words, the overall density of galaxies is not really relevant to determining the clustering parameters of groups; it is the galaxy number density within groups themselves that fixes the linking length. As a further check we calculated the mean sizes of Eke et al. groups. We found that the mean size of Eke groups increases considerably with the distance, i.e. the group catalogue is not homogeneous. To summarise, these tests indicated that both extreme linking methods have drawbacks, and a new linking method is needed.

In the search for a better method we decided to use real groups to study the scaling of group properties with distance. First we created a 2dF group catalogue

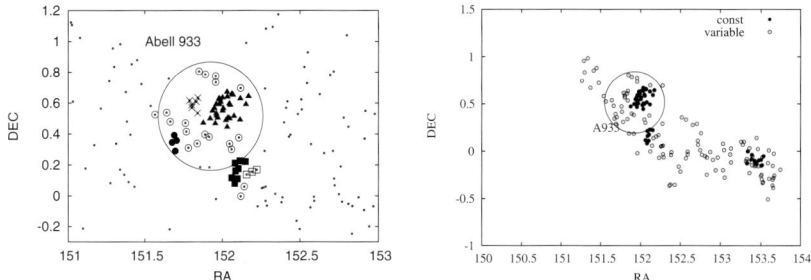

Fig. 7.11 The distribution of galaxies in and around the Abell cluster A933, located in the core of the supercluster SCL82 in the Sloan Great Wall. The circle shows the size of the Abell cluster. In the left panel galaxies that belong to different groups, found by using a constant linking length, are plotted with different symbols. Open circles show single galaxies or pairs of galaxies. All large symbols form a single group according to Eke et al. (2004). Small dots show field galaxies in the vicinity of the cluster. The right panel shows an extended region around the same cluster using FoF with a constant and a variable linking length, which increases with distance as the mean number-density of galaxies in the cluster decreases (Tago et al., 2006).

with constant linking length, selected in the nearby volume ($d < 100\,h^{-1}$ Mpc), all rich groups, and determined for group galaxies their absolute magnitudes and radial velocities. Then we shifted the groups progressively to larger distances, and calculated new k-corrections and apparent magnitudes for the group members. As with increasing distance more and more fainter members of groups fall outside the observational window of apparent magnitudes, the group membership changes. We calculated the minimum FoF linking length necessary to keep the group together at this distance. To determine that, we built the minimal spanning tree for the group (Martínez & Saar, 2002), and found the maximum length of the MST links. The average linking length increases from redshift 0 to 0.2 by a factor of about 2. Using this scaling we found the 2dF group catalogue (Tago et al., 2006).

We regard every galaxy as a visible member of a group or cluster within the visible range of absolute magnitudes, corresponding to the observational window of apparent magnitudes at the distance of the galaxy. To calculate total luminosities of groups we have to find the estimated total luminosity per one visible galaxy, taking into account galaxies outside of the visibility window. For this purpose the observed luminosity of a galaxy is to be multiplied by the distance dependent weight:

$$W_L = \frac{\int_0^\infty L\,F(L)\,\mathrm{d}L}{\int_{L_1}^{L_2} L\,F(L)\,\mathrm{d}L}, \qquad (7.2)$$

the ratio of the expected total luminosity to the expected luminosity in the visibility window. Here $F(L)$ is the luminosity function, and L_1 and L_2 are the lower and upper limit of the luminosity window, respectively. In our first group catalogues we used the Schechter luminosity function (Schechter, 1976); in our last catalogues we used the double power-law function with smooth transition:

$$F(L)\mathrm{d}L \propto (L/L^*)^\alpha (1 + (L/L^*)^\gamma)^{(\delta-\alpha)/\gamma} \mathrm{d}(L/L^*), \qquad (7.3)$$

where α is the exponent at low luminosities $(L/L^*) \ll 1$, δ is the exponent at high luminosities $(L/L^*) \gg 1$, γ is a parameter that determines the speed of transition between the two power laws, and L^* is the characteristic luminosity of the transition, similar to the characteristic luminosity of the Schechter function. The double power-law luminosity function was in general use in the 1970's (Abell, 1977), however with a sharp transition from one to the other power law.

To check the correctness of our weighting procedure in the estimation of total luminosities we again shifted nearby groups to larger distances. Now we selected two subsamples of clusters at different true distances from the observer. The first subsample was chosen in the nearby region with distances $100 \leq d < 200\,h^{-1}$ Mpc, and the other sample in the distance interval $200 \leq d < 300\,h^{-1}$ Mpc. In both cases the number of visible galaxies in groups was chosen, $N_{\mathrm{gal}} \geq 10$. Next the clusters were shifted to progressively larger distances, galaxy apparent magnitudes were calculated, and galaxies inside the visibility window selected. The number of galaxies inside the visibility window for shifted clusters decreases; the mean number of galaxies in shifted clusters is shown in the upper panel of Fig. 7.12. We see that the mean number decreases almost linearly in the $\log N - d$ diagram. At the far side of our survey the mean number of remaining galaxies in clusters is between 1 and 2; some groups disappear, only the main galaxy remains.

The expected total luminosity of clusters, calculated on the basis of galaxies inside the visibility window, and using the procedure outlined above, is shown on the lower panel of Fig. 7.12. We see that the mean values of restored total luminosity of clusters are almost identical with the true luminosity at the initial distance. The restored luminosities of individual clusters have a scatter that increases with the distance.

A similar approach was used in the compilation of the group catalogues for the Sloan Survey galaxies. For the Data Release 5 of the Sloan Survey the group catalogue was published by Tago et al. (2008), and for the Data Release 8 by Tempel et al. (2012c). Analysis of the dependence of group properties on the mean distance from us shows that main properties (the mean size and expected total luminosity) do not depend on the distance, i.e. our group catalogues are homogeneous. Of course, the mean number of galaxies in groups decreases with distance, since only brighter group members remain in the observational window of apparent magnitudes.

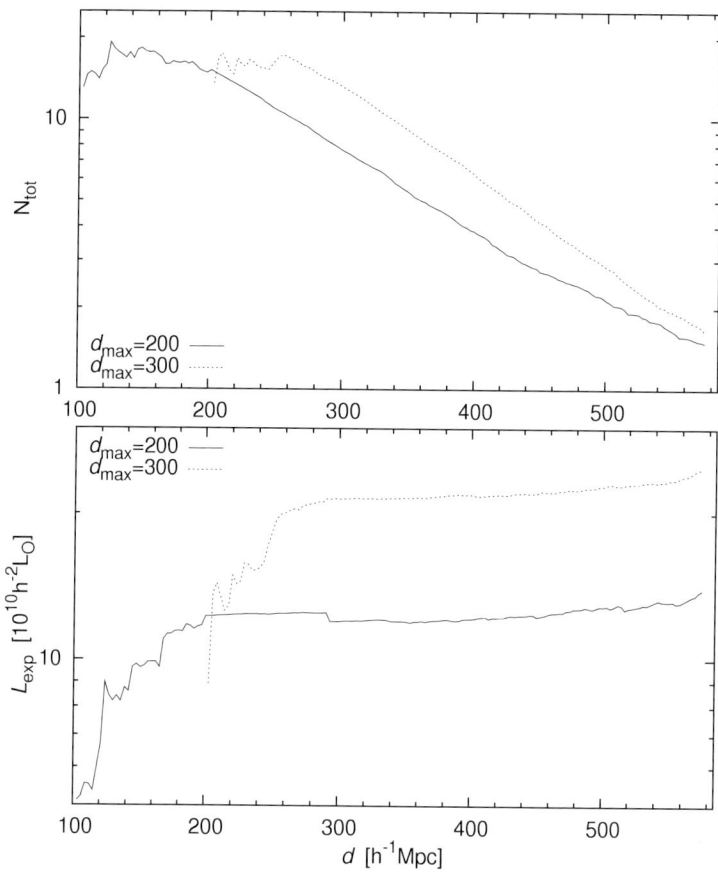

Fig. 7.12 Upper panel: the mean number of galaxies in shifted clusters as a function of distance for the 2dF. Lower panel: restored mean total luminosities of shifted clusters. Solid line shows results for clusters located initially at distances $100 \leq d < 200\,h^{-1}$ Mpc, dashed line is for clusters of initial distance $200 \leq d < 300\,h^{-1}$ Mpc (Tempel et al., 2009).

7.2.3 *Catalogues of superclusters*

To investigate large-scale properties of the cosmic web more distant markers of the web are needed. The best markers of the web are rich clusters of galaxies. In earlier studies superclusters were defined using clusters of galaxies. George Abell (1958) in his catalogue found evidence for second-order clustering, and called these systems of clusters as clusters of clusters of galaxies. Soon the term "superclusters" was established, but as before, only systems consisting of several clusters were considered as superclusters. This definition does not take into account the property of the relatively rich nearby Virgo supercluster, where there is only one rich cluster surrounded by galaxy sheets and filaments.

In our first study of superclusters we compiled a list of ten probable superclusters (Jõeveer et al., 1978), using mostly Abell clusters as supercluster tracers. The Abell (1958) catalogue of rich clusters covers the Northern sky up to declination -27 degrees, and contains 2,712 clusters. When I visited George Abell in 1982, we discussed with him and his collaborator Harold Corwin the extension of the catalogue to the Southern sky. The main purpose of the catalogue is to find high-density knots in the cosmic web. Here the richness limit 30 galaxies in the cluster area is not so important, because less rich clusters also mark knots of the web. For this reason I suggested including all clusters found during the search into the catalogue, the poorer clusters being added as a Supplement to the main catalogue. So the Southern Survey adds 1,361 rich clusters to the main catalogue, bringing the total number of rich clusters to 4,076, and the Supplement has 1,174 Southern clusters which were not rich enough to be included into the main catalogue (Abell et al., 1989).

To use the Abell catalogue for the study of the cosmic web, distances to clusters are needed. For most nearby rich clusters redshifts were found by special observing programs. However, the number of clusters with known distances was still too low. Thus Erik Tago in collaboration with Heinz Andernach started to collect redshifts of galaxies belonging to Abell clusters, using all available galaxy redshift sources.

In our early studies of the distribution of galaxies and clusters (Jõeveer & Einasto, 1978; Jõeveer et al., 1978; Einasto et al., 1980a) we noticed another problem in the definition of superclusters — superclusters are connected by filaments of galaxies, groups and clusters, forming a continuous cosmic web. Thus it was not evident where a particular supercluster ends and the another one begins.

Following Zeldovich's ideas we expected that the evolution of density perturbations starts from pancaking to form flat sheets (Zeldovich, 1970, 1978). According to this scenario filaments form by the flow of pre-galactic matter towards the crossing of sheets, and clusters by the flow of matter towards the corners of the cellular network. Thus we considered the possibility of identifying one cell wall with surrounding cluster chains as a supercluster (Einasto et al., 1980a). When more redshift data arrived we noticed that cell walls do not form continuous surfaces (Zeldovich et al., 1982; Einasto et al., 1983b; Tago et al., 1984), thus cell walls cannot be used to define superclusters. We concluded that it is reasonable to consider superclusters as high-density regions of the cosmic web, using in their identification both clusters and galaxies.

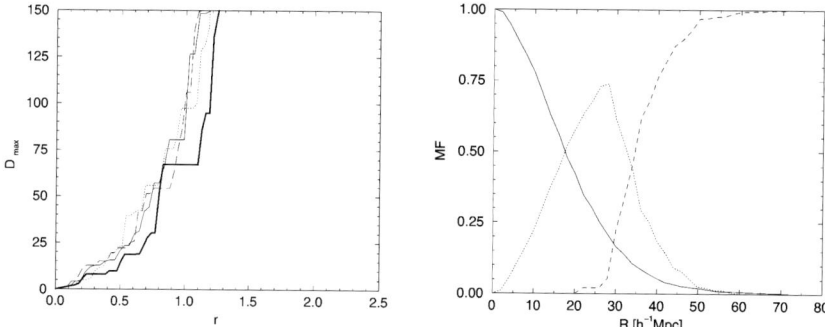

Fig. 7.13 Left: the length of the largest supercluster, $D_{\max}$ in h^{-1} Mpc, as a function of the neighborhood radius r in dimensionless Poisson units. Solid line shows all clusters, dotted line shows Abell clusters of richness $R = 0$, and dashed line shows clusters of richness $R \geq 1$. Heavy solid line shows the Rosat Bright X-ray Survey sample. Right: multiplicity functions of systems of Abell clusters as a function of the neighbourhood radius R. The solid line shows the fraction of isolated clusters, the short-dashed line shows the fraction of clusters in superclusters of multiplicity 2 to 31, and the dashed line shows the fraction of clusters in merged superclusters with at least 32 member clusters (Einasto et al., 2001).

In the early 1990's the Tago and Andernach collection of Abell cluster redshifts was complete enough to use it to compile new catalogues of superclusters. Our first supercluster catalogue based on all near Abell clusters was published by Maret Einasto et al. (1994b). Here again the formal definition of a supercluster was essential. Maret constructed a number of supercluster catalogues for varying neighbourhood radii, calculated sizes of all superclusters, and found the length of the largest supercluster, see the left panel of Fig. 7.13. Further she calculated for various neighbourhood radii multiplicity functions of systems of various richness, separately for isolated clusters, for superclusters of multiplicity $2\ldots 32$, and for merged superclusters with multiplicity at least 32 clusters, see the right panel of Fig. 7.13. If we understand superclusters as *the largest non-percolating high-density regions in the cosmic web*, then it is reasonable to use for their selection the neighbourhood radius just below the radius where superclusters start to merge. Using data shown in Fig. 7.13 Maret chose $R = 24\, h^{-1}$ Mpc to select superclusters of Abell clusters.

The redshift collection by Tago and Andernach was improving, so it was possible to improve also the supercluster catalogues (Einasto et al., 1997b, 2001). The last catalogue contains superclusters up to redshift $z = 0.13$ and contains in addition to Abell clusters also X-ray clusters, using the Rosat Bright X-ray

Survey sample. In this redshift interval the mean spatial density of Abell clusters is practically distance independent, thus the supercluster sample is volume limited.

In the early 2000's data on deep redshift surveys became available, which covered large areas of the sky. First of such surveys was the Las Campanas Redshift Survey; the Two-degree Field (2dF) Survey and the Sloan Digital Sky Survey (SDSS) followed. We used all these surveys to compile catalogues of groups of galaxies and superclusters.

In earlier supercluster catalogues we used the FoF method, applied to clusters. Now data on individual galaxies were known, and a more effective method was needed to make use of all the data. We found that for supercluster searches the luminosity density field of galaxies can be used. But there is a problem here: galaxies have random velocities in groups and clusters, thus, in order to use redshifts as distance indicators, peculiar velocities must be eliminated. To do this, catalogues of groups of galaxies are needed, which contain information on peculiar velocities.

The key element in our scheme is the restoration of the expected total luminosity of superclusters as accurately as possible. This goal can be achieved using the weights for galaxies and groups in the calculation of the density field; for details see the previous section.

The final step in the selection of superclusters is the proper choice of the threshold density to separate high and low-density galaxy systems. We compiled supercluster catalogues in a wide range of threshold densities. For each threshold density we found a list of superclusters, calculated the number of superclusters, and the maximal diameter of the largest system. At a low threshold density the largest systems span the whole observational volume. At a very high threshold density only the densest parts of the density field are considered as superclusters. The optimal choice is in the medium range of threshold density, which yields about 100–150 h^{-1} Mpc for the size of largest superclusters, as known from the study of nearby rich superclusters.

The first supercluster catalogue based on the luminosity density field was compiled from the Las Campanas Redshift Survey. This survey covered six 1.5×80 degree slices: three slices located in the Northern Galactic cap, and three slices in the Southern Galactic cap. The slices are so thin that only 2-dimensional luminosity density fields can be found. We calculated high-resolution and low-resolution density fields using Gaussian filters with smoothing scale 0.8 h^{-1} Mpc and 10 h^{-1} Mpc, respectively (Einasto et al., 2003a). A high-resolution field was used to find density field clusters, which are some equivalent of Abell clusters. A low-resolution field was used to find superclusters. The density threshold to define superclusters was chosen so that the largest superclusters do not form percolating systems.

Next we used the 2dF Galaxy Redshift Survey data to compile catalogues of superclusters for the Northern and Southern regions of the 2dFGRS. Altogether 543 superclusters at redshifts $0.009 \leq z \leq 0.2$ were found (Einasto et al., 2007b). We analysed methods of compiling supercluster catalogues using complete flux-limited galaxy catalogues. Results of the Millennium Simulation (Springel et al., 2005) were used to investigate possible selection effects and errors. We found that the most effective method is the density field method using smoothing with an Epanechnikov kernel of radius $8\,h^{-1}$ Mpc. Superclusters were defined as galaxy systems larger than groups and clusters which have a certain minimal overdensity of the smoothed luminosity density field, but are still non-percolating. Superclusters form intermediate-scale galaxy systems between groups and filaments, and the whole cosmic web. Similar procedures were applied in the preparation of the supercluster catalogue for the SDSS Data Release 4 (Einasto et al., 2006), and a preliminary catalogue for SDSS Data Release 7.

The group catalogue for SDSS DR8 was compiled by Tempel et al. (2012c), and the supercluster catalogue for SDSS DR7 by Juhan Liivamägi et al. (2012). Here a number of improvements in the calculation of the density field and the supercluster search were made. In the calculation of the density field instead of the Epanechnikov kernel the B_3 kernel was used (see Martínez & Saar (2002)). The SDSS density field of the main galaxy sample was calculated with the kernel size $8\,h^{-1}$ Mpc, the field for the Luminous Red Giant (LRG) sample with the kernel of size $16\,h^{-1}$ Mpc. Superclusters were found for a series of density thresholds to find the best way to define superclusters. It was found that it is not possible to use a fixed threshold for the whole sample, since superclusters in different regions of space have different mean densities.

To see the problem we plot in Fig. 7.14 the high-resolution luminosity density field for a spherical layer at a distance $240\,h^{-1}$ Mpc from us. The thickness of the shell is $10\,h^{-1}$ Mpc. In the plotting we used SDSS coordinates η, λ. Because we use a spherical shell at a fixed distance, distance dependent selection effects are excluded. We see that in the lower part of the Figure a huge complex of several superclusters is located — this complex is called the Sloan Great Wall; actually it consists of several very rich superclusters. Here the overall luminosity density is higher, and galaxy and cluster chains joining superclusters to the cosmic web have much higher luminosities than in other regions.

As a solution of the problem Juhan prepared two sets of supercluster catalogues, one set with a number of fixed threshold densities, and the second set with adaptive local threshold densities chosen individually for each supercluster. The idea is to follow the growth of individual superclusters from a compact volume around its centre, by lowering the density level and observing the supercluster

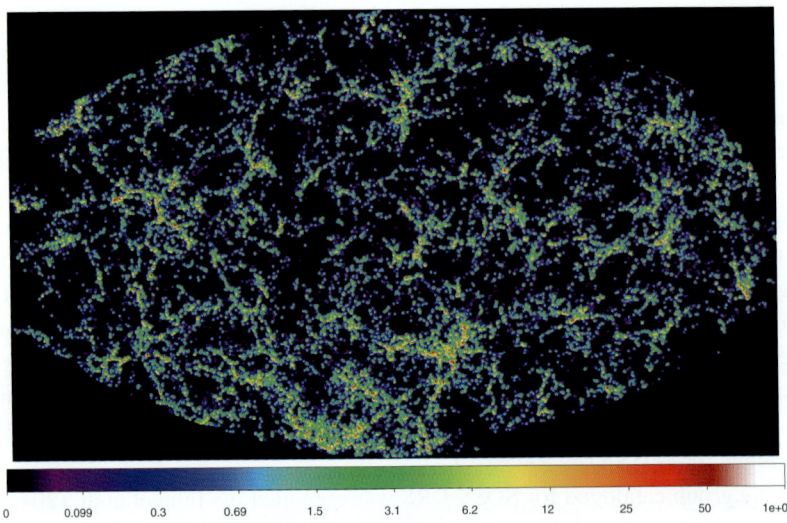

Fig. 7.14 The luminosity density field of the SDSS in a spherical shell of $10\,h^{-1}$ Mpc thickness at a distance of $240\,h^{-1}$ Mpc. To enhance the faint filaments in voids between the superclusters, the density scale is logarithmic, in units of the mean luminosity density for the whole DR7. The rich complex in the lower area of the picture is part of the Sloan Great Wall; it consists of two very rich superclusters, SCL 111 and SCL 126 in the list by Maret Einasto et al. (2001) (Suhhonenko et al., 2011).

mergers. By defining a supercluster as the volume within the density contour until the first merger, we can break the large-scale structure into a collection of compact components. Every component (supercluster) then has its own limiting density level D_a.

Figure 7.15 gives an example of how supercluster diameters and luminosities change during mergers when lowering the density level and, while still growing, remain relatively stable in between. Supercluster SCl 24 in this Figure is a part of the Sloan Great Wall. At densities $D < 4.7$ it actually includes all of the SGW superclusters (Einasto et al., 2010).

The Liivamägi et al. (2012) supercluster catalogues contain 982 and 1313 superclusters in the fixed and adaptive threshold catalogues of the main galaxy sample, respectively. The LRG supercluster catalogues contain 3761 and 2701 superclusters in the fixed and adaptive catalogues. Analysis of supercluster catalogues shows that the main supercluster properties do not depend on the distance from the observer, i.e. selection effects have been corrected properly. The authors find that superclusters are well-defined systems, and the properties of the superclusters of the main and LRG samples are similar. The Millennium simulation galaxy catalogue provides similar superclusters to those observed.

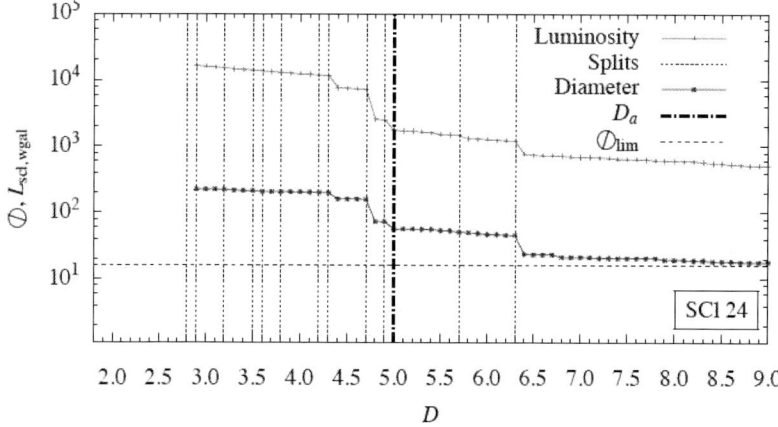

Fig. 7.15 An example of the dependence of the supercluster diameter and the total luminosity on the density level D. Vertical thin dashed lines show splitting/merger events, and the thick dashed line shows the adaptive density threshold. The minimum diameter limit $16\,h^{-1}$ Mpc is also shown. The lines begin at the level where the object separates from the larger structure (Liivamägi et al., 2012).

7.3 Elements of the cosmic web

7.3.1 *Galaxies in different environments*

Already in the early stages of the study of galactic populations I found that there exists a close correlation between physical and dynamical properties of populations: colour and mass–luminosity relation (Einasto & Kaasik, 1973), and a relation between dynamic properties and ages of galaxies (Einasto, 1973, 1974b). These relations suggested the presence of fundamental relations between various properties of galaxies. About ten years earlier we derived the fundamental plane of stars (Tiit & Einasto, 1964), and made preparations to apply this method to galaxies by compiling a bibliography of galaxies (Brosche et al., 1974). However, in the early 1970's we had so many programs running simultaneously that this study was postponed. A few years later the problem was investigated in much more detail by Faber & Jackson (1976). Sandra used her own determinations of velocity dispersions in elliptical galaxies, and found the Faber–Jackson fundamental plane for elliptical galaxies.

In the further study of galaxies and their environment we found that giant galaxies are surrounded by dark matter halos and dwarf satellite galaxies, and that satellite galaxies are morphologically segregated (Einasto et al., 1974c, 1975b). Dark halos form the largest and the most massive populations of the central giant galaxies. Satellite galaxies move inside the halo of the maternal galaxy. As noted earlier, to avoid confusion (a dwarf galaxy inside a giant galaxy), we called such systems hypergalaxies (Einasto et al., 1974a). Hypergalaxies are actually small groups of galaxies with one concentration center. In terms of DM simulations hypergalaxies are equivalent to halos, and dwarf satellite galaxies are equivalent to subhalos. Einasto et al. (1974a) assumed that hypergalaxies are the primary sites of galaxy formation, similar to stars which form in stellar associations and star-forming regions (Ambartsumian, 1958). Further Einasto et al. (1974a) assumed that large multicomponent groups and clusters are formed by merging of hypergalaxies.

The morphological segregation of satellite galaxies is actually an early hint to the presence of the density–morphology relation, discussed in detail by Dressler (1980). Einasto & Einasto (1987) showed that the density–morphology relation is valid not only in clusters, but also in the environment of clusters up to distance ~ 10–$15\, h^{-1}$ Mpc.

Einasto et al. (1976d) investigated in detail the dynamics of aggregates of galaxies and their relation to the dynamics of member galaxies, in particular to the dynamics of main galaxies. We calculated the mean velocity dispersion of members of aggregates for various distances from the primary (main) galaxy. Results of our calculations are shown in Fig. 7.16. We see that in the inner regions of aggregates the velocity dispersion of galaxies is higher than the mean dispersion. This suggests

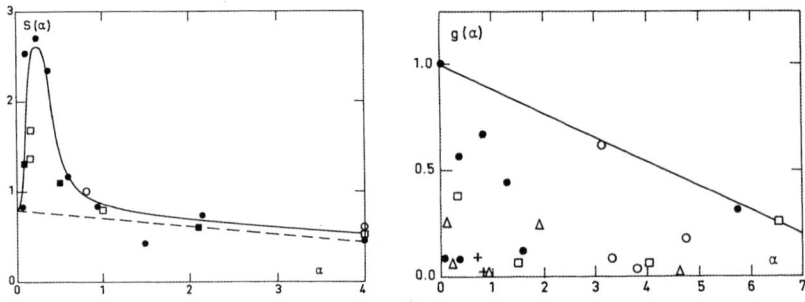

Fig. 7.16 Left: the velocity dispersions of companion galaxies, $s(\alpha) = [\sigma_r(\alpha)/\sigma_{\mathrm{comp}}]^2$, as a function of the distance from the main galaxy, $\alpha = R/a_0$, where a_0 is the harmonic mean radius of the group or cluster. The full line represents the adopted mean $s(\alpha)$ while the broken line is the upper envelope of the internal velocity dispersion of companion galaxies $g(\alpha)$, shown in the right panel. Filled circles show data on companion galaxies in the Coma cluster, filled squares are mean data on groups with a giant E galaxy as the main galaxy, open spheres show mean data on groups with a giant S galaxy, and open squares with a medium bright S galaxy as the main galaxy. Right: internal velocity dispersion of companion galaxies, $g(\alpha) = (\sigma/\sigma_{\mathrm{main}})^2$ (in units of the dispersion of the main galaxy), at various distance from the main galaxy. The straight line is the upper envelope. Symbols are: filled circles — M87 and other members of the Virgo cluster; open circles — members of the M94 group; squares — members of the M81 group; triangles — members of the M31 subgroup (hypergalaxy); crosses — members of the Galaxy subgroup (hypergalaxy) (Einasto et al., 1976d).

that galaxies in this region are in their peri-cluster parts of orbits around the center of the system. With increasing distance the velocity dispersion decreases slowly. Internal velocity dispersions of satellite galaxies also decrease with increasing distance from the system center, see the right panel of Fig. 7.16.

Data available in 1975 suggested that velocity dispersions of main galaxies are approximately equal to the mean velocity dispersions of galaxies in the system. More accurate velocity dispersion measurements by Faber & Jackson (1976) showed that velocity dispersions of main galaxies are actually lower. This result is expected since otherwise stars of the main galaxy would have orbits covering the whole cluster. More accurate data suggest that the broken line in the left panel of Fig. 7.16 should lie lower than shown in the Figure.

The close link between dynamical properties of galaxies and systems they belong to suggests a common origin of galaxies in groups/clusters.

One of the principal functions which characterises physical properties and the evolutionary stage of galaxies is the luminosity function. Einasto et al. (1974a) studied the luminosity function of galaxies in systems of various richness from hypergalaxies (groups with one bright galaxy and dwarf companion galaxies) to rich clusters of galaxies. We found that main galaxies of hypergalaxies have a definite *lower* limit of the luminosity. Member galaxies can be divided into dwarf and giant galaxies, and there is a gap in the luminosity between giant and dwarf members. These data suggest that the formation mechanism of giant and dwarf galaxies can be different. The reason for such a difference was investigated in detail by Dekel & Silk (1986).

In the mid 1970's data were available only for nearby groups and clusters, thus the above results can be considered as preliminary. Recently large and deep redshift data became available, and we resumed the study of the luminosity function of galaxies. First we investigated the luminosity function of the 2dF Galaxy Redshift Survey. I made the preliminary analysis myself, while the principal author of the final version is our young collaborator Elmo Tempel (Tempel et al., 2009).

Here we had several problems to elaborate, both technical and principal. The technical problem was to calculate total luminosities of groups of galaxies on the basis of data available in the visibility window of the survey. As discussed above, we corrected the observed luminosities to take into account galaxies which are too faint to be visible in the observational window of apparent magnitudes. Our check indicates that mean corrected luminosities of groups are independent of the distance from the observer. This shows that statistically our method to find expected total luminosities of groups is correct.

One of the main goals of the study of the luminosity function was to find its dependence on galaxy properties and on the large-scale environment of galaxies.

Earlier studies of many authors, including our own earlier work (Einasto & Einasto, 1987; Einasto, 1991), have shown that luminosities and morphological properties of galaxies depend on the environment. Using new and more complete data Tempel et al. (2009) calculated luminosity functions (LF) of 2dF galaxies for four different environment, defined by the global luminosity density found by smoothing with a kernel of radius $8\,h^{-1}$ Mpc. Void, filament, supercluster, and supercluster core environments were defined by threshold densities 1.5, 4.6, and 7, and were designated as D1, D2, D3, and D4.

The second main goal of our study was to understand the nature of galaxies in various local (group/cluster) environments. For this purpose luminosity functions were found separately for first-ranked galaxies, second-ranked galaxies, all satellite galaxies, and isolated galaxies.

Our main results are shown in Fig. 7.17. The Figure shows that in voids, the bright end of LFs of all galaxy populations is shifted toward lower luminosities. LFs in filament and supercluster environments are rather similar; actually our supercluster environment corresponds to poor superclusters. The luminosity functions

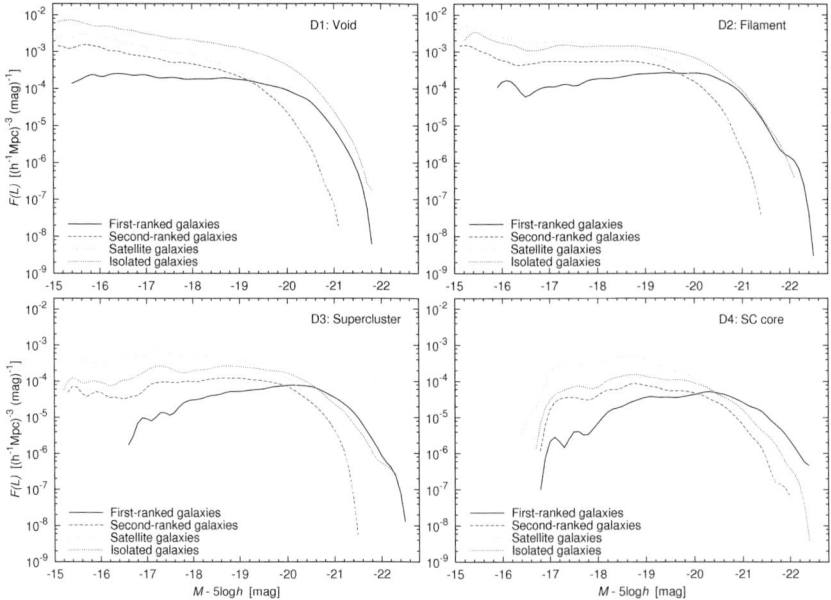

Fig. 7.17 Differential luminosity functions of the 2dF galaxy samples in different environments and for different galaxy populations. Top-left panel — void environment D1; top-right panel — filament environment D2; bottom-left panel — supercluster environment D3; bottom-right panel — supercluster core environment D4. Solid line shows first-ranked galaxies; dashed line — second-ranked galaxies; short-dashed line — satellite galaxies; dotted line — isolated galaxies (Tempel et al., 2009).

for supercluster cores are different from functions in all other environments: here all LFs have a well-defined lower luminosity limit, about 17 mag, which for first-ranked galaxies had been seen already in supercluster environment. Also, in supercluster cores the brightest first ranked galaxies are more luminous than the brightest first-ranked galaxies in other environments. This shows that the densest environment (supercluster cores) is different from other environments. Differences in less dense environments for galaxies of various types (local environment) are much smaller.

Of special interest is the LF of isolated galaxies. Figure 7.17 suggests that isolated galaxies may be a superposition of two populations: the bright end of their LF is close to that of the first-ranked galaxy LF, and the faint end of the LF is similar to the LF of satellite galaxies. This is compatible with the assumption that the brightest isolated galaxies in the sample are actually the brightest galaxies of invisible groups. This assumption is supported by results of our group shifting procedure: at large distance only the brightest member of the group falls into the visibility window of the survey.

To examine this possibility, Tempel et al. (2009) made the following test. A group has only one galaxy in the visibility window if its second-ranked galaxy (and all fainter group galaxies) are fainter than the faint limit of the luminosity window at the distance of the galaxy. Thus we calculated for each isolated galaxy the magnitude difference between the galaxy and the faint limit of the absolute magnitude of the sample at the distance of the galaxy. The distribution of magnitude differences was compared with the distribution of the actual magnitude differences between the first-ranked and second-ranked group galaxies. The distributions look rather similar.

Tempel et al. (2011) studied the morphology, luminosity, and environment in the SDSS Data Release 7 galaxy sample. The authors constructed the LFs separately for galaxies of different morphology (spiral and elliptical) and of different colours (red and blue) using data from the Sloan Digital Sky Survey (SDSS), correcting the luminosities for intrinsic absorption.

The results suggest that the evolution of spiral galaxies is slightly different for different types (colours) of spirals. A possible interpretation of our results may lie in the fragility of spiral galaxies: they form and survive only in specific conditions (e.g. the preservation of the gas, the absence of major mergers) which are typical in low density regions, but to some extent can be present also in high density regions.

The derived LF of elliptical galaxies can be reconciled with hierarchical galaxy formation through mergers. The denser the environment, the brighter the galaxies that should reside there because of the increased merger rate. The difference between the LFs of elliptical galaxies in different environments is more notable

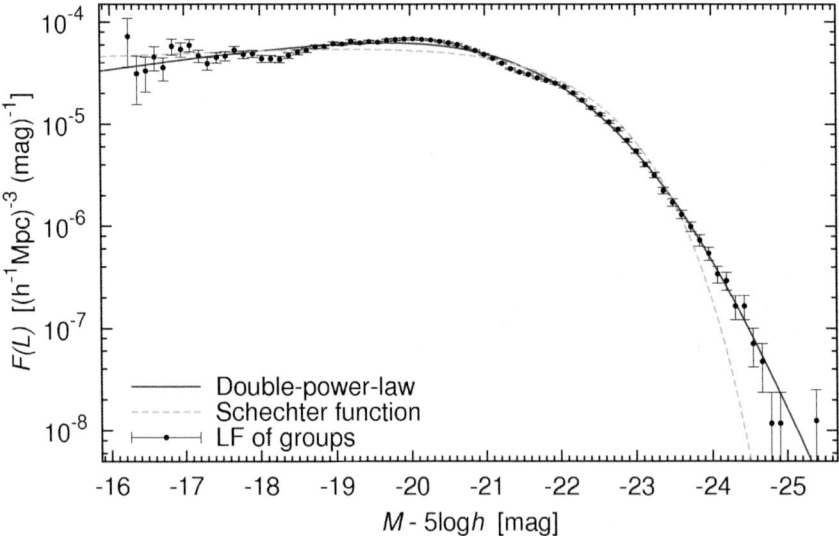

Fig. 7.18 The differential luminosity function of 2dF groups (shown by points). The solid line is the double-power-law fit and the dashed line is the Schechter function (Tempel et al., 2009).

for red galaxies, in accordance with their supposed merger origin. This interpretation agrees well also with the picture of hierarchical formation of galaxies: for blue galaxies, the evolution is more quiescent and major mergers are not so important; for red ellipticals, merging is the dominant factor of galaxy evolution. Since blue ellipticals are most likely S0-s or late-type ellipticals, they still have some gas available for star formation and therefore the evolution of blue ellipticals is closer to the evolution of spiral galaxies — the global environment is less important.

Tempel et al. (2009) calculated also the LF for groups of galaxies, using the Tago et al. (2006) group catalogue. The differential LF of 2dF groups is shown in Fig. 7.18. For comparison Tempel et al. (2009) found also the best-fit model luminosity functions, applying the conventional Schechter (1976) function and the double-power-law function. A similar comparison was made for 2dF and SDSS galaxy luminosity functions. We note that the double-power-law function was used in the early 1970's by many authors. In earlier applications a sharp transition between two power laws was made; in our case we made the transition smooth. Our analysis suggests that the double-power-law function represents galaxy and group luminosity functions better than the Schechter function, especially at the high-luminosity end.

7.3.2 Groups and clusters of galaxies

There exists a large number of studies of the structure and properties of groups and clusters of galaxies. We studied group properties in the 1970's in relation to the dark matter problem, as discussed above. In recent years we resumed the study of groups of galaxies. We needed group catalogues to eliminate the Finger of God effect in the cosmic web, and to calculate the luminosity density field. As a byproduct we had the possibility of studying group and cluster properties, since they are principal building blocks of the web.

Here I shall discuss briefly only our recent study of morphological properties of clusters in our list of SDSS DR8 groups of galaxies by Tempel et al. (2012c). This list includes also rich clusters if found by our group finding algorithm. Einasto et al. (2012) searched for the presence of substructure, the non-Gaussian, asymmetrical velocity distribution of galaxies, and large peculiar velocities of the main galaxies in clusters with at least 50 member galaxies, drawn from this SDSS DR8 group catalogue; in total we found 109 clusters. Using various tests Einasto et al. (2012) found that 70–80% of the clusters in the sample have multiple components. The authors find that in about half of the multicomponent clusters the peculiar velocity of the main galaxy is larger than 250 km/s (0.5 of the normalised peculiar velocities). There is a clear difference between the median values of the peculiar velocities and the normalised peculiar velocities of the one-component and multicomponent clusters: these velocities are much larger in the multicomponent clusters.

Figure 7.19 shows the distance of the main galaxy from the cluster centre, both for the multicomponent and one-component clusters (in the plane of the sky). It is easily seen that in multicomponent clusters a large fraction of main galaxies are located far away from the cluster centre (grey dotted line). However, when looking at the components found by the 3D normal mixture modelling, we see that the main galaxies of clusters are preferentially located close to the centre of one of the components (solid line). The distribution of distances from the component centre for the brightest galaxies in the components shows that these galaxies are also located preferentially close to the component centre (dashed line).

Figure 7.19 (dark long-dashed line) shows that in one-component clusters in most cases the main galaxy lies close to the cluster centre. But there is also a substantial number of main galaxies, which are further away from the cluster centre. We calculated the minimum distance from the cluster centre for the three brightest galaxies in the one component clusters. As seen from Fig. 7.19 (light dotted-dashed line), one of the three brightest galaxies in clusters is always located close to the cluster centre. This shows that the central galaxy of a cluster is typically one of the most luminous galaxies, but not always the most luminous one.

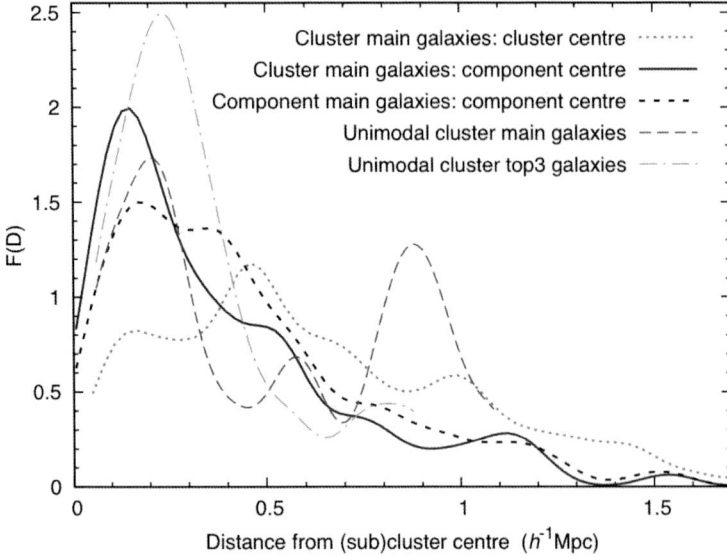

Fig. 7.19 Distribution of the distance of the main galaxy from the cluster (subcluster or component) centre for various subsamples of galaxies. Multicomponent clusters: grey dotted line — the distance of the main galaxy from the cluster centre; solid line — the distance of the main galaxy from the (nearest) component centre; dashed line — the distance of the brightest galaxy in a component from the component centre. One-component clusters: dark long-dashed line — the distance from the cluster centre; light dotted-dashed line — the minimum distance of one of the three brightest galaxies from the cluster centre (Einasto et al., 2012).

The basic conclusion from this study is that the presence of substructure, large distances of main galaxies from the cluster centre, and their large peculiar velocities are signs of mergers and/or infall. This suggests that most clusters in our sample are not yet in dynamical equilibrium. The high frequency of such clusters indicates that mergers between groups and clusters are common — galaxy groups continue to grow and are still assembling. Unimodal clusters are examples of clusters which are probably already in dynamical equilibrium.

7.3.3 Chains, strings and filaments

Jõeveer & Einasto (1978) noticed that the Perseus–Pisces supercluster contains rich Abell clusters, less rich clusters and groups, and galaxies. Clusters, groups and most galaxies form a network of chains. The main chain of the supercluster has 7 high-density knots. 3 of them are Abell clusters, and all are Zwicky near clusters. Knots are fairly regularly spaced as pearls in a chain. One knot contains only one very bright peculiar galaxy NGC 315; it lies exactly on the place where a cluster should be. We discussed with Mihkel if this object should be included as a member of the chain. Then we decided that this very bright peculiar galaxy has probably 'eaten' all other members of the group via galaxy mergers (Toomre & Toomre, 1972; Toomre, 1977). This is an example of a fossil group, as they were later called.

Also we noticed that chains can be very different in richness. The main chain of the Perseus–Pisces supercluster contains clusters and groups as members. Some less rich chains join various superclusters to a connected network; such chains were mentioned by Jõeveer & Einasto (1978); Jõeveer et al. (1978) and in our more detailed analysis by Einasto et al. (1980a, 1984) and Tago et al. (1984). Even fainter chains consisting of galaxies only cross large voids; they are seen in our wedge diagrams Fig. 5.5 and 5.4. A detailed analysis of the distribution of galaxies around the main chain of the Perseus–Pisces supercluster and in chains crossing the large void between the Local, the Coma and the Hercules supercluster suggested that galaxy and cluster chains are essentially one-dimensional structures, surrounded by empty regions devoid of galaxies.

Zeldovich et al. (1982), Einasto & Miller (1983), and Einasto et al. (1984) analysed in more detail the shape of various systems of galaxies in and around the Virgo and Coma superclusters, see Fig. 7.20. Thin slices in supergalactic coordinates

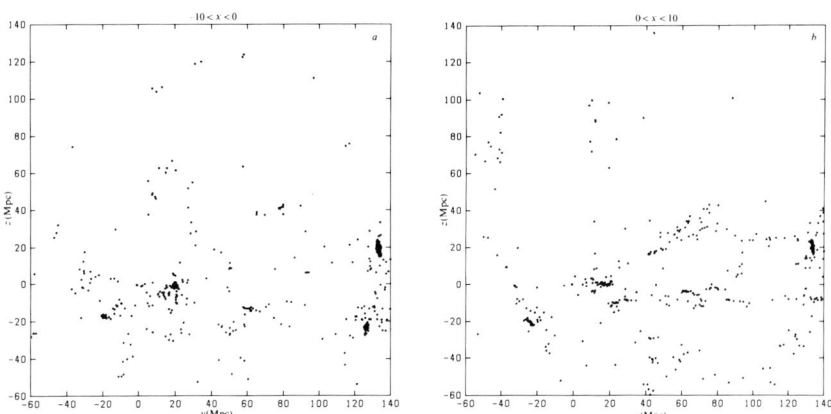

Fig. 7.20 Slices through the Virgo and Coma superclusters in supergalactic rectangular coordinates. The Galaxy is at the centre, $x = y = 0$, the Virgo cluster at $y = 20$ Mpc, the Coma cluster at $y = 120$ Mpc, and the other rich cluster in the Coma supercluster, A1367, at $y = 130$, $z = -10$ Mpc. Galaxies brighter than -19.5 absolute magnitude have been plotted. All coordinates correspond to the Hubble constant $h = 0.5$ (Zeldovich et al., 1982).

indicate that galaxy chains between the Local and the Coma superclusters are very narrow. For further analysis galaxies were collected into systems using the clustering Friends-of-Friends method. At small neighbourhood radius all systems are round; at this radius only high-density cores of clusters and groups are combined into systems. At medium neighbourhood radii most systems become elongated and are mostly tri-axial. When one axis exceeds considerably two other axes, these systems can be called strings or filaments. If two shorter axes are approximately equal, they can be called sheets of galaxies. One sheet forms the main body of the central part of the Virgo supercluster, see Fig. 7.1. At still larger neighbourhood radii galaxy strings merge into a connected network, as seen in Figs. 5.5, 5.4, and 7.20. The respective radius is called the percolation radius, as discussed above.

Our analysis was questioned by Geller (1987) in her talk at the Vatican Conference on Theory and Observational Limits in Cosmology, July 1985. She cites our studies on this problem (Jõeveer et al., 1978; Einasto et al., 1980a; Zeldovich et al., 1982), but disagrees with our conclusion on the filamentary character of the distribution. On the basis of the first results the Second CfA redshift survey by de Lapparent et al. (1986) she argues that galaxies are located in shells which surround voids, and that filaments are just slices through walls in relatively thin observational wedges. According to her interpretation observations strongly favour the Ostriker & Cowie (1981) explosive scenario of structure formation.

Reading this conference talk and the original paper by de Lapparent et al. (1986) I was rather surprised. John Huchra showed me his model during my first visit to Harvard, based on the First CfA Redshift Survey, where the filamentary distribution of galaxies was clearly seen. Our analysis, both qualitative and quantitative, indicated the filamentary character of the distribution. In Harvard I discussed the filaments using our latest data, as shown in Fig. 7.20, where it was clearly seen that filaments are very thin and are not cross-sections through larger shells.

The de Lapparent et al. (1986) paper and Geller (1987) talk had no quantitative tests. However, after the popular description of the structure by Geller & Huchra (1989) the term "walls" was widely used to describe the distribution of galaxies. When our "Nature" paper (Einasto et al., 1997a) was reported in "Physics Today", the editors also used the term "walls" instead of "superclusters". The editors sent the preliminary version of their report to me, and I explained that the term "superclusters" had already been in use for many years, starting from de Vaucouleurs (1953) and Abell (1958), and there is no need to replace it with "walls", taking into account the filamentary character of the cosmic web. Walls have a completely different meaning as we see below.

New high-resolution simulations indicate that tenuous sheets really surround voids, but the density of these sheets is too low for galaxy formation — they consist of pre-galactic matter, see Einasto et al. (1986a) and the review by van de Weygaert (2002).

Filaments differ not only by their richness, but also by their topological behaviour. Most rich filaments form sections of the continuous web network, i.e. they are connected at both ends with the network. Examples are filaments which connect neighbouring superclusters. The filament found by Tago et al. (1986) is different. One of its tips joins the filament with the Local supercluster, but the other end just fades inside the Local supervoid, and has no connection with the Hercules supercluster on the other side of the void.

Several teams performed in last few years a more detailed study of the structure of filaments, including their objective definition. Aragon-Calvo et al. (2006); Aragón-Calvo et al. (2007b,a) used the Multiscale Morphology Filter (MMF) to automatically segment cosmic structure into basic components: clusters, filaments, and walls. The density field is calculated using the Delaunay Tessellation Field Estimator. Thereafter a series of morphology filters are applied that identify particular kinds of structure in the data. The method is referred to as the Multiscale Morphology Filter (MMF). To check the method it has been applied to a simple Voronoi model. Structural elements are defined as *field* if particles are located in the interior of Voronoi cells (these particles can be called void particles), as *wall* if particles lie within and around the Voronoi walls, as *filament* if particles are within and around the Voronoi edges, and as *blobs* if particles are within and around the Voronoi vertices. Aragón-Calvo et al. (2007b) applied the MMF technique to find the spin alignment of dark matter halos in filaments and walls.

Stoica et al. (2010) suggested using a marked point process to find filaments. This method does not use the density estimation step. Filaments are defined as cylinders filled with particles, and located between two clusters. Tempel et al. (2013) applied this filament search method to study the alignment of spiral and elliptical galaxies in filaments.

7.3.4 *Walls*

The term *walls* has been used in three different meanings: as walls in the cellular network, as a complex of clusters and superclusters between supervoids, and as rich superclusters with their neighbouring filaments.

In N-body and Voronoi models walls had the meaning of surfaces between cell interiors surrounded by cell edges. As shown by Einasto et al. (1986a) on the basis of a ΛCDM model, such walls exist only in unbiased models where all dark matter particles are present. The density of these walls is so low that here no galaxy formation takes place, thus in real galaxy samples there are no walls in this meaning.

In the early 1980's we used the term 'wall' in another context, as a complex of rich clusters or superclusters which form a flattened disk between large low-density regions (supervoids). My own experience with the existence of such complexes goes back to the early 1980's. When working in ESO as a visiting scientist in 1981 I used the plotting facilities of ESO to make plots of galaxies and clusters in sequence at various coordinate systems to understand better the spatial distribution of these objects. One product of this work was the preparation of a movie together with Dick Miller, discussed elsewhere.

First I used equatorial coordinates to plot galaxies and clusters. But soon I noticed that for relatively near objects it is better to use supergalactic coordinates, introduced by Gerard de Vaucouleurs. I plotted near rich clusters in two distance intervals, $37.5 \leq r < 75\, h^{-1}$ Mpc, and $75 \leq r < 125\, h^{-1}$ Mpc. To my surprise I found that in the first plot all rich clusters form a relatively narrow belt close to the supergalactic plane. In the plot of more distant clusters the distribution is uniform.

Then I remembered our first results of the distribution of galaxies and clusters by Jõeveer et al. (1977); Jõeveer & Einasto (1978), which showed the presence of a large void between the Virgo, the Coma and the Hercules superclusters. Thus I assumed that the belt of near clusters is a complex of clusters and superclusters which form one sidewall of this large void. However, this conclusion was only tentative. Since for Southern clusters I did not have data, the picture was not full.

At this time George Abell with his collaborator Harold Corwin had started to inspect the Southern sky to extend the Abell cluster catalogue to the South. So I wrote a letter to Harold asking if he can find in the Southern sky all nearby rich clusters, up to approximately redshift 12,500 km s^{-1}. Soon I got the list of nearby rich clusters of galaxies. Thereafter I wrote letters to John Huchra and Massimo Tarenghi, who were engaged in redshift determinations of clusters of galaxies. John sent me redshifts from his redshift compilation, and Massimo was able to measure a number of new redshifts for these clusters.

To my joy the new Southern data confirmed my expectations — all close Southern clusters formed a continuation of the Northern part of the belt close to the supergalactic plane, and more distant clusters were evenly distributed. Time was pressing; it was spring 1982, and soon the next IAU General Assembly would took place in Athens, as well as a Symposium in Grete. So I prepared quickly short reports of our fresh results (Einasto & Miller, 1983; Einasto et al., 1983a). These papers contain Figures which show the location of clusters and superclusters, and show the presence of a disk, which forms a wall between two low-density regions, see Fig. 7.24. A figure with the distribution of galaxies and rich clusters in supergalactic z-coordinate was included into the review paper by Zeldovich et al. (1982)

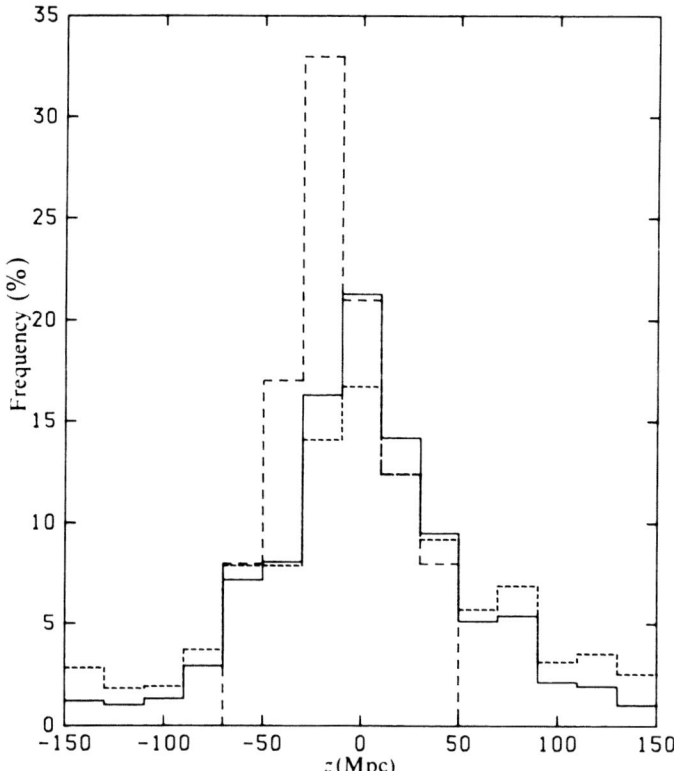

Fig. 7.21 Distribution of galaxies and rich clusters of galaxies in supergalactic z coordinate. Solid and dotted lines are for galaxy distributions uncorrected and corrected for incompleteness, respectively; the dashed line is for rich cluster distribution (Zeldovich et al., 1982).

(Fig. 7.21). The main conclusions of the study were formulated as follows (Einasto & Miller, 1983; Einasto et al., 1983a):

(1) The distribution of rich clusters suggests the presence of two local cells, the Northern Local Cell and the Southern Local Cell, each about $100\,h^{-1}$ Mpc in diameter;
(2) The Virgo, Coma, Perseus–Pisces, Lynx–Ursa Major, Hydra–Centaurus and Pavo–Corona Australes superclusters form a disk between the Northern and the Southern Local Cells; its diameter is about $125\,h^{-1}$ Mpc, and thickness about $25\,h^{-1}$ Mpc;
(3) Superclusters located at distances $75\ldots 125\,h^{-1}$ Mpc (Hercules, Ursa Majoris–Leo, and several Southern superclusters) form side-walls of Local

Cells. The Hercules and Ursa Majoris–Leo superclusters form a wall between the Northern Local Cell and the Bootes Cell; the Perseus–Pisces supercluster is located in a wall between the Northern Local Cell and a cell beyond that supercluster, seen in Figs. 5.7, 5.8 and 5.9.

The concentration of nearby Abell clusters close to the Supergalactic plane was confirmed by Brent Tully & Fisher (1987) and Peter Shaver (1991), see the discussion by Jim Peebles (2012). Brent also discovered independently local voids, the Northern Local Void is often called Tully's Void. Our earlier papers on local voids and the concentration of Abell clusters between these voids were published only in proceedings of IAU (Einasto & Miller, 1983; Einasto et al., 1983a). Evidently this was not enough to be noticed by the community.

Our new study confirmed our earlier results that the principal structural elements in cell walls are galaxy and cluster chains. Within superclusters chains are rich and consist mainly of clusters. In cell interiors there are several chains, consisting of galaxies and poor clusters. Galaxy chains connect neighbouring superclusters into a single lattice, and there are no large sheets of galaxies, uniformly filled with galaxies. Small sheets surround some clusters (Virgo).

In the 1990's more data on Abell clusters became available, and Einasto et al. (1994b, 1997b) compiled catalogues of superclusters. Maret found that very rich superclusters are concentrated to a Dominant Supercluster Plane which is situated at a right angle with respect to the plane of the Local Supercluster and adjacent nearby superclusters. Einasto et al. (1997a) found that rich clusters of galaxies, located in rich superclusters, form a quasi-regular lattice. Part of this lattice coincides with the Dominant Supercluster Plane, see Fig. 7.29 below.

de Lapparent et al. (1986) applied the term "Great Wall" to designate the Coma supercluster and its extension towards the Hercules supercluster, a rich filament of galaxies, as seen in our absolute magnitude limited distribution of galaxies across the Coma and Hercules superclusters, shown in Fig. 7.7 (Einasto et al., 1999a). Praton et al. (1997) investigated the "wall" phenomenon in detail using numerical simulations. The authors showed that structures perpendicular to the line of sight are enhanced in redshift space, due to the Finger-of-God effect and distortions caused by infall. The effect is enhanced in apparent-magnitude-limited samples of galaxies near the maximum of the sensitivity of the sample. The authors call this the bull's-eye effect.

The Sloan Great Wall (Vogeley et al., 2004) is a complex of several very rich superclusters. Its structure has been recently investigated by Einasto et al. (2010). In front and beyond the Sloan Great Wall there are large under-dense regions, as shown by Liivamägi et al. (2012). The richest supercluster in this complex is SCL126 from the list by Einasto et al. (2001). In redshift-slices of apparent-magnitude-limited samples the Sloan Great Wall is enhanced due to the bull's-eye effect. If corrected for incompleteness, the region remains a very rich one (see Fig. 7.14), but not so extreme as in apparent-magnitude-limited slices.

7.3.5 Superclusters

The largest elements of the cosmic web are superclusters of galaxies. Their properties can be divided into physical and general properties, similarly to properties of smaller elements of the web. Physical properties characterise the internal structure and physical nature of supercluster populations. The most important general properties of superclusters are their luminosity and richness.

Einasto et al. (2006) used two independent parameters to quantitatively characterise the richness of superclusters: the multiplicity and the total luminosity. These parameters were calculated for the 2dF and SDSS Data Release 4 superclusters, and for comparison superclusters found for the Millennium Simulation galaxy samples by Croton et al. (2006). The colour systems of our various samples are different: r in SDSS and Mill.A8, b_j in 2dFGRS and g in Mill.F8.

To characterise the multiplicity of superclusters we used density field (DF) clusters, defined as high-density peaks of the density field, smoothed on a scale of $8\,h^{-1}$ Mpc. We defined the multiplicity of a supercluster by the number of DF-clusters in it. The spatial density of DF-clusters in our observational samples is 62 per million cubic h^{-1} Mpc, about twice the spatial density of Abell clusters, 25 per million cubic h^{-1} Mpc (Einasto et al., 1997b). Thus the expected multiplicity of superclusters is about two times higher than the multiplicity of Abell superclusters of Einasto et al. (2001).

The other integral parameter of a supercluster is its total luminosity, determined by summing luminosities of all galaxies and groups of galaxies inside the threshold isodensity contour, which was used in the definition of superclusters. We are interested in the fraction of rich and very rich superclusters relative to poor superclusters. To avoid complications due to the use of different color systems and mean luminosities, we defined relative luminosities as the luminosity in terms of the mean luminosity of poor superclusters, i.e. superclusters that contain only one DF-cluster or Abell-cluster, and hence are classified as richness class 1.

Fig. 7.22 shows the relative luminosity functions and multiplicity functions for the observational and model samples. The most striking feature of the figure is the demonstration of the presence of numerous very luminous superclusters in observational samples, and the absence of such systems in simulated samples. This

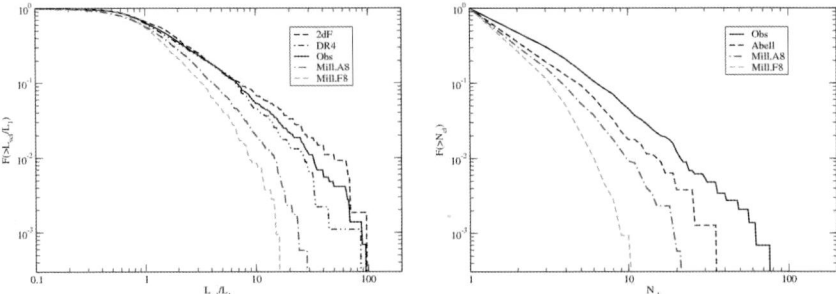

Fig. 7.22 A comparison of relative luminosity functions and multiplicity functions of observational and model supercluster samples in the left and right panels, respectively. In the left panel we show relative luminosity functions for observational samples SDSS DR4, 2dF, and the combined sample Obs; in the right panel we use the combined observational sample Obs and the Abell supercluster sample (here multiplicity is defined by the number of Abell clusters; isolated Abell clusters are considered as richness class 1 superclusters) (Einasto et al., 2006).

difference between real and simulated supercluster richness is well seen using both richness criteria, the multiplicity and luminosity functions.

When comparing models with observations we used the sample Mill.F8 by Einasto et al. (2007b), which is formed using similar selection criteria as observational samples. The most luminous simulated superclusters of the Mill.F8 sample have a relative luminosity of about 15 in terms of the mean luminosity of richness class 1 superclusters. The most luminous superclusters of real samples have a relative luminosity of about 100, i.e. they are about 6 times more luminous. The richest model superclusters of the sample Mill.F8 have a multiplicity of 10, whereas the richest real superclusters have DF cluster multiplicities over 70. The number of Abell clusters in the richest Abell supercluster is 34 (Einasto et al., 2001). The differences between real and simulated samples are observed not only in the region of most luminous superclusters: over the whole richness scale the number of DF-clusters in simulated samples is smaller than in samples of real superclusters, and the relative luminosity function lies lower.

The absence of very rich superclusters in simulated samples can be explained either as a real deviation of ΛCDM models from reality or by too small volume of the model. This problem has been studied by many astronomers; the most recent and perhaps most important contribution came from the analysis by Park et al. (2012).

Changbom Park with collaborators used the largest so far simulation of a ΛCDM universe, the so-called Horizon Run 2 (HR2) (Kim et al., 2011). This simulation was made for a box of side length 10 Gpc, it contains 6000^3 particles and allows to extract dark matter subhalos of the minimum mass 5.2×10^{12} $M_\odot$. The mean subhalo separation is 12.5 h^{-1} Mpc, equal to that of the SDSS volume limited galaxy sample used by the authors. Park et al. (2012) made 200 SDSS-like surveys of simulated galaxies for the present epoch on the basis of HR2, and analysed the properties of these mock survey samples in exactly the same way as the observational data. Out of the 200 mock samples, 137 samples contain rich supercluster complexes similar to or richer than the Sloan Great Wall (SGW). The authors conclude that the SGW type structures can be easily found in surveys like the SDSS in the ΛCDM universe. This analysis shows that the absence of very rich systems of galaxies in previous simulations was the result of a too small size of the simulation.

To study the internal structure of superclusters a number of methods have been used; an overview of such methods is given by Martínez & Saar (2002). In particular, Minkowski functionals can be used to describe overall structural properties of superclusters; they describe the volume, the surface, the mean curvature, and the integrated Gaussian curvature of the system. Einasto et al. (2007d) derived these parameters for the richest systems of the 2dF supercluster catalogue by Einasto et al. (2007a,b). Sahni et al. (1998) and Shandarin et al. (2004) introduced shapefinders, a set of combinations of Minkowski functionals: H_1 (thickness), H_2 (width), and H_3 (length). These quantities have dimensions of length, and are normalized to give $H_i = R$ for a sphere of radius R. Additionally Sahni et al. (1998) defined their combinations — shapefinders K_1 (planarity) and K_2 (filamentarity). Einasto et al. (2007d) found that the information about the shapes of superclusters can be best described by their morphological signature, i.e. the path in the shapefinder K_1–K_2 plane for varying mass fraction, calculated for various threshold densities to derive the supercluster.

The results of these calculations of Minkowski functionals and shapefinders are presented in Fig. 7.23. For morphological study we used volume-limited galaxy samples; this makes our results insensitive to selection corrections. As the argument labeling the isodensity surfaces, we chose the mass fraction m_f — the ratio of the mass in regions with density *lower* than the density at the surface to the total mass of the supercluster. When this ratio runs from 0 to 1, the isosurfaces move from the outer limiting boundary into the center of the supercluster, i.e. the fraction $m_f = 0$ corresponds to the whole supercluster, and $m_f = 1$ to its highest density peak. At small mass fractions the isodensity surface includes the whole supercluster. As we move to higher mass fractions, the iso-density surfaces include only higher density parts of superclusters, and their volumes and areas get smaller. At very high mass fractions only the highest density clumps in superclusters give their contribution to the supercluster. Individual high density regions in a supercluster, which at low

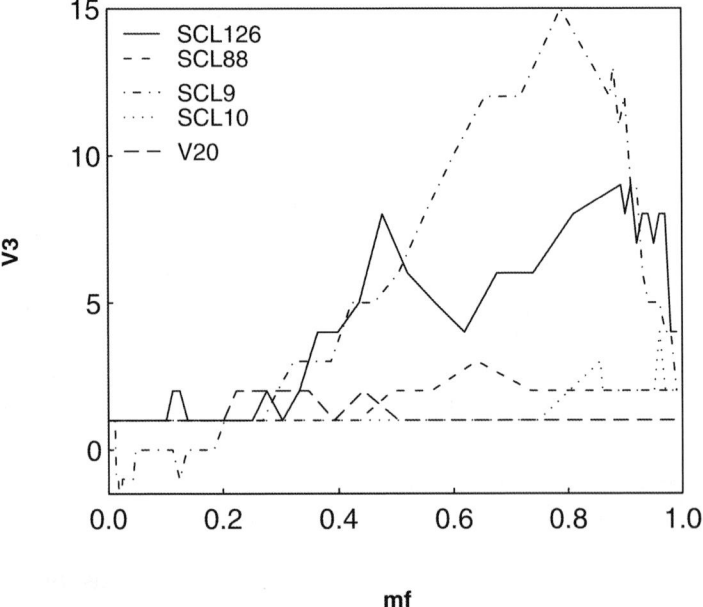

Fig. 7.23 The Minkowski functional V_3 (the Euler characteristic) for the observed superclusters according to list by Einasto et al. (2001) (Einasto et al., 2007d).

mass fraction are joined together into one system, begin to separate from each other, and the value of the fourth MF (V_3) increases. At a certain density contrast (mass fraction) V_3 has a maximum showing the largest number of isolated clumps in a given supercluster at the spatial resolution determined by the smoothing kernel. At still higher density contrasts only the highest density peaks contribute to the supercluster.

Figure 7.23 shows that at the mass fraction value of about 0.2, the value of V_3 for the supercluster SCL9 begins to increase and reaches a maximum value at the mass fraction $m_f \approx 0.7$. Then the value of V_3 begins to decrease. This indicates that the overall morphology of the supercluster SCL9 is clumpy; this supercluster consists of a large number of clumps or cores connected by relatively thin filaments, in which the density of galaxies is too low to contribute to the supercluster, starting at certain mass fraction values. The maximum value of the fourth Minkowski functional V_3 shows that the supercluster SCL9 has the largest number of isolated clumps in it.

The second richest and largest supercluster among the observed superclusters in the Sloan region of sky is the supercluster SCL126, a member of the Sloan

Great Wall. The V_3 curve for the supercluster SCL126 shows several peaks at a high mass fraction, $m_f > 0.95$. This indicates the presence of a very high density core region with several individual clumps in it — this is the main core region of the supercluster with several Abell clusters, which are also X-ray clusters.

Einasto et al. (2011c,d,e, 2012) investigated morphological properties of Sloan Digital Sky Survey rich superclusters, using the fourth Minkowski functional V_3, the morphological signature (the curve in the shapefinders $K_1 - K_2$ plane) and the shape parameter (the ratio of the shapefinders K_1/K_2). The superclusters in our samples form three chains of superclusters; one of them is the Sloan Great Wall. Most superclusters have filament-like overall shapes. Superclusters can be divided into two sets: more elongated superclusters are more luminous, richer, have larger diameters and a more complex fine structure than less elongated superclusters. The fine structure of superclusters can be divided into four main morphological types: spiders, multispiders, filaments, and multibranching filaments.

The analysis demonstrates that almost all superclusters in our sample of rich superclusters are elongated; they have larger filamentarities K_2 than planarities K_1. The two most elongated superclusters are the richest and the most luminous. Almost all superclusters studied are elongated and have filamentarities that are larger than their planarities. More elongated superclusters are also more luminous, have larger diameters and contain a larger number of rich clusters. The values of the fourth Minkowski functional V_3 show that they also have a more complicated inner morphology than less elongated superclusters.

7.3.6 *Voids and supervoids*

Already early studies of the distribution of galaxies and clusters showed that voids defined by objects of different type and luminosity have different sizes, see Figs. 5.5 and 5.4. The largest voids are determined by clusters and superclusters of galaxies, as seen from the distribution of near Abell clusters and superclusters in Fig. 7.24. Lindner et al. (1995) used the term "supervoids" for voids defined by superclusters of galaxies. As discussed above, supervoids are not empty; they are crossed by chains of galaxies and poor clusters or groups, as seen in Figs. 5.5 and 5.4.

Fig. 7.25 shows void diameters as a function of the size L of galaxy and Zwicky near cluster samples (Einasto et al., 1989). All galaxy samples are volume (absolute magnitude) limited; the limit corresponds to the apparent magnitude limit of the sample, $m = 14.5$, of the CfA first redshift survey, at the far side of the sample. The Figure shows an almost linear growth of void diameters with distance (limiting luminosity) in log–log scale.

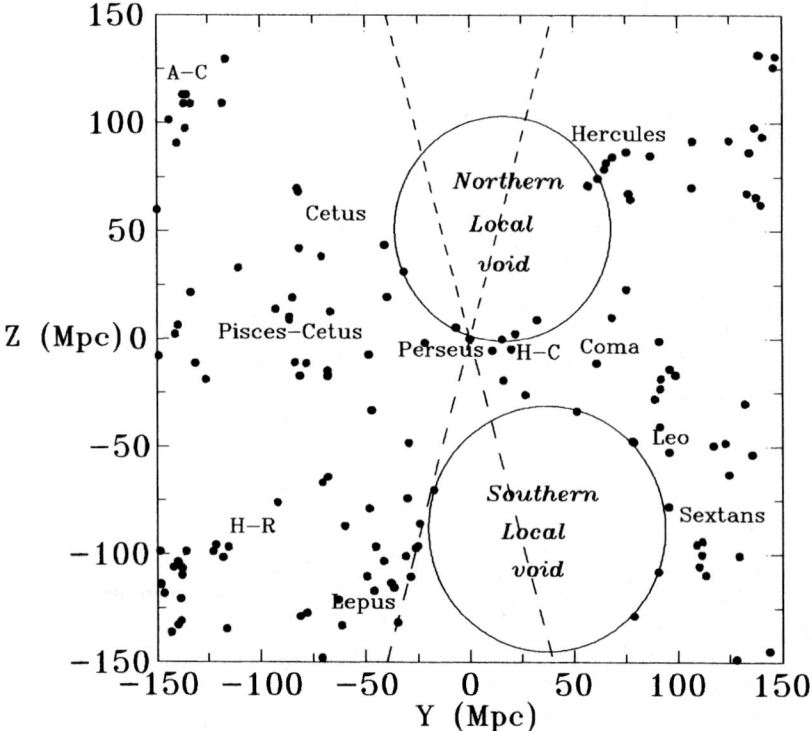

Fig. 7.24 The distribution of Abell clusters in rectangular supergalactic coordinates in the interval $X = -75 \ldots 50\, h^{-1}$ Mpc. The zone of avoidance is shown by dashed lines, outer contours of Local Voids by circles. Supercluster names are marked. H–R, H–C and A–C are for the Horologium–Reticulum, Hydra–Centaurus and Aquarius–Capricornus superclusters, respectively (Einasto et al., 1994b).

The right panel of the same Figure shows correlation lengths of the same samples as a function of the sample size (and the absolute magnitude limit). Here also in a log–log diagram the correlation length increases with sample size almost linearly. This is the result of biasing, with increasing luminosity limit the correlation length increases, as discussed above.

Another method to characterise void sizes is to use the void probability function VPF, introduced by White (1979). A number of authors have used this function. Einasto et al. (1991) derived the void probability function for a number of observational and model samples of various sizes, as used earlier to derive the distribution of void diameters (Einasto et al., 1989). The results show that both methods describe the void properties well. For the CfA2 sample the void probability function was

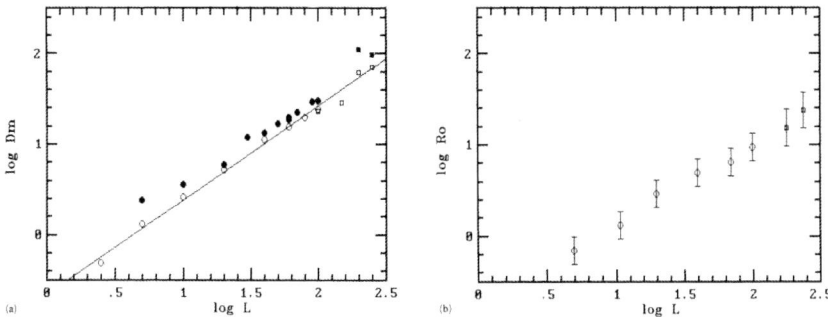

Fig. 7.25 The panel (a) shows the void diameters, D_m, as a function of the sample size, L, both expressed in units h^{-1} Mpc. Open circles mark diameters of galaxy samples, found with the beam method. Filled circles are for diameters found with empty sphere method; squares are for cluster samples. Panel (b) gives the correlation length R_0 versus the sample size for the same galaxy and cluster samples (Einasto et al., 1989).

found by Vogeley et al. (1994a). The VPF analysis of the CfA2 sample confirmed the principal results obtained from a similar analysis of the CfA1 sample by Einasto et al. (1991).

Lindner et al. (1995, 1996, 1997) investigated the structure of voids in more detail, in particular the distribution of faint galaxies in large voids defined by rich clusters and superclusters. He found the distribution of void diameters for galaxies of different absolute magnitude limits for various environments. The environment was defined using the luminosity density field, smoothed with a Gaussian kernel of radius $8\,h^{-1}$ Mpc. The dependence of void sizes on the absolute magnitude limit is shown in Fig. 7.26. As expected, brighter galaxies form larger voids. In larger voids the central smoothed density is lower, and void galaxies are fainter.

In subsequent years there has been a lot of progress in the study of voids. In 2006 a special workshop on voids was organised in Aspen, and later in Amsterdam. At this workshop a program to compare void finder programs was initiated, and results were published by Colberg et al. (2008). One of the most effective void finders is the cosmic watershed method suggested by Platen et al. (2007).

In March 2012 Rien van de Weygaert visited Tartu Observatory; in November 2012 his previous collaborator Miguel Aragon-Calvo also visited us. Rien works in the Kapteyn Institute of the Groningen University. The Groningen team has concentrated their efforts on the detailed study of the cosmic web and voids. What impressed me in the work of the Groningen team was the study of the same phenomena we did many years earlier, but now the study was made much more methodically, using novel methods of analysis.

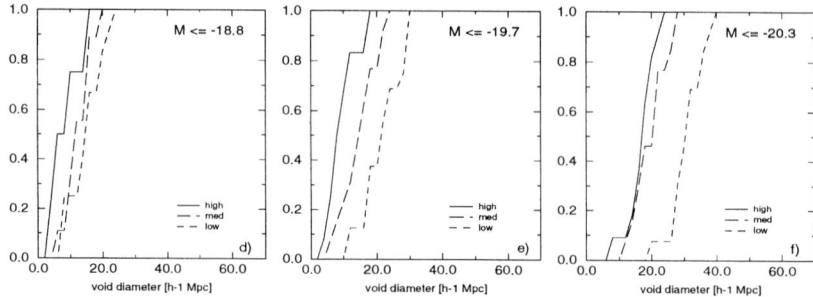

Fig. 7.26 The cumulative distributions of void diameters for various absolute magnitude limits in different environments. The environment is defined by the density field, smoothed with Gaussian dispersion $8\,h^{-1}$ Mpc; high, medium and low density is defined using threshold density levels 2 and 1 in mean density units (Lindner et al., 1995).

Both Rien and Miguel had seminar talks in our Observatory. In his seminar talk ("The Alpha and Betti of the Universe") Rien described his recent work in the study of the cosmic web. Together with his collaborators he initiated the void galaxy survey (van de Weygaert et al., 2011). Void galaxies are of great interest in understanding the evolution of galaxies, since in a low-density environment one can observe galaxies which have had little influence from interactions with other galaxies. Rien also described his results of the study of the void phenomenon using different techniques and galaxies (van de Weygaert & van Kampen, 1993; van de Weygaert & Platen, 2009; van de Weygaert et al., 2009, 2010; Aragon-Calvo et al., 2010b,c,a).

7.3.7 Cosmic web — cells and the cosmic foam

The largest natural units of the cosmic web are cells, defined by rich clusters and superclusters, as suggested already by Jõeveer et al. (1977); Jõeveer & Einasto (1978). The simplest characteristics of the cosmic web are the mean distances between superclusters, and the nearest neighbour distribution. Both characterise the size of cells of the cosmic web.

Previous analysis has shown that the sizes of voids determined by superclusters of different richness are rather close. This result, and the absence of a randomly located population of rich clusters in voids, suggest that practically all rich clusters of galaxies are located in void walls, and the overall distribution of superclusters of different richness is rather similar. Einasto et al. (1997b) studied the distribution of superclusters of different richness in void walls. For that they calculated for each

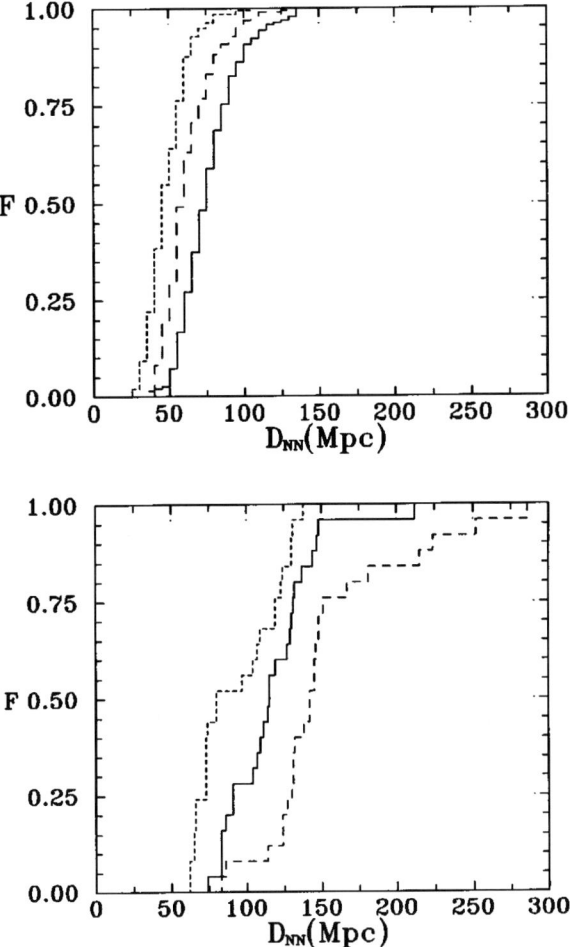

Fig. 7.27 The distribution of distances between centres of superclusters. Upper panel shows the distributions for poor and medium rich superclusters, lower panel the distributions for very rich superclusters. Curves correspond to the first (line with short dashes), second (line with long dashes) and third (solid line) neighbour (in the lower panel the last two lines are interchanged) (Einasto et al., 1997b).

supercluster centre the distances to the centres of the three nearest superclusters, separately for poor, medium rich, and very rich superclusters (see Fig. 7.27).

On the upper panel of Fig. 7.27 these distributions are given for poor and medium rich superclusters. We see firstly, that these distances are small, and secondly, that these distributions are smooth and do not show the presence of

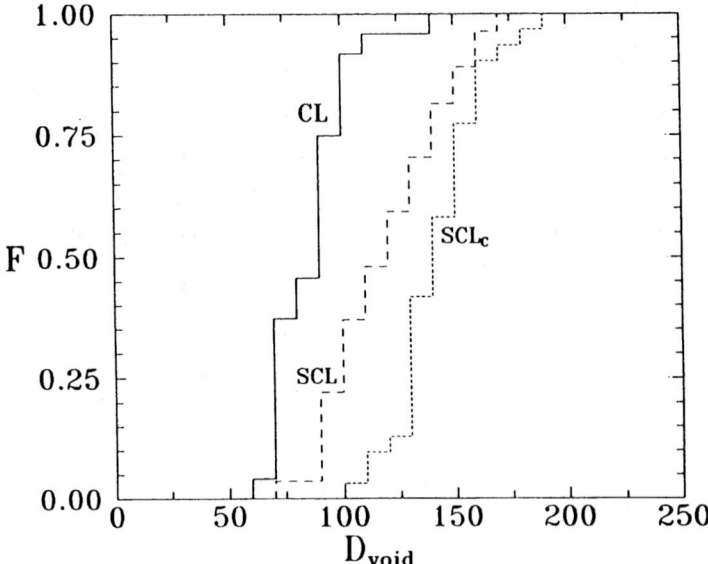

Fig. 7.28 The cumulative distribution of void diameters. The solid line shows diameters of voids determined by all clusters (CL), the long-dashed line shows diameters of voids defined by clusters that belong to superclusters (SCL), and the dotted line shows diameter of voids defined by centres of superclusters (SCL_C) (Einasto et al., 1994b).

any preferred distance between superclusters. That would be seen as peaks in the distance distribution.

The distributions of distances between very rich superclusters (Fig. 7.27 lower panel) are different. None of these distributions is as smooth as in the upper panel. The most important feature in this Figure is the presence of a peak in the distribution of distances of the second and the third neighbour in the interval $110 < D_{NN2;3} < 150\,h^{-1}$ Mpc — over 75% of very rich superclusters have a second or third neighbour at this distance interval. The median distances to the second and third neighbours are, correspondingly, $D_{NN2} = 115$ and $D_{NN3} = 142\,h^{-1}$ Mpc.

Einasto et al. (1994b) analysed the distribution of void diameters as defined by Abell clusters of galaxies. Distributions were found for three cases: voids defined by all clusters, voids defined by clusters located in superclusters, and voids defined by supercluster centres, see Fig. 7.28. The median diameter of voids defined by clusters in superclusters is $120\,h^{-1}$ Mpc, and that of centres of superclusters is

$145\,h^{-1}$ Mpc. These values characterise the scale of cells of the Universe and the mean sizes of voids between superclusters.

van de Weygaert (2002) characterized the cosmic web as a "cosmic foam" — a tenuous space-filling frothy network permeating the interior of the Universe. Sheth & van de Weygaert (2004) investigated the hierarchy of voids in more detail.

Miguel in his seminar talk in Tartu described his algorithm to simulate the evolution of the cosmic web, the Multum In Parvo (MIP) constrained ensemble simulation, which allows a very high resolution model with minimal computer resources (Aragon-Calvo, 2012). He used these high-resolution simulations to investigate the internal structure of voids. His simulations confirmed that the cosmic web has properties of a cellular distribution. He compared the cosmic cells with other cellular systems in Nature on very different scales, from molecular to cosmic scales. He found that all cellular systems have similar properties, depending on the number of neighbouring cells. If the number of neighbouring cells is small, then during the evolution the cell shrinks and disappears (Sheth & van de Weygaert, 2004). If the number of neighbours is large, then the cell expands. The most stable configuration is a cell with 8 neighbouring cells. Such cells have the structure of a honeycomb.

Miguel's high-resolution simulations also showed in more detail the hierarchical nature of the cellular and void structure. Within a large cell (void) there are sub-cells (sub-voids), within sub-cells there are sub-sub-cells (sub-sub-voids) etc. The highest level in this hierarchy have super-cells (super-voids), surrounded by superclusters, which are well visible both in simulations and in Nature. The next level is also seen both in Nature and in simulations. Lower levels in this hierarchy are located inside sub-cells (sub-voids). Here the density of void surfaces is so low that in most cases no galaxy formation can take place here. These substructures are seen only in high-resolution simulations of dark matter. But their existence is highly probable taking into account the similarity of the cellular structure in other natural phenomena of different scale. Thus the cosmic web has properties of a "cosmic foam", as already suggested by Rien ten years ago.

When we introduced the term "cellular structure of the Universe" in the late 1970's, we did not guess that this term could have such a deep physical meaning.

7.3.8 Regularity of the cosmic web

Already our first pictures of the wedges shown in Fig. 5.5 and 5.4 show that there is some regularity in the distribution of rich superclusters. The relatively rich Perseus–Pisces supercluster is located at a distance of about $50\,h^{-1}$ Mpc. The Local supercluster is fairly poor; the closest rich supercluster is the Coma supercluster almost behind the Local one. The distance between the Coma and the Perseus–Pisces superclusters is about $120\,h^{-1}$ Mpc. Behind the Perseus–Pisces supercluster there is a relatively empty region; the next region containing several rich clusters is located at a distance about $150\,h^{-1}$ Mpc, i.e. the distance between the Perseus–Pisces supercluster and the system of superclusters behind is about $100\,h^{-1}$ Mpc.

The Bootes void found by Kirshner et al. (1981) has a diameter of about $100\,h^{-1}$ Mpc. It is surrounded by several rich superclusters, one of which is the Hercules supercluster. In our terminology "cellular structure" we had in mind just large low-density regions surrounded by rich superclusters.

Maret Einasto compiled catalogues of superclusters formed by Abell clusters (Einasto et al., 1994b, 1997b, 2001). She noticed that in the nearest neighbour distribution of superclusters there are peaks: the first neighbour is located at a distance of about $60\,h^{-1}$ Mpc, this is usually on the same side of the wall between two adjacent supervoids. But the second and third nearest neighbours are located, as a rule, across the supervoid, and have a mean distance of about $100\,h^{-1}$ Mpc.

This issue became topical after the discovery by Broadhurst et al. (1990) that the distribution of high-density regions of galaxies may be quasi-regular or periodic. Broadhurst *et al.* found such a regularity in the direction of the North and South Galactic poles.

Thus we started to investigate the distribution of rich clusters and superclusters in more detail. Our study confirmed that high-density regions marked by rich superclusters form a quasi-regular lattice (Einasto et al., 1997b,a), see Fig. 7.29.

Initially we believed that this quasi-regularity is due to the presence of a bump on the corresponding scale in the power spectrum of density fluctuations. To check this possibility I made a series of numerical simulations in Fermilab during my visit in 2000. My results were negative.

Then I started to think of other reasons for such quasi-regularity. I discussed the problem with Enn. It was clear that the present density field is the sum of density perturbations of various scales which have various amplitudes and phases. The power spectrum characterises only the amplitudes of density waves of different scales. But phases of density perturbations also play an important role in the formation of the cosmic web. So far the phase information has been almost ignored by the astronomical community. In the next Chapter I shall describe our efforts to understand the evolution of the cosmic web, taking into account the phase information of density perturbations.

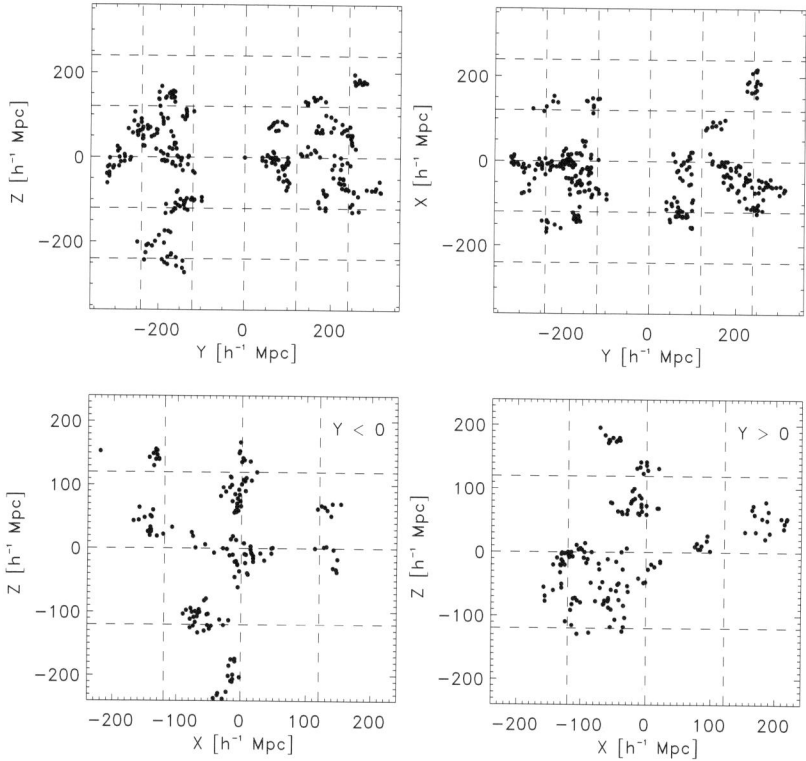

Fig. 7.29 The distribution of 319 clusters in 25 very rich superclusters with at least 8 members (including 58 clusters with photometric distance estimates) illustrates the network in the cluster distribution in supergalactic coordinates. In the lower panels clusters in the northern and southern Galactic hemispheres are plotted separately. The supergalactic $Y = 0$ plane coincides almost exactly with the galactic equatorial plane, i.e. with the zone of avoidance due to galactic absorption. The grid with step-size $120\,h^{-1}$ Mpc corresponds approximately to distances between high-density regions across voids. In the two upper panels and in the lower right panel several superclusters overlap due to projection but are actually well-separated in space (Einasto et al., 1997a).

7.3.9 Baryonic acoustic oscillations

The early Universe consisted of a hot plasma of electrons, baryons and photons, and dark matter. The plasma was so hot that electrons, baryons and photons were tightly coupled. Overdensities of matter (baryons and dark matter) attract matter gravitationally towards the center of the overdensity, while the heat of photon–matter interactions creates outward pressure. Competing forces of gravity and pressure create oscillations, similar to sound waves in air. In this way the pressure forms a spherical sound wave of baryons and photons around each overdense region. These sound waves move outwards from the overdensity. The dark matter does not interact with the hot plasma, and stays at the center of the sound wave at the origin of the overdensity.

After the decoupling (recombination) the photons no longer interact with the baryonic matter and diffuse away. The pressure vanishes and the shell of baryonic matter is left at a fixed radius, called the sound horizon. The shell continues to attract matter and galaxies formed in a similar pattern in a shell surrounding a high-density peak. The number of peaks is large, thus actually there are many overlapping shells. It is difficult to see all the overlapping shells individually; they can be detected statistically by looking at the separations of a large number of galaxies.

These acoustic waves in the tightly coupled baryon–photon fluid prior to the epoch of recombination will lead to the characteristic maxima and minima in the post-recombination matter power spectrum. The same mechanism is responsible for the peak structure in the CMB angular power spectrum (Sunyaev & Zeldovich, 1970; Peebles & Yu, 1970). The scale of these features reflects the size of the sound horizon, which is fully determined by the physical densities $\Omega_b h^2$ and $\Omega_m h^2$.

Peaks in the CMB angular power spectrum have been discovered by the WMAP satellite, see the next Chapter. Recently similar features have been found also in the distribution of galaxies of the SDSS Luminous Red Giant (LRG) sample by Eisenstein et al. (2005), using the correlation function. The same sample was analysed by Hütsi (2006) using the power spectrum of LRG galaxies. Hütsi has found evidence for a full series of acoustic features down to the scales of $k \sim 0.2\ h$ Mpc^{-1}. This corresponds up to the 7th peak in the CMB angular power spectrum. The acoustic scale derived, $105.4 \pm 2.3\ h^{-1}$ Mpc, agrees very well with the "concordance" model prediction and also with the one determined via the analysis of the spatial two-point correlation function by Eisenstein et al. (2005). Figure 7.30 shows the power spectrum of the SDSS LRG sample. This work is the first determination of the power spectrum of this sample. Acoustic features in the power spectrum are clearly visible, and are in good agreement with features expected from the model spectrum.

Hütsi (2006) emphasised that the ability to observe baryonic features in the low redshift galaxy power spectrum demands a rather high baryonic to total matter density ratio. Blanchard et al. (2003) suggested an Einstein–de Sitter type model with zero cosmological constant, and emphasised that with such a model it is possible to fit a large body of observational data, if one adopts a low value for the Hubble parameter. In light of the results obtained by Hütsi (2006) these models are

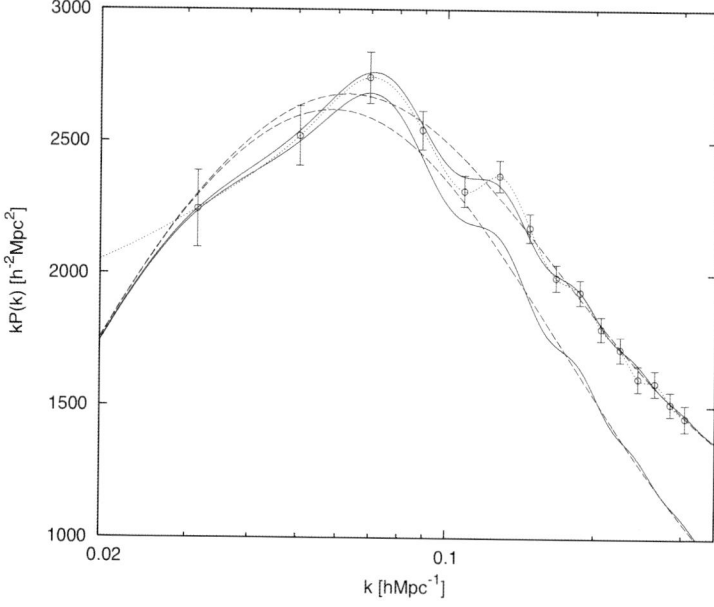

Fig. 7.30 Power spectrum of the SDSS LRG sample with bin width $\Delta k \approx 0.02\ h$ Mpc^{-1}. The upper solid line shows the best fitting model spectrum and the lower one corresponds to the linearly evolved matter power spectrum of the "concordance" cosmological model multiplied by the square of the bias parameter $b = 1.95$. Both of the spectra are convolved with a survey window. The dashed lines represent the "smoothed-out" versions of the above model spectra. The dotted line is the cubic spline fit to the data points. (Hütsi, 2006).

certainly disfavored due to the fact that the high dark matter density completely damps the baryonic features. And finally, purely baryonic models are also ruled out since for them the expected acoustic scale would be roughly two times larger than observed here. So the data seems to demand a weakly interacting nonrelativistic matter component.

The effect of baryonic acoustic oscillations (BAO) had previously been detected using correlation functions and power spectra of the galaxy distribution. Arnalte-Mur et al. (2012) presented a new method to detect the real-space structures associated with BAO. Baryon acoustic structures are spherical shells of relatively small density contrast, surrounding high density central regions. The authors designed a specific wavelet adapted to search for shells, and applied this method to detect shells surrounding high-density peaks of the SDSS density field. Peaks were found using the LRG sample of galaxies; to find shells around peaks the main galaxy sample of SDSS was used. To enhance shells they were stacked around high-density peaks.

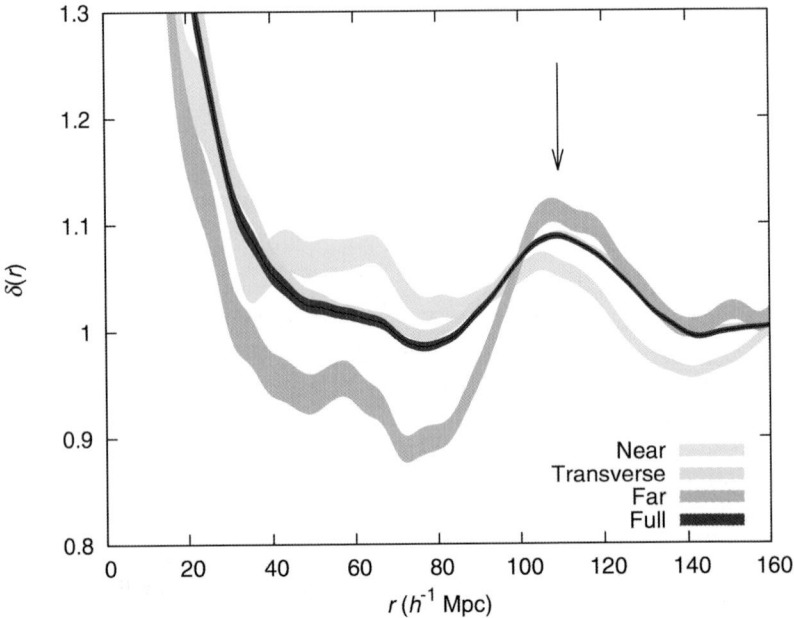

Fig. 7.31 Radial density profiles averaged over high-density centres. Density is expressed in units of the mean average density of the sample. The continuous black line with an error band shows the radial profile for the full sphere. The arrow signals the location of the maximum, $r_{max} = 109.5\,h^{-1}$ Mpc (Arnalte-Mur et al., 2012).

Radial density profiles averaged over high-density centres are shown in Fig. 7.31. The lines in different shades of gray mark the profiles of various subsamples; the black line shows the radial profile for the full sphere. The resulting profile has a high bump on short scales, corresponding to the central peak, and a clear maximum on the acoustic scale, $r_{max} = 109.5 \pm 3.9\,h^{-1}$ Mpc. This study shows that BAO shells are real spatial structures, and that the BAO phenomenon can be studied in detail by examining those shells.

Presently it is not clear whether there is a clear connection between the distribution of rich clusters and superclusters, discussed in the previous subsection, and the baryonic acoustic oscillations. The physics of these phenomena is different. The large-scale distribution of clusters and superclusters is probably given by the initial density perturbation field, generated during or after the inflation period of the evolution of the Universe. Baryonic acoustic oscillations are generated by sound waves in the epoch just before recombination. The connection between both effects is given by the fact that the BAO phenomenon is generated by the initial density perturbation field. To understand the possible relations between these two effects more detailed studies are needed.

7.4 Tartu Observatory in the 1990's

7.4.1 *Estonian path to independence*

In the Soviet period we often had guests who spent their summer vacation in the Observatory's guesthouse. Among them were Yakov Zeldovich, Josif Shklovsky and several other prominent astronomers. During these visits we had a lot of time to discuss various topics on cosmology as well as Zeldovich's experience in the secret work. We used our sauna also for receptions of guests. One of the most memorable sauna-visits was with Remo Ruffini and his students. It was winter, the pond next to sauna was covered with ice, and a hole was made in the ice to jump into the cold water from the sauna. For Italians this was a rather unexpected experience.

Our sauna was in some aspects similar to ancient thermae in Rome — we used the sauna to discuss various topics from science to everyday life. So, in the late 1980's an informal "sauna-club" formed which started to discuss problems related to the transition from a command-economy in the Soviet system to a free market-economy. At this time it was already clear that the Soviet system was near to a collapse, thus it was time to think about the transition. This club had its discussions confidentially, so its activity was not known to the whole community of the Observatory. The club developed, among other topics, the main principles to be used in privatisation of the economy.

Elections for the last Estonian Supreme Soviet (Parliament in present sense) were held in March 1990. The elections were practically free and a number of intellectuals were elected to perform the reforms needed for the transition of the society. Among other intellectuals one representative was elected from the Tartu Observatory: Liia Hänni, a stellar astrophysicist and member of our secret sauna-club. Popular Front participated in elections and won the majority of seats. In contrast, the Estonian Citizens Committee declared that the Supreme Soviet is an organ of occupation and refused to participate in its activities.

The Estonian Supreme Soviet formulated as its main goal the making of all necessary preparations to restore the independence of Estonia. As a first formal juridical act on March 30, 1990, it openly declared 'a transition period' to restore the Estonian Republic. In this way cooperation with the Soviet Government was still possible. In contrast, Lithuania declared its independence already on March 11, 1990, but a Soviet blockade followed, and no country recognised Lithuanian independence. On June 12, 1990, the Russian Federation declared its sovereignty, and elected Boris Yeltsin as the Russian President.

In October 1990 the Estonian Supreme Soviet accepted the law of economic borders, and many checkpoints were organized at the main crossing points to check documents. This was the first step in fixing the borders of the country. Similar laws were accepted in Latvia and Lithuania. At this time the Soviet Union had Special Purpose Militia Units, called OMON. Estonia did not form such units, but in Latvia and Lithuania they were formed. At the end of 1990 these units refused to obey the orders of Latvian and Lithuanian Interior Ministries. On January 20, 1991, the Riga OMON attacked Latvia's Interior Ministry, killing six people. On June 14, 1991, the OMON attacked the Estonian crossing point between Latvia and Estonia, wounding three people. One of them was my relative Enn Vilbaste. His uncle Juhan Vilbaste was my close friend, an entomologist and a "Finnish Boy" during WWII.

In January 1991 the Soviet army together with OMON units tried to suppress the independence movements in Lithuania, Latvia and Estonia. The first actions were in Vilnius. Lithuanians defended the TV-tower and many civilians were killed. In Latvia the Soviet army attacked demonstrators in Riga. To avoid this in Estonia, our leaders phoned Yeltsin and invited him for negotiations at Tallinn. In this meeting Yeltsin signed a "Treaty on the Basis of Interstate Relations Between the Russian Federal Socialist Republic and the republic of Estonia"; similar treaties were signed also with representatives of Latvia and Lithuania. Yeltsin also made an appeal to the Soviet army condemning attacks against civilians. This appeal prevented Soviet military actions in Estonia. A small remark — all these documents were typed in Lippmaa's office.

A union-wide referendum was performed on March 17, 1991 which restricted the rights of the Soviet republics. Estonia and Latvia boycotted this referendum and organised their own preventive referendums; Lithuania had already declared its independence. The Estonian referendum was held on March 3. The overwhelming majority supported the restoration of an independent Estonian state.

This development made conservative forces in the Soviet Union very anxious and they formed the State Committee of Emergency, which organised the August Coup. In the early morning of August 19 over the state radio and television a broadcast was transmitted stating the Emergency. I was on this day in Tallinn to buy air tickets to attend a summer school in Erice, Italy, and heard of the Coup while standing in the queue for tickets.

The tickets purchased, I went to Toompea (Dome Hill) where our government and Estonian Supreme Soviet had their offices. I searched for Liia Hänni, our representative in the Soviet. I found her in a small office. To my question of what the Soviet is doing, Liia said that they were discussing the next law project. I replied that all other items are of secondary importance, and now the only major issue was to declare Estonian independence, as it might be impossible later. Liia answered that she has the same opinion, but the Estonian Committee is strongly opposing this, because in their view the Supreme Soviet is an illegal organ of occupation.

In my opinion the actual reason was that both leading forces, the Popular Front and the Estonian Committee, wanted sole credit for the declaration of independence. The Estonian Committee did not understand that neither the Soviet Union nor any foreign country would recognise an independence declaration made by an informal organisation which did not have actual control of the country. To overcome this rivalry the chairman of the Supreme Soviet Arnold Rüütel argued that a compromise between the Popular Front and the Estonian Committee was needed. In further discussions with Liia we both found that the major issue is to find the compromise.

The next day two other astronomers visited Tallinn and also spoke to Liia Hänni urging just the same. To achieve the compromise, on the next day, August 20, a mass meeting was organised by Popular Front in the central square of Tallinn (Vabaduse plats — the Square of Freedom). Soviet tanks were already in Tallinn to take over the TV tower and other important buildings, so it was high time for action. The people's message in the mass meeting to our leaders was clear — "Take it now!"

Two scientists, Liia Hänni and Marju Lauristin (both were members of Supreme Soviet as well as Estonian Congress), played a major role in achieving a compromise between the Estonian Congress and the Supreme Soviet. The idea of the compromise was to declare the Estonian Republic as a continuation of the pre-war Republic, and to form the Estonian Constitutional Assembly to prepare a new constitution of Estonia. It should consist of equal numbers of members from the Estonian Congress and the Supreme Soviet. That very evening the compromise was achieved, and the Supreme Soviet proclaimed the restoration of the independent state of Estonia.

The real events of the Coup were in Moscow. Already on August 19 citizens of Moscow began to gather around the White House, the Parliament of the Russian Federation. A tank battalion was sent to guard the White House, and the chief of the tank battalion declared its loyalty to the Russian Federation. The president of the Russian Federation Yeltsin climbed one of the tanks and addressed the meeting, declaring the Coup as illegal. On August 22 the Coup leaders understood that the Coup had failed. The real power was in the hands of the President of the Russian Federation.

On August 24 many thousands of Moscow citizens took part in the funerals of three men killed during the Coup. Rüütel was in Moscow to meet Yeltsin. After the funerals Yeltsin invited Rüütel to his office, and Yeltsin signed the document of recognition of Estonia as an independent state. In the document an appeal to the Soviet Union and foreign countries was included inviting them to recognise Estonian independence. In the following days almost all European countries followed this appeal. But some countries waited until the State Council of USSR recognised Estonian independence on September 6, 1991. On December 21, 1991, representatives of all Soviet republics except the three Baltic states and Georgia declared the end of the Soviet Union.

After the declaration of independence the economical situation quickly worsened, because almost all imports were made on the basis of international prices, but the value of the ruble was falling very rapidly. In June 1992 the Estonian currency, kroon, was introduced, and in September 1992 the new Estonian Parliament was elected. Our currency was pegged to the Deutsche Mark, the most stable currency in Europe, which gave confidence to our economy. On January 1, 2011, the Euro replaced our kroon as the official currency.

After the introduction of the kroon the economic situation started to improve. The new government formed after the election of the Parliament was dominated by the nationalist party who started radical economic reforms. The minister for Privatisation was Liia Hänni, the minister of Social Affairs was Marju Lauristin, both members of the Moderate Party, which was later renamed as the Social Democratic Party.

But the Russian Army was still there. It took several years of difficult negotiations to withdraw all Russian troops. Here our allies in the West helped: the U.S. Senate threatened in July 1994 to halt all aid to Russia if the forces were not withdrawn by the end of August. Final withdrawal was really completed on August 31, 1994.

For us, only now has World War II ended.

7.4.2 *Science reform*

In the first year of restored independence the value of the currency, ruble, fell very rapidly, and the prices for imported goods became international. Thus the budget of the Observatory was sufficient only to buy fuel for central heating, and a bit was left over for salaries. This led to a drastic decrease in the number of staff — almost the whole technical staff left and formed new private enterprises. Thus in the early 1990's on Observatory territory there were about ten private enterprises ranging from a computer company and typography to a center of education. A number of scientists also left, mostly young men and women. Most of them organised their own enterprises and went to work there, some left to join politics like Liia Hänni, and some found a better place abroad. Among them were our best young people in the cosmology department, Lev Kofman and Dimitry Pogosyan.

During the Soviet period most scientific studies were made in institutes of the Estonian Academy of Sciences. In the early 1990's the science and university system was reformed. Most former institutes of the Academy were included into universities — Tartu University (the only classical university in Estonia), Tallinn Technical University, Estonian University of Life Sciences, or Tallinn University (former Tallinn Pedagogical University). In most cases these institutes dissolved as scientific units as scientists were distributed between many subunits within universities. To avoid such a situation we succeeded in keeping Tartu Observatory as an independent institution which collaborates closely with Tartu University. So most astronomers teach courses in the University or serve students supervisors.

In the early 1990's the Soros Foundation helped us to get some equipment; most urgently we needed modern personal computers. In the mid 1990's in Estonia a grant system for science was introduced. Grants are given on the basis of scientific excellence, and in most cases foreign referees are used to estimate the scientific level of the grant. Also collaboration with astronomers from other countries was easier.

The general level of science financing was in the 1990's still rather low. So a lot of young men and women entering universities chose soft sciences such as financing, law, project managers etc instead of physics, mathematics or astronomy. Only in the early 2000's did the financial situation improve, and now we have a number of young investigators in our staff.

7.4.3 *Participation in international organisations*

Soon after the restoration of Estonian independence and its recognition by Western countries I got a letter from George Contopoulos. He was the Editor-in-Chief of the European Journal "Astronomy and Astrophysics" (A&A). This is a journal for all European astronomers, created by merging of all previous European astronomical journals, and coordinated by the European Southern Observatory. Contopoulos suggested that Tartu Observatory should make an official appeal to the Journal and ask for membership, first as associate member-state, and later as a full member-state. He asked me to write the letter promptly, so that he could make necessary arrangements during his period as Editor-in-Chief. The director of the Observatory wrote the letter, and soon I got an invitation to participate in the next annual Meeting of the Board of Directors of the Journal.

This time the Meeting was in Budapest; just recently Hungary's membership in A&A had been accepted. When I entered the room where the Board had its Meeting, there was a burst of applause. I knew most of the members of the Board, and now I was congratulated because Estonia has regained its role in the European community. But a question immediately followed: *"Will you publish papers on cosmology in our journal?"* In mainland Europe cosmology was not as well developed as in the USA and Great Britain, and the Journal was interested in our publications. I answered that certainly we shall publish our future cosmology papers also in A&A; so far we published our papers mostly in "Monthly Notices of the Royal Astronomical Society". Now we really use for publications mostly A&A. I participated a few times in Meetings of the Board over next few years, but soon I proposed that the Director of Tartu Observatory, Laurits Leedjärv, should take over the role of the Estonian representative in the Board. My suggestion was accepted. Now Estonia is already for many years a full member of the Journal. Once the Meeting of the Board was held in Tartu; members of the Board made a visit to our Observatory in Tõravere.

Soon Estonian membership in the International Astronomical Union was restored. The first IAU General Assembly after the restoration of our independence was in The Hague in August 1994. We participated with a fairly large delegation. Initially we planned to drive there in two personal cars, but one car broke down, so a group of four astronomers used my car, while the rest came in buses and trains — there were no flights still in those years. I felt proud to drive in the city with an Estonian number plate. At the opening ceremony I discovered that the Estonian flag was upside down. I hurried to the organisers, and during the break the flag was placed correctly. At the Assembly banquet around our table there were astronomers congratulating us. I had a longer talk with Harry van der Laan, the former Director General of ESO. During my visits to ESO we had discussions on our freedom movement. Now he had a small blue-black-white badge on his breast, which I had presented him in our discussions some years ago.

The next IAU General Assembly was held in Kyoto, the old capital of Japan. I was the representative of Estonia, so I participated. This was my only trip to Japan. Japan has old traditions and culture, a bit similar to the culture of China. We had a chance to visit the old Emperor's palace and many other places of interest. In the opening ceremony of the Assembly Japan's Emperor, Akihito, participated. It was very interesting to see all these old traditions still alive.

During the Assembly the American delegation organised a reception, and I had the honour to be invited. I came a bit earlier, and found only the head of the US delegation, Vera Rubin, awaiting guests. So we had plenty of time for a very friendly discussion. I thanked her for her popular paper in Scientific American on dark matter, where she gave a detailed overview of our work on the subject, always calling us Estonian or Tartu astronomers. She smiled and answered that our enthusiasm for Tartu and Estonia was so evident that everybody knew which observatory and country we are from.

Chapter 8

Cosmic inflation, dark energy and the evolution of the Universe

In this chapter I shall discuss some aspects of the evolution of the Universe. In the 1980's our cosmology group welcomed 4 young cosmologists: Lev Kofman, Dmitri Pogosyan, Maret Einasto and Mirt Gramann. Lev was a postdoc in Moscow in the early 1980's; his supervisor was Alexei Starobinsky. He studied the theory of inflation with Alexei Starobinsky and Andrei Linde, and models dominated by the cosmological term with Alexei Starobinsky. Kofman initiated the Tartu Cosmology Seminars; the first Seminar was held in May 1982, the Second in June 1985 (Kofman et al., 1986). In both seminars the main topics were the nature of dark matter, the evolution of the Universe, and inflation theory.

These theoretical studies started in the early 1980's. Later theoretical studies were complemented with numerical simulations of cosmic evolution to understand some particular aspects of the evolution — the quasi-regularity of the structure, the understanding of the void problem, and the early evolution of the cosmic web. Our team is rather small and our computational possibilities modest, thus we were not able to study all aspects of the cosmic evolution. We concentrated on problems which for us seemed interesting, and where we had ideas on how to solve the problems.

8.1 The birth of the Universe and inflation

8.1.1 *The classical inflation theory*

Astronomers have direct data on the past by observing galaxies, clusters, quasars and other objects. The most distant objects found so far have a redshift about 6–8. The Universe was at this time rather young, about 1 billion years old. However, direct observational evidence is available for earlier epochs: CMB radiation was emitted when the Universe was about 350 thousand years old, and data on the nucleosynthesis of light chemical elements tell us what properties our Universe had when it was only a few minutes old. What happened in earlier moments we do not know so well.

According to the presently accepted Big Bang model the Universe started from a singularity. But "singularity" is a mathematical term. Big Bang theory says nothing about the physics of the primordial explosion. The theory of inflation is a physical description of the bang itself. It tries to answers a number of questions which could not be explained in the framework of the classical Big Bang model.

The first problem is the flatness problem. During the evolution the Universe expands so much that any deviation from exact critical density would increase during the expansion. In order to have an approximately critical density today, it must have critical density in the early epoch, at the time of nucleosynthesis, with an accuracy of at least 15 decimal places, i.e. it must be very accurately tuned.

The second problem is the homogeneity of the Universe. Data from the COBE satellite indicate that the gas in opposite directions of the sky have the same temperature with an accuracy of one part in 100,000 at the recombination epoch. Such high accuracy is possible only if these different regions have communicated. However, this was impossible at the time of recombination, since such communication would be possible only with a speed roughly 100 times the speed of light. In other words, the identical temperature must be achieved much earlier when the Universe was more compact.

These and some other difficulties of the classical Big Bang theory can be avoided if in the very early phase of the evolution of the Universe there was a period of very rapid expansion by a factor of at least 10^{26}. This rapid expansion is called inflation. The inflation scenario was suggested by Aleksei Starobinsky (1980, 1982, 1985) and independently by Alan Guth (1981).

This classical inflation model solves the flatness problem. During the inflationary period the Universe is driven very accurately towards the critical mass density. The model also solves the problem of homogeneity. The presently visible Universe has a radius of about 15 billion light years. Since the expansion factor is at least 10^{25} times, the present Universe was before the inflation so small that there was plenty of time for it to come to a uniform temperature. So in the inflationary model, the uniform temperature was established before the inflation took place, in an extremely small region.

8.1.2 *The new inflation theory and the birth of the Universe*

The above classical version of the inflation model is called "old". It provides a simple solution to several crucial problems, but does not answer some other fundamental questions: Why does the Universe have elementary particle parameters as they are? What happened before inflation started? To answer these questions Andrei Linde suggested a scenario of chaotic inflation (Linde, 1982, 1983).

Andrei Linde describes his first contacts with cosmologists from Western countries in discussing his new inflationary model (Linde, 2002b). In 1981 there was a conference on Quantum Gravity in Moscow. This was the first conference where Andrei gave a talk on the new inflation scenario. After the conference one of the participants, Stephen Hawking, was invited to give a talk at the Sternberg Astronomy Institute. Zeldovich asked Andrei to translate. At that time Stephen did not have his computer, so his talks usually were given by his students. Stephen said one word, his student repeated the word, and Andrei translated it. Because Andrei knew the subject, he started adding lengthy explanations in Russian. In the second part of the lecture Stephen said that recently Andrei Linde had suggested an interesting way to solve the problems of the old inflationary theory. But then Stephen said that the new inflationary scenario cannot work. When the talk was over Andrei said that he translated but cannot agree, and explained why. Thereafter Andrei and Stephen discussed the issue privately. In Summer 1982 Stephen organised a workshop in Cambridge dedicated to the new inflation theory. As Andrei writes, this was the best and most productive workshop he ever attended.

A small remark. About the same time another Cambridge astronomer, Donald Lynden-Bell, visited Moscow and gave a talk at the Sternberg Institute. Zeldovich asked me to be the translator. Similarly to Andrei, I knew the paper, and added detailed explanations. It happened several times that I was ahead with my explanations, so when Donald finished a step in his talk, I had to say that I had already talked about this, accompanied with a laugh from the audience.

Linde (2002a) explains the problems which led him to his new inflation scenario as follows.

"Most of the parameters of elementary particles look more like a collection of random numbers than a unique manifestation of some hidden harmony of Nature. For example, the mass of the electron is 3 orders of magnitude smaller than the mass of the proton, which is 2 orders of magnitude smaller than the mass of the W-boson, which is 17 orders of magnitude smaller than the Planck mass M_p. Meanwhile, it was pointed out long ago that a minor change (by a factor of two or three) in the mass of the electron, the fine-structure constant, the strong-interaction constant, or the gravitational constant would lead to a universe in which life as we know it could never have arisen. These facts, as well as a number of other observations, lie at the foundation of the so-called anthropic principle. According to this principle, we observe the universe to be as it is because only in such a universe could observers like ourselves exist."

Andrei continues:

"One can consider different universes with different laws of physics in each of them. This does not necessarily require introduction of quantum cosmology. It is sufficient to consider an extended action represented by a sum of all possible actions of all possible theories in all possible universes. One may call this structure a 'multiverse.'"

Similar arguments were discussed by Martin Rees (2000) on the meaning of six numbers which determine the essential properties of our Universe. Martin considers as fundamental numbers the following: $\mathcal{N} \approx 10^{36}$ — the ratio of the strength of electrical forces that hold atoms together to the force of gravity; $\epsilon = 0.007$ — the effectivity of the nuclear burning of hydrogen to helium; $\Omega = 0.28$ — the amount of matter in the Universe, in units of the critical density; $\Lambda = 0.72$ — the amount of dark energy in the Universe, also in units of the critical density; $\mathcal{Q} \approx 10^{-5}$ — the ratio of the energy needed to disperse large structures (superclusters) to their internal rest mass energy (mc^2). The final important number is the number of spatial dimensions in our world, $\mathcal{D} = 3$. Martin shows that if these numbers were a bit different from their actual values, our Universe in such form as we know it would be impossible. The question is: Why do these number have exactly the values needed for the existence of our Universe and the life in it, including ourselves?

Martin discussed the possible explanations for the presence of just these values of fundamental numbers. One simple solution is favoured by theologists — this was the will of a Creator. The other possibility is that during the formation of the universe all possible values of these numbers were possible. It is clear that we can live only in an universe where all these numbers take 'proper' values. This leads us again to the concept of the multiverse.

Martin writes that *"the multiverse concept lies within the province of science as a tentative hypothesis. This hypothesis allows to map what questions must be addressed in order to put the concept on a more credible footing"*.

Trimble (2008) describes the status of the multiverse concept as follows: *"The core multiverse concept is that our universe (the 4-dimensional spacetime with which we are or could be connected and all its contents) is one of many, perhaps infinitely many, probably with different values of the constants of nature and other physical differences, which cannot communicate with ours even in principle. Such ensembles are predicted by some versions of inflation, string and M-theory. The anthropic principle is the idea that our universe has (or even must have) the structure, physics, chemistry and all required for me to be writing this and you to be reading it (editors are optional).*

On previous occasions, Martin Rees has said that he has enough confidence in the multiverse to bet his dogs life on it, while Andrei Linde said he would bet his own life. Weinberg concludes his contribution by saying that he has just enough confidence in the multiverse to bet the lives of both Andrei Linde and Martin Rees's dog.

The problem with the multiverse concept is that it cannot be proved nor disproved, because there is no possibility in principle to contact other universes. On the other hand, it reflects the nature of science to try find answers to ultimate questions like: Why we are here?

Now back to our activities. Together with Kofman and Linde we discussed cosmic voids (bubbles) as remnants from inflation (Kofman et al., 1987). Kofman & Shandarin (1988) developed the adhesion model to find the skeleton of the present cosmic web from density peaks in the early universe just after the inflation stage. Calculations show that the seeds of the present cosmic web were created already in the very early Universe. Since the expansion of the Universe was in the early phase very rapid, all perturbations exceeding some scale (about 140 Mpc) were outside the horizon and could not grow or have mutual contact after inflation and before recombination (CMB radiation epoch).

8.2 Structure formation in hot, cold and lambda models

8.2.1 *Initial conditions*

In numerical simulation of the evolution of structure two issues are of crucial importance: (1) the method to evolve the ensemble of particles, and (2) initial conditions for calculations.

In the first numerical simulations of the evolution the authors applied the direct integration of equations of motion under the influence of mutual gravitational interaction of particles. As particles usually galaxies were considered. Pioneering numerical simulations of the evolution of the structure of the Universe using direct integration method were made in the 1970's by Peebles (1970, 1971b, 1974b), Aarseth (1971b,a); Aarseth et al. (1979), Miller (1978) and others. The number of particles was in the interval from ~ 300 to ~ 1000. As to initial conditions, mostly particles were put at random locations with either zero or random initial velocities. Several values of the density parameter of the Universe were used: $\Omega = 1$, and $\Omega \simeq 0.1$ for the present epoch. To compare results of simulations with observations usually the correlation function was used. Also plots of the distribution of particles were compared with similar plots found for actual galaxies. According to the dominant ideas of the early 1970's the first objects to form are globular cluster sized systems (Peebles & Yu, 1970), which by clustering form larger objects, such as galaxies and clusters of galaxies. This hierarchical clustering model is sometimes called the bottom-up scenario.

Yakov Zeldovich and his team in Moscow used a completely different approach to study the evolution of the structure of the Universe. First of all, Zeldovich started from the fact that the primordial matter is a continuous medium — the primordial gas. For this reason the evolution of the medium must be treated as a hydrodynamical problem. Since the early Universe was almost uniform, it is convenient to use the perturbations of coordinates (displacements from the uniform state) and the perturbations of velocities (departures from the Hubble velocities) with respect to the uniform (unperturbed) state. Initial perturbations were small and the evolution was in the linear regime.

Zeldovich found one crucial aspect of the early cosmic evolution. In the linear regime the gravitational instability amplifies a particular combination of coordinate and velocity perturbations. This combination is called the growing mode. In the growing mode the initial displacements in the medium and its initial velocities are proportional to each other. Therefore, it requires only one function to describe both displacements. The other combination of the displacements and velocities comprises the decreasing mode that decays in the course of the evolution and can be neglected.

To evolve the ensemble of particles Zeldovich's team applied the "cloud-in-cell" (CIC) method which makes it possible to study the collective effects of a large number of particles, while suppressing two-body effects (Hockney & Eastwood, 1981). In this case the medium is considered as a continuous one, i.e. a fluid, and particles are used only as markers or test objects to show the evolution of the medium. Particle masses are distributed over a finite volume (cloud), their movement is followed in a mesh. On this mesh Poisson equations are solved using the fast Fourier transform (FFT) with periodic boundary conditions. Since the FFT works very fast, it is possible to use much more particles than in the direct integration method to simulate the evolution of the Universe.

Further Zeldovich (1970) investigated how far the linear theory can be applied. He found that the linear theory can be used also for further stages of the evolution, this approach is called the "Zeldovich approximation". Using this approximation the density and velocity are calculated for a continuous medium, not for a finite number of discrete point masses, as in the CIC method. For practical purposes it is convenient to place test particles in a regular grid, and calculate perturbations for each particle. Perturbations correspond to density waves of different scale and phase, and have a certain power spectrum. Zeldovich assumed that perturbations consist of common motion of photons and baryons. These perturbations conserve entropy and therefore are called "adiabatic".

Initial perturbations are random, thus at any point the perturbation tensor is three-axial, i.e. in one direction the growth of perturbations is faster. The essential aspect of the Zeldovich approximation is the understanding that in this direction the growth goes to the non-linear regime faster and leads to the formation of flat structures — pancakes. This result is independent of the scale of perturbations: both on large and on small scales pancakes do form.

The amplitude of fluctuations on different length scales is described by the power spectrum. The primordial power spectrum is usually assumed to have a power law dependence on scale: $P(k) \sim k^n$, where k is the wavenumber. A popular choice is the scale-invariant spectrum with spectral index $n = 1$, proposed by Zeldovich and Harrison. In this case, fluctuations on different length scales correspond to the same amplitude of fluctuation in the gravitational potential. If the matter is baryonic, as was believed in the early 1970's, then during the radiation dominated era of the evolution small-scale fluctuations are damped due to photon viscosity (Silk damping). For this reason the power spectrum contains only waves larger than a critical length. First large-scale density enhancements grow, and thereafter they fragment into smaller units. This scenario is thus called top-dawn.

The first calculations of the evolution of inhomogeneities in the non-linear regime using the Zeldovich approximation were done by Doroshkevich et al. (1973); Doroshkevich & Shandarin (1973). Tens of thousands of particles were moved according to the growing mode of the Zeldovich approximation until the formation of pancakes, i.e. into a highly non-linear regime. A relatively small set of particles (about two dozen) was selected for checking self consistency of their motion. The particles had 'twins'. The original particles moved as other particles — i.e. according to the Zeldovich approximation. But the trajectories of the twin particles were obtained by integration of their paths in the gravitational field of all particles. Gravitational forces on these twin particles were computed by direct summation of the Newton gravitational forces from all the particles. Comparing the two sets of trajectories of the selected particles with their twins allowed the authors to estimate the accuracy of the Zeldovich approximation. The differences between the two sets of particles were small, which indicated the good accuracy of the Zeldovich approximation.

The distribution of particles shown in Fig. 5.3 was calculated by Sergei Shandarin using only the Zeldovich approximation. The first true N-body simulations (Zeldovich approximation was applied only in the early phase of the evolution) were done by Doroshkevich et al. (1980) in 2-dimensions (64^2 mesh) and by Klypin & Shandarin (1983) in 3-dimensions (32^3 mesh). Both simulations were done using power spectra cut on small scales. At this time the non-baryonic nature of the dark matter was also a serious possibility to consider, and the first natural candidate was massive neutrinos. Massive neutrinos move with very high speed, thus density perturbations on small scales are damped. For this reason simulations of the neutrino dominated Universe were similar to previous simulations of the baryonic matter, where also spectra were cut on small scales. In all simulations the formation of a cellular structure of particles was clearly visible, as seen in Fig. 5.3.

The first simulations in Western countries using the Zeldovich approximation for initial conditions were made by Melott (1983) in 2-dimensions (100^2 mesh), and by Centrella & Melott (1983) in 3-dimensions (32^3 mesh, in each cell up to 27 particles). In both cases the Zeldovich method was applied to calculate initial positions and the evolution. The use of a larger number of particles allowed the evolution of both over-dense and under-dense regions to be better seen. Calculations by Centrella & Melott (1983) confirmed the formation of the cellular structure. Voids at isodensity level $\varrho/\varrho_{\mathrm{mean}} = 0.5$ are isolated from each other, suggesting a "Swiss-cheese" topology. The distribution of particles and their velocities are similar to the distribution found earlier by Doroshkevich et al. (1980) and Klypin & Shandarin (1983).

Frenk et al. (1983) and White et al. (1983) also used the Zeldovich method to determine initial conditions, but applied the direct integration method to calculate the evolution of 1000 particles. The formation of a system of filaments was confirmed. However, as shown by White et al. (1983), the structure forms too late. This contradicts observations, which suggest an early formation of galaxies. In other words, the conventional neutrino-dominated cosmology is not possible. Problems with the neutrino-dominated Universe were discussed earlier by Zeldovich et al. (1982), see Ch. 6. These are the first papers where Efstathiou and his colleagues used the Zeldovich approximation to set initial conditions.

I discussed recently the development of numerical simulation methods with Sergei Shandarin. He wrote to me that cosmologists in the West started to use the Zeldovich approximation only after reading papers of the Zeldovich team cited above, and listening to Sergei's talk at the Erice Summer School in 1981, where Simon White was present. Sergei noticed that Frenk et al. (1983) cited the Klypin & Shandarin (1983) paper several times, but not in the context of setting initial conditions which was actually the core of the Zeldovich method.

I also asked Sergei how he would explain the fact that astronomers in Western countries were more than ten years behind Moscow scientists in the understanding of the evolution of the Universe. Sergei's answer was that one possible reason may be the difference in education. In most American and British universities hydrodynamics is not in the curriculum of astronomy students. In contrast, to be accepted by the physics community in Moscow, every student must take the "Theoretical Minimum" exam conducted by Landau or Lifshitz in person. The basis of the exam was the Landau–Lifshitz series of books covering all aspects of theoretical physics. Our collaborator Lev Kofman also wanted to be accepted into the Moscow community of physicists, and took Lifshitz's exam, in spite of the fact that he had already graduated from Tartu University in theoretical physics.

In connection to this I remember one winter school in the Caucasus in the early 1980's, where Andrei Linde gave an overview of the inflation theory. All major astrophysicists attended, among them Josif Shklovsky. After the talk Shklovsky said that he had not understood anything about what Andrei said. Andrei's answer was simple: *"This is new physics, it is not better nor worse than the old physics, it is just different"*.

8.2.2 *HDM and CDM simulations*

Numerical simulations of structure evolution for both hot and cold dark matter were made by Melott et al. (1983), and by White et al. (1987) (HDM and CDM models with density parameter $\Omega_m = 1$). Both dark matter scenarios were already discussed in Chapters 5 and 6, thus here I repeat only the main results.

In contrast to the HDM model, in the CDM scenario the structure formation starts at an early epoch. Superclusters consist of a network of filaments of DM halos which can be identified with galaxies, groups, and clusters of galaxies, similar to the observed distribution of galaxies. Thus CDM simulations reproduce quite well the observed structure with clusters, filaments and voids, including quantitative characteristics: the correlation function, the percolation or connectivity, and the multiplicity distribution of systems of galaxies (Melott et al., 1983). The Melott *et al.* paper ends with a statement: *"we see here strong support for the structure formation process in an axion-, gravitino-, or photino-dominated universe. Galaxy formation proceeds from collapse of small-scale perturbations, as in the Hierarchical Clustering theory, but large-scale coherent structure forms as in the Adiabatic theory"*.

The CDM numerical simulation by Adrian Melott was basically made to explore the viability of the CDM scenario. The CDM power spectrum was modeled by two power-law segments with $n = 1$ on large scales and $n = -3$ on smaller scales with a sharp bend. In spite of this simplification the model behaved surprisingly well also on small scales. This model and additional models calculated by Adrian and his group with higher resolution were used by our team to compare the observed distribution of galaxies with the model.

An extensive series of numerical simulations was made by Marc Davis, George Efstathiou, Carlos Frenk, and Simon White, nicknamed the "Gang of Four". First Efstathiou et al. (1985) compared various simulation methods: direct integration, particle–mesh (PM), and particle–particle/particle–mesh (P^3M) codes. The last two methods use the CIC technique to calculate the influence of all particles. The P^3M method uses additionally direct integration (particle–particle) to take into account gravitational interactions of nearby particles, thus the resolution of simulations is much higher than in the simple CIC or PM code. The authors used various softening parameters in the calculation of gravitational force, as well various density parameters, $\Omega_m = 1$ and $\Omega_m = 0.2$. The number of particles in simulations was 32^3 in a grid of 64^3 cells in most cases.

Davis et al. (1985) described the evolution of large-scale structure in a CDM dominated Universe, and White et al. (1987) analysed the distribution of clusters, filaments, and voids in a CDM dominated Universe. Several simulation runs were made using a grid with 64^3 cells and 32^3 or 60^3 particles, in cubic boxes of comoving size $L = 280$ and $L = 360\,h^{-1}$ Mpc, respectively. In all cases $\Omega_m = 1$ was assumed. The modeled structures are in good agreement with observed structures. However, the comparison of models with observations was made only in a qualitative manner.

This series of models is considered as the first comprehensive study of the CDM dominated Universe.

8.2.3 Simulations with cosmological constant

A flat cosmological model with $\Omega_{\rm tot} = 1$ is theoretically preferred (arguments which led to the formulation of the inflation theory). On the other hand, direct observational data suggested that the density of matter, including dark matter in galaxies and clusters, yields a value $\Omega_{\rm m} \approx 0.2$ (Einasto et al., 1974b; Ostriker et al., 1974). Thus the reminder must lie in some smoothly distributed background. The most suitable candidate for such a background is the cosmological term, $\Omega_\Lambda = \Omega_{\rm tot} - \Omega_{\rm m} \approx 0.8$. Additional arguments in favour of a flat cosmological model are data on the Hubble constant and on the age of the Universe. A flat cosmological model with Λ-term was discussed by Gunn & Tinsley (1975), Turner et al. (1984), Kofman & Starobinsky (1985), as well as during the Second Tartu Cosmology Seminar (Kofman et al., 1986).

In the 1970's and early 1980's our team compared observation with simulations using models calculated by the Zeldovich team and later by Adrian Melott. In the mid 1980's it was clear that we needed our own models suited for our particular tasks. Our student Mirt Gramann started to write the program for numerical simulations. As a starting point she took the program developed by Anatoly Klypin, the particle–mesh code. Following the suggestion of Enn, Mirt Gramann calculated models with the cosmological term. The first trial calculations in 1984 were done with a small number of particles. Soon Tartu Observatory got its first UNIX computer with 2 MB of core memory, so it was possible to simulate 3-dimensional models with 64^3 particles in a 64^3 mesh. One run took about a month — not too much in those days. Several simulations were made with cube sizes $40\,h^{-1}$ Mpc and $80\,h^{-1}$ Mpc. Some models were calculated also for higher values of the matter density, $\Omega_m = 0.4$ and $\Omega_m = 1.0$, the latter model corresponds to the "standard" CDM model of those days (Gramann, 1987, 1988, 1990). The first application of the model with cosmological constant was in our topology paper by Einasto et al. (1986a).

In 1988 I received an invitation from the Institute of Astronomy of Cambridge University. At this time it was possible to take with me our post-graduate student Mirt Gramann. Our basic goal was to compare her ΛCDM models with other models and observational data. Carlos Frenk, a member of the US–British team modelling the structure evolution, gave us a copy of data of their standard CDM model with density parameter $\Omega_m = 1$ (White et al., 1987). This comparison suggested that the structure of the cosmic web is similar in ΛCDM and standard CDM models. However, in order to get the correct amplitude of density fluctuations, the evolution of the standard CDM model has to be stopped at an earlier epoch (Gramann, 1990). In ΛCDM there are no problems with timing — all model properties fit observational quantities well at just the right time.

We compared our ΛCDM models with observations applying various quantitative methods. The first check was always the correlation function. Additional tests included the connectivity or percolation, the multiplicity of connected systems, the filling factor of both filled and empty regions using various density levels, the void diameter distribution, and the void probability distribution. All quantitative checks suggested that the ΛCDM model represents real galaxy samples very well (Einasto & Einasto, 1989; Gramann, 1990; Einasto et al., 1991). Our ΛCDM model was probably the first one with the presently popular cosmological term.

Independent evidence favouring a CDM model with the cosmological term was found by Efstathiou et al. (1990). Efstathiou made a series of N-body simulations with the "standard" $\Omega = 1$ CDM model, and several spatially flat ΛCDM models with matter density $\Omega_m = 0.2$, for comoving box sizes 50, 150, $300\,h^{-1}$ Mpc. For all models angular correlation functions were calculated, and compared with the correlation functions found for the galaxy survey made with the APM (Automatic Plate Measuring machine in the Institute of Astronomy of Cambridge University).

The authors find that the "standard" model lacks power on large scales, whereas the ΛCDM model fits the observed correlation function well.

An interesting remark. Together with Mirt Gramann we visited the Institute of Astronomy in Cambridge in 1988, 1989, and 1990, and had a lot of discussions with Cambridge astronomers on numerical simulations of the structure. Our comparison of ΛCDM and CDM models with observations was made basically in Cambridge, thus George Efstathiou and his colleagues were aware of our ΛCDM models. Their paper Efstathiou et al. (1990) contains, however, no citations to our papers on this subject.

An essential property of ΛCDM models is an accelerating expansion speed of the Universe. Experimental proof for this effect came from the comparison of distant and nearby supernovae, see below.

8.2.4 Modern cosmological simulations

In the following years the technique to calculate numerical simulations to follow the evolution of the structure of the Universe and to see how galaxies were formed progressed a lot. A detailed description of this development is, however, outside the scope of this review. I only mention one of the largest simulations of the evolution of the structure — *the Millennium Simulation*. It was made in the Max-Planck Institute for Astrophysics in Garching by Volker Springel and collaborators (Springel et al., 2005; Gao et al., 2005a; Springel et al., 2006). The simulation assumes the ΛCDM initial power spectrum. A cube of the comoving size of $500\,h^{-1}$ Mpc was simulated using about 10 billion dark matter particles that allowed the evolution of small-scale features in galaxies to be followed. Using a semi-analytic model the formation and evolution of galaxies was also analysed (Di Matteo et al., 2005; Gao et al., 2005b; Croton et al., 2006). For simulated galaxies photometric properties, masses, luminosities and sizes of principal components (bulge, disk) were found. The comparison of this simulated galaxy catalogue with observations shows that the simulation was very successful.

The results of the Millennium Simulation are frequently used as a starting point for further more detailed simulations of the evolution of single galaxies. We used the Millennium Simulation to construct simulated supercluster catalogues and to compare properties of real and simulated superclusters (Einasto et al., 2007a,c,d). One difference was evident: there are more very rich superclusters than expected from simulations, see Fig. 7.22 (Einasto et al., 2006). One possible explanation for the large difference between the distribution of luminosities of real and simulated samples is that the role of very large density perturbations is underestimated. The other feasible explanation is the presence of some unknown processes in the very early Universe which give rise to the formation of extremely luminous and massive superclusters. The size of the simulation box, $L = 500\,h^{-1}$ Mpc, is, however, not sufficient to conclude which of these possibilities is correct.

One difficulty with the original pancake scenario by Zeldovich was the shape of objects formed during the collapse. It was assumed that forming systems are flat pancake-like objects, whereas the dominant features of the cosmic web are filaments (Einasto et al., 1980a, 1983b). This discrepancy was explained by Bond et al. (1996). They showed that, due to tidal forces, in most cases only essentially one-dimensional structures, i.e. filaments form.

The ΛCDM model of structure formation and evolution combines all essential aspects of the original structure formation models, the pancake and the hierarchical clustering scenario. First structures form at very early epochs soon after recombination in places where the primordial matter has the highest density. This occurs in the central regions of future superclusters. The first objects to form are small dwarf galaxies, which grow by infall of primordial matter and other small galaxies. Thus, soon after the formation of the central galaxy other galaxies fall into the gravitational potential well of the supercluster. These clusters have had many merger events and have "eaten" all its nearby companions. During each merger event the cluster suffers a slight shift of its position. As in central regions of superclusters merger galaxies come from all directions, the cluster settles more and more accurately into the centre of the gravitational well of the supercluster. This explains the fact that very rich clusters have almost no residual motion with respect to the smooth Hubble flow. According to the old paradigm galaxies and clusters form by random hierarchical clustering and could have slow motions only in a very low-density universe — an argument against the presence of a large amount of dark matter by Materne & Tammann (1976). However, the low random velocity of central galaxies of clusters is valid only for the richest clusters near the centres of superclusters, see the discussion above on the morphology of clusters in superclusters.

8.3 The formation and evolution of the cosmic web

Numerical simulations of the evolution of the Universe applying the Zeldovich approach to set initial conditions reproduce the present structure of the cosmic web well. I was interested in understanding how the evolution actually works — How one gets from a random field of density perturbations of various scale the cosmic web with filaments, clusters, superclusters and voids? In other words, What processes are involved in this transformation from a random field to a quasi-regular cosmic web? One of our goals was to understand what causes the formation of superclusters, in particular the quasi-regularity of the location of very rich clusters and superclusters, discussed in the previous Chapter (Einasto et al., 1997a).

First I assumed that the observed quasi-regularity may be caused by a strong feature in the power spectrum of density perturbations. During a visit to Fermilab in 2000 I calculated a number of N-body models with an artificial strong peak at the wavelength $120\,h^{-1}$ Mpc, as found from observations by Einasto et al. (1997a). This attempt failed. Even a narrow peak at this scale, exceeding tenfold the amplitude of the mean spectrum around the peak, did not create the observed quasi-regularity of rich clusters and superclusters. Evidently the reason of this phenomenon must lie somewhere else.

8.3.1 *The luminosity density field of the SDSS*

I discussed the problem with Enn Saar, who had applied various methods to study the density field. We came to the conclusion that we had to start the study with a more detailed analysis of the real luminosity density field. As an example of the real density field we chose the equatorial slice of the Sloan Digital Sky Survey (SDSS). The density field is usually described by the amplitudes of Fourier components of the field, expressed by the power spectrum of density perturbations. However, the spatial structure of the field depends both on the amplitudes and on the phases of the density field. Thus both aspects of the field must be investigated.

To study the role of phase information in more detail, we extracted a 2-dimensional rectangular region of size 512 Mpc, calculated for the Hubble constant $h = 0.8$, from the density field of the SDSS Northern equatorial wedge of 2.5 degrees thickness up to redshift $z = 0.2$. This region is shown in the upper left panel of Fig. 8.1. The observer is located at the lower left corner. The colour-coded density levels used in plotting are in the interval from 0 to 10 in mean density units with white corresponding to the highest value using the SAO Image program DS9 colour palette SLS. Then we Fourier transformed the 2D density field, randomised phases of all Fourier components, and thereafter Fourier transformed it back to see the resulting density field.

The modified field has the same amplitudes of all wave-numbers k as the original field; only the phases of waves are different. The results are shown in the upper right panel of the Fig. 8.1. With shifting phases of density waves some densities become negative, thus in this case colour codes are in the density interval ± 3.5. We see that the whole structure of superclusters, filaments and voids has disappeared; the field is fully covered by tiny randomly spaced density enhancements. There are even no clusters of galaxies in this picture, comparable in luminosity to real clusters of galaxies.

This simple test shows the importance of phase information in the understanding of the structure of the web. Coles & Chiang (2000) came to the same conclusion by randomising phases of a simulated filamentary network.

To study the influence of waves of different scales Enn developed programs where only waves in a specified scale interval are randomised. I tried this program using again the 2D slice of the SDSS Northern equatorial wedge. To our dissatisfaction, we did not find clear evidence of the influence of randomising phases of waves of different scales.

Next we tried the wavelet method. It is well known that the Fourier space is not sensitive to the location of particular high-density features in real space, such as filaments, clusters, and superclusters. To have a better understanding of the texture of the cosmic web, the web must be studied in real space. For this purpose wavelet analysis can be applied, which analyses properties of waves of various scales in real space (Jones, 2009).

In the wavelet analysis the density field is decomposed into several frequency bands as follows. The high-resolution (zero level) density field is calculated with the kernel of width, equal to the size of one cell of the field. Every next field is calculated with the twice larger kernel. Wavelets are found by subtracting higher level density fields from the previous level fields. In such a way each wavelet band contains waves twice longer the previous band, in the range $\pm\sqrt{2}$ times the mean (central) wave (Martínez & Saar, 2002).

Figure 8.1 shows wavelets 7 to 4 of the Sloan Northern rectangular region. These wavelets characterise waves of length about 256, 128, 64, and 32 Mpc, respectively. In wavelet figures both under- and over-densities are shown. Extreme levels were chosen so that mean features of the structure are well visible.

The middle left panel of Fig. 8.1 shows the waves of length about 256 Mpc. In its highest density regions there are three very rich superclusters: N20 from the list by Einasto et al. (2003b), located in the upper part of the Figure, supercluster N13 (SCL126 from the list by Einasto et al. (2001) in the Sloan Great Wall) near the centre, and supercluster N02 (SCL82) in the lower right part of the panel. Supercluster numbers are shown in the middle right panel w6.

The next panel shows waves of about 128 Mpc. Here the most prominent features are superclusters N13 (SCL126) and N02 (SCL82). The supercluster N23 (SCL155) in the upper left part of the panel is also fairly strong, seen as weak density peak already in the previous panel. In addition we see the supercluster N15 just above N13 near the minimum of the wave of the 256 Mpc scale, and a number of poorer superclusters located mostly in voids defined by waves of larger size.

The lower left panel plots waves of scales about 64 Mpc. Here all superclusters seen on larger scales are also visible. A large fraction of density enhancements are

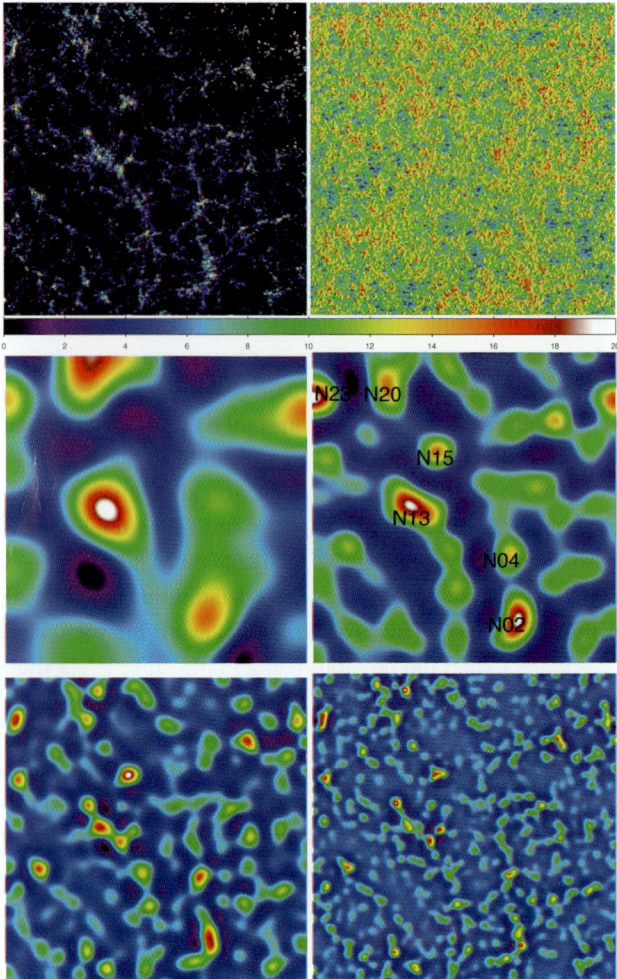

Fig. 8.1 The 2-dimensional rectangular region of size 512×512 Mpc of the luminosity density field of the SDSS. Upper panels show the actual density field, and the field with phases randomly shifted. Middle and lower panels show the wavelets w7, w6, w5, and w4 of the field (Einasto et al., 2011a).

either situated in the middle of low-density regions of the previous panel, or they divide massive superclusters into smaller subunits. This property is repeated in the next panel. Here the highest peaks are substructures of rich superclusters, and there are numerous smaller density enhancements (clusters) between the peaks of the previous panel.

When we compare density waves of all scales, then we come to the conclusion that superclusters form in regions where large density waves of various scales combine in *similar over-density phases*. The larger the scale of the wave where this coincidence takes place, the richer the superclusters are.

Similarly voids form in regions where density waves of medium and large scales combine in *similar under-density phases*. In large voids medium-scale perturbations generate a web of filamentary structures with knots.

This simple analysis demonstrates very clearly the role of phase coupling (synchronisation) of density waves of different scales in the formation of the supercluster–void network.

8.3.2 *The role of density waves of various scales*

Our next goal was to investigate in more detail how structure has formed and what role is played by density perturbations of various scale. I discussed the problem again with Enn. After some thoughts and trial calculations we found that the best way to understand the role of density perturbations of different scales in the formation of the cosmic web is to use models with a varying large-scale cutoff of the power spectrum of initial perturbations. So far the influence of small-scale waves was studied by cutting small-scale perturbations; one such example is the HDM model where small-scale perturbations are absent. Now we decided to do it otherwise. Enn developed a variant of the generation of initial conditions of the N-body simulations, where the amplitude of density perturbations in a given scale interval is put to zero.

Thus we performed several series of simulations. All simulations of a given series had identical initial conditions with 'random' initial positions and velocities of test particles, but the amplitudes of all perturbations on a scale exceeding a given scale were forced to be zero. In this way, we can follow how systems of galaxies grow under the influence of perturbations of various scales.

The first trial models were calculated by Enn already in 2003 with a rather small resolution, with 128^3 particles and cells. As our computer park evolved we made new series with higher resolution. We also used simulations made in Potsdam in collaboration with Volker Müller, and in Tartu University computer center. The last simulations were made by our young collaborator Ivan Suhhonenko. Preliminary results of our calculations were reported at various conferences: "Bernard60" conference in Valencia 2006, void conference in Amsterdam 2006, Zeldovich memorial conference in Minsk 2009. The final results of this study were published only recently by Einasto et al. (2011b,a); Suhhonenko et al. (2011).

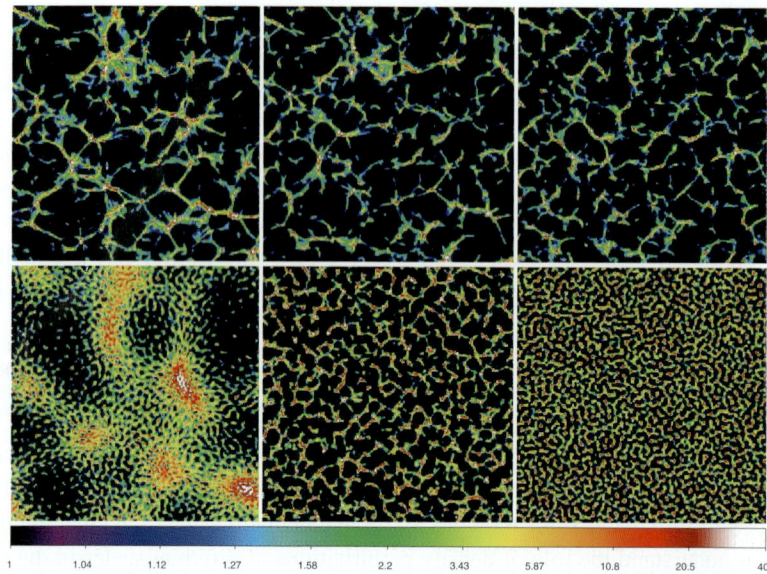

Fig. 8.2 Density fields for the models of the M256 series. The upper panels show the high-resolution fields for the models M256.256, M256.064, and M256.032, the lower panels for the models M256.864, M256.016, and M256.008 (from left to right). The densities are shown for a layer of $6\,h^{-1}$ Mpc thickness at the $k = 75$ coordinate. All fields correspond to the present epoch $z = 0$. The densities are expressed on a logarithmic scale to enhance the low-density regions; only the overdensity regions are shown (Suhhonenko et al., 2011).

To have both a high spatial resolution and the presence of density perturbations in a larger scale interval we used simulations in boxes of sizes varying from $64\,h^{-1}$ Mpc to $768\,h^{-1}$ Mpc, and resolutions $N_{\mathrm{grid}}^3 = 256^3$ and $N_{\mathrm{grid}}^3 = 512^3$; our basic simulations were made in boxes of sizes 100 and $256\,h^{-1}$ Mpc (Einasto et al., 2011b,a). The notations for our models are: the first characters M and L designate models with resolutions of $N_{\mathrm{grid}} = 256$ and $N_{\mathrm{grid}} = 512$, respectively. The following number gives the size of the simulation box, L, in h^{-1} Mpc; the subsequent number indicates the maximum wavelength used in the simulation, also in h^{-1} Mpc. Some simulations were made in boxes of sizes 64, 512, and 768 h^{-1} Mpc, to understand the role of the smallest and the largest density perturbations.

Fig. 8.2 shows the density fields of models of the series with cube length $L = 256\,h^{-1}$ Mpc, M256, at the present epoch $z = 0$. In the model M256.864 the amplitude of perturbations between wavelength $8\,h^{-1}$ Mpc and $64\,h^{-1}$ Mpc has been put to zero, to see the influence (actually the absence) of density perturbations of medium scales.

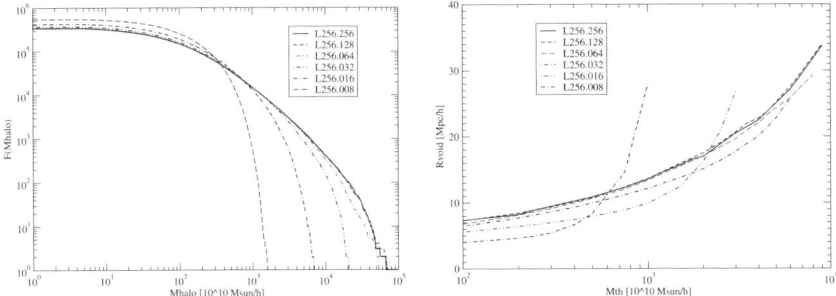

Fig. 8.3 The left panel show the cumulative mass functions of the density field clusters for the models with various cutoff scales. The right panel show the mean radii of voids, defined by the DF clusters for different threshold masses, M_{th}, and for various cutoff scales (Suhhonenko et al., 2011).

The Figure demonstrates very clearly the role of density perturbations of various scales in the formation of the cosmic web. In the absence of large-scale perturbations, systems larger than the cutoff scale do not form. The removing of only the medium-scale perturbations is of particular interest. In such a model there are no filaments, both within superclusters and between them. The distribution of small-scale systems is more or less random, and there are no compact systems of galaxies such as clusters — the compact systems are rather small. Superclusters are present, but they lack the fine structure. This simple exercise shows that for the formation of the cosmic web the combined influence of density waves of all scales is important.

To characterise quantitatively the dependence of the cosmic web on the scale of perturbations Suhhonenko et al. (2011) used two tests, the cumulative mass function of density field clusters and the void radius distribution, as determined by density field (DF) clusters of different mass. Density field clusters were defined as peaks of the high-resolution density field. In the search for maxima of the field we used a minimum density threshold, $D_0 = 2$, and minimal mass, $D_p = 5$. The mass was calculated as a sum of density values within ± 3 cells from the central one; the mass is given in Solar mass units.

The cumulative mass functions of the DF clusters for the models of the series L256 are shown in Fig. 8.3. We see that DF cluster masses strongly depend on the scale of the power spectrum cutoff. All models with a cutoff on the scale $8\,h^{-1}$ Mpc have DF cluster mass distributions with rather sharp decreases on the high mass side. The maximum masses of the DF clusters in these models are $\simeq 1.5 \times 10^{13}\,M_\odot$. With increasing spectrum cutoff scale, the mass distribution rapidly shifts to higher masses. This rapid increase in the maximum DF cluster mass continues up to the cutoff scale of $64\,h^{-1}$ Mpc. This increase stops at the cutoff scale $128\,h^{-1}$ Mpc.

Analysis of models in larger boxes of sizes 512 and 768 h^{-1} Mpc confirms this result.

Next we studied the distribution of radii of voids — regions of space devoid of certain kinds of objects. Different types of objects define voids of different size. Large voids are determined by rich clusters and are crossed by filaments of faint galaxies. Almost all systems of galaxies contain outlying faint members (see Fig. 7.14 for the luminosity density field of a spherical shell of the Sloan Digital Sky Survey). Dwarf galaxies define much smaller voids than giant ones.

To find the distribution of void radii, we used a simple void finder suggested by Einasto et al. (1989). The mean void radii were found for a broad range of the DF cluster mass thresholds, $M_{\rm th}$. This variation imitates the various luminosity limit in real galaxy and cluster samples. The highest threshold used was selected so that the volume density of the DF clusters in the sample is approximately equal to the volume density of the Abell clusters, about 25×10^{-6} $(h^{-1}$ Mpc$)^3$ (Einasto et al., 2006).

The results of our calculations are shown in Fig. 8.3. We see that for a low DF cluster mass threshold, the void radii are almost independent of the mass threshold $M_{\rm th}$. This means that in this DF cluster mass interval the clusters are located in identical filaments. If the DF cluster mass threshold increases further, the void radii start to increase. This means that some filaments are fainter than the respective mass threshold limit, and do not contribute to the void definition. In the case of models L256.008 and L256.016 with a high mass threshold the void radii increase very rapidly until clusters disappear. This effect is due to the very sharp decrease in the number of DF clusters of a high mass. These rare clusters define very large voids which are not characteristic of the overall cosmic web pattern of the particular model.

Fig. 8.3 shows that the R_v versus $M_{\rm th}$ curves for the models M256.032, M256.064, M256.128, and M256.256 are almost identical. This means that the scale of the cosmic web is determined essentially by density perturbations of a scale up to $32\,h^{-1}$ Mpc. Some small differences in the R_v versus $M_{\rm th}$ curves remain between the models M256.032 and M256.064. The higher cutoff models are practically identical in this void size test.

Thus, the void analysis confirms our results from the mass distribution of the DF clusters, that density perturbations of large scales have little effect on the pattern of the cosmic web as characterised by void sizes. This analysis suggests that the cosmic web with filamentary superclusters and voids is formed by the combined action of all perturbations up to the scale $\sim 100\,h^{-1}$ Mpc. The largest perturbations in this range determine the scale of the supercluster–void network. Perturbations of the largest scales $>100\,h^{-1}$ Mpc modulate the richness of galaxy systems from clusters to superclusters, and make voids emptier, but do not change the pattern of the web.

8.3.3 *The phase coupling of density perturbations of various scale*

Our next task was to look at how the phase coupling or synchronisation may arise. To clarify this problem numerical simulations are needed, where the role of density perturbations of various scales can be investigated. Ryden & Gramann (1991) studied the phase shifts in the evolving density field, and showed that small-scale phase information is lost during the evolution. Our aim was to find the evolution of the phases of large and medium scales which determine the basic elements of the cosmic web.

If the hypothesis of primordial Gaussianity is correct, then density waves of different scales began with random and uncorrelated spatial phases. As the density waves evolve, they interact with others in a non-linear way. This interaction leads to the generation of non-random and correlated phases which form a spatial pattern of the present cosmic web. How is this process in action?

To see this process we used again the wavelet decomposition of the density field. Figure 8.4 shows the high-resolution density fields of the full model M256 at four redshifts: $z = 10, 5, 1, 0$. Wavelet decompositions at levels w6, w5, and w4 for the same redshifts are shown in the same Figure. Colour-coding of wavelets at different redshifts is chosen so that a certain colour corresponds approximately to the density level, corrected by the linear growth factor for that redshift. Blue wavelet colours correspond to under-dense regions of density waves, green colours to slightly over-dense regions, and red and white colours to highly over-dense regions.

We see that the pattern of the cosmic web on the scale of the wavelet w6 is almost identical at all redshifts; only the amplitude of the density waves is increasing approximately in proportion to the linear growth factor. This linear growth is expected for density waves of large scales, which are in the linear stage of growth. The pattern of the web of the wavelet w5 changes little, but the growth of the amplitude of density waves is more rapid. The pattern of the wavelet w4 changes much more during the evolution, and the amplitude of density waves increases more rapidly, but essential features remain unchanged, i.e. the locations of high-density peaks and low-density depressions are almost independent of the epoch.

Now let us now compare the evolution of the locations of high-density peaks of wavelets of various scale. Fig. 8.4 shows that at all redshifts high-density peaks of wavelets of medium and large scales almost coincide. In other words, *density*

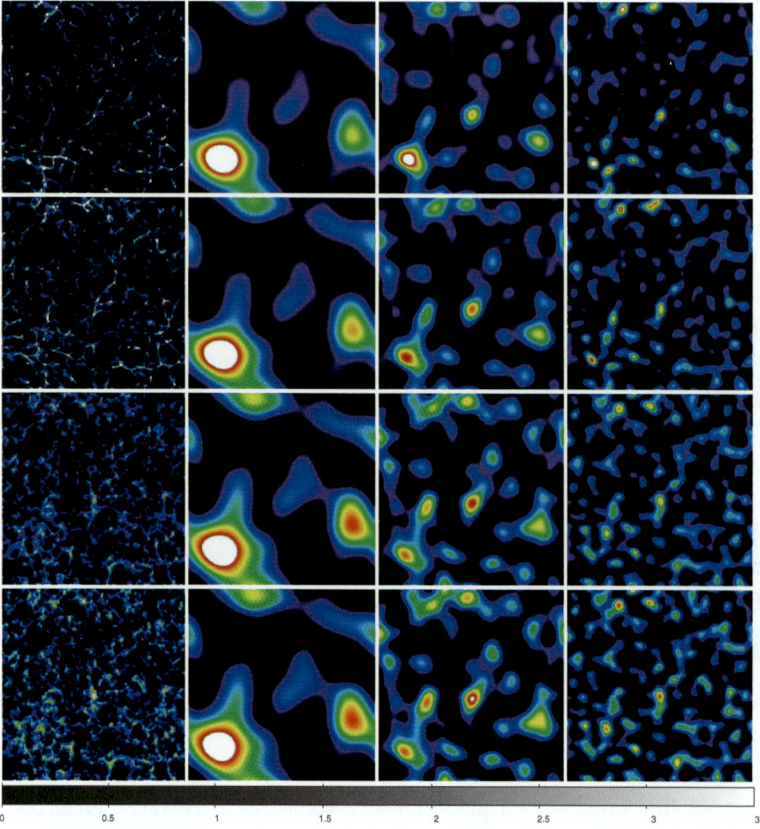

Fig. 8.4 The high-resolution density field of the full model M256 (where density perturbations of all scales are present) is shown in the left column, at $k = 153$ coordinate. The second, third, and fourth columns shows the wavelet w6, w5, and w4 decompositions at the same k, respectively. The upper row gives data for present epoch, $z = 0$, the second row for redshift $z = 1$, the third row for redshift $z = 5$, and the last row for redshift $z = 10$. Densities are expressed in linear scale (Einasto et al., 2011a).

perturbations of medium and large scales have a tendency of phase coupling or synchronisation at peak positions. Figure 8.4 shows that the synchronisation of medium and large scales applies also to underdense regions.

This analysis suggests that the synchronisation of peak positions of wavelets of *various scales* represents a general property of the evolution of the density field of the Universe. How this occurs will be discussed in the next section.

As shown by Kofman & Shandarin (1988), the skeleton of the supercluster–void network is created already at a very early stage of the evolution of the Universe, in the post-inflation phase. Thereafter the structure is frozen as respective waves move out of the horizon. Scales larger than the sound horizon at recombination, ≈ 146 Mpc, were outside the horizon most of the time. This scale ($105\,h^{-1}$ Mpc for the presently accepted Hubble constant $h = 0.72$) is surprisingly close to the characteristic scale of the supercluster–void network (Einasto et al., 1997a, 2001), and to the scale of the baryon acoustic oscillations peak, seen in the correlation function of main galaxies of the Sloan survey by Eisenstein et al. (2005) and of LRG-galaxies by Hütsi (2006), as well as in the distribution of galaxies as demonstrated by Arnalte-Mur et al. (2012).

8.3.4 The fine structure of the cosmic web

So far we looked at how the general pattern of the cosmic web forms. Our next question is: How to explain the evolution of the fine structure of the web, in particular the formation of the filamentary web? It is well known that the structure of the cosmic web is hierarchical. Supervoids are not empty, but are crossed by filaments of galaxies and groups of various strength (Jõeveer & Einasto, 1978; Einasto et al., 1980a; Einasto & Einasto, 1989; van de Weygaert & van Kampen, 1993; Lindner et al., 1995). We have to understand how all this evolves in a natural way. To follow the evolution of the fine structure of the cosmic web we use our high-resolution models of the series L100; for more details see Einasto et al. (2011b).

To follow the evolution of the fine structure of the density field we compared high-resolution density fields of models of the L100 series at various time-steps and wavelength cuts. Fig. 8.5 shows slices at the coordinate $k = 51$ of the full model L100.100, and of the strongly cut model L100.016 with $\lambda_\text{cut} = 16\,h^{-1}$ Mpc, at epochs $z = 0$ and $z = 2$. The k coordinate is chosen so that the slice of the model L100.100 crosses a large under-dense region between a rich supercluster and several rich clusters. To see the differences between the density fields of models L100.100 and L100.016 better we show in Fig. 8.5 only the zoom-in of the central $50 \times 50\,h^{-1}$ Mpc (256×256 pixels) region. To compare the present field with the initial density field, we use the smallest scale wavelet $w1$ for both models at the redshift $z = 10$, shown as a zoom-in plot in Fig. 8.6, which is similar to the plot for the present epoch in Fig. 8.5.

The power spectrum of density perturbations has the highest power at small scales. Accordingly, the influence of small-scale perturbations relative to large-scale perturbations is strongest in the early period of structure evolution. For this reason the density fields and wavelets $w1$ at early epochs are almost identical for the full model L100.100, and for the model cut at small scales, $\lambda_\text{t} = 8\,h^{-1}$ Mpc, L100.008, as seen in Fig. 8.6. Eventually, perturbations of larger scale start to affect the evolution. These perturbations amplify small-scale perturbations near maxima

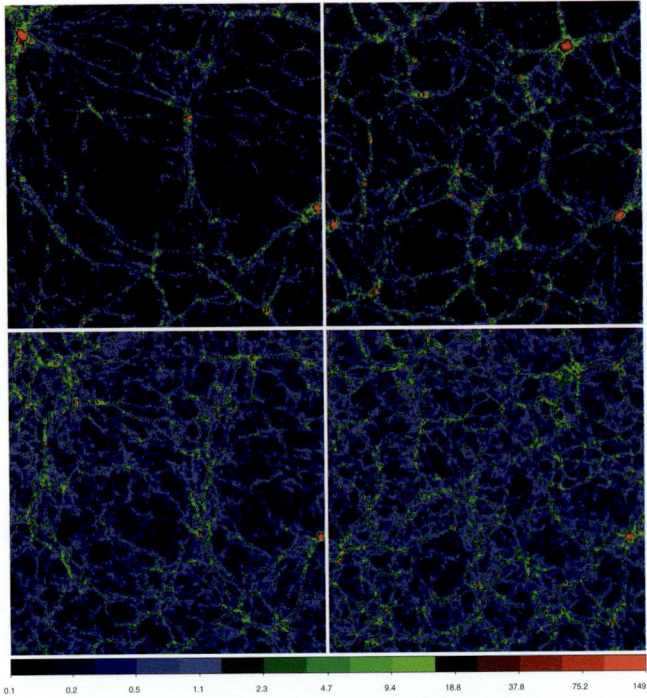

Fig. 8.5 Zoom-ins to the high-resolution density fields of the models L100.100 and L100.016, left and right columns, respectively. Zoom factor is 2; central $50 \times 50\,h^{-1}$ Mpc (256×256 pixels) of all models are shown. Upper panels are for the present epoch $z = 0$, lower panels for the epoch $z = 2$. All panels are at the $k = 51$ coordinate. Cross sections (beams) at coordinates $j = 222$ and $k = 51$ for both models are shown in Fig. 8.7 for three redshifts to see the evolution of the density field and its wavelets. In the upper left corner of the Figure there is a rich supercluster in the model L100.100, absent in the model L100.016. Both models have at the right edge of the Figure a rich cluster. This cluster is well seen in Fig. 8.7 at the $i = 390$ coordinate. Densities are expressed in the logarithmic scale, identical lower and upper limits for plotting with the SAO DS9 package are used. The border between the light blue and the dark green colours corresponds to the critical density $D_{\mathrm{loc}} = 1.6$, which separates low-density halos and halos collapsed during the Hubble time (Kaiser, 1984; Bardeen et al., 1986). Note that in both models and simulation epochs the majority of filaments in voids have densities below the critical density (Einasto et al., 2011b).

and suppress small-scale perturbations near minima. In this way the growth of small-scale perturbations becomes non-linear. Thereafter still larger perturbations amplify smaller perturbations near their maxima, and suppress smaller perturbations near their minima, and so on. The largest amplification (non-linearity) occurs in regions where the maxima of perturbations of all scales happen to coincide. In such a way the synchronisation of phases of waves of different scales occurs as a natural process.

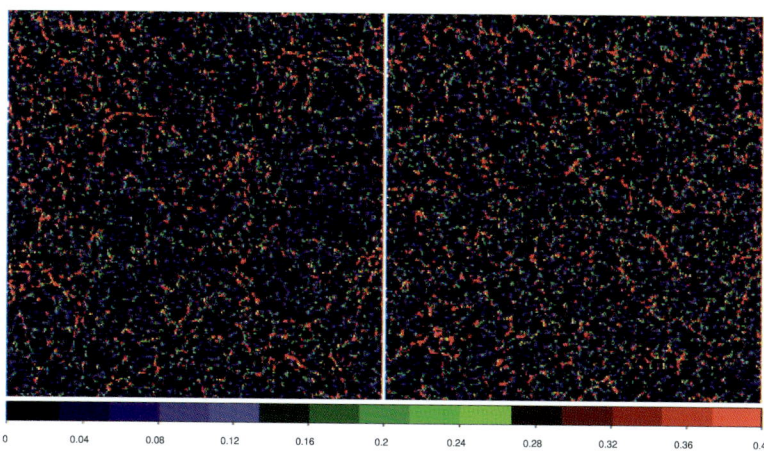

Fig. 8.6 Wavelets w1 of models L100.100 and L100.008 at redshift $z = 10$ are shown in the left and right panels, respectively, at coordinate $k = 51$. Densities are expressed on a linear scale; only over-dense regions are shown. As in Fig. 8.5, the central 256×256 pixels of the full models are shown. The characteristic scale of density perturbations corresponding to this wavelet is $0.4\,h^{-1}$ Mpc. Note the weakening of peak densities of the model L100.100 in the region of the future large under-dense region, seen in Fig. 8.5 (Einasto et al., 2011b).

The differences between the models L100.100 and L100.016 at the present epoch $z = 0$ are very well seen in Fig. 8.5. In the model L100.100, between the supercluster at the left corner and the cluster at the right edge there is a large low-density region. This region is crossed near the centre by a filament having several knots in the green and red colour. In the model L100.016 there are no rich superclusters; the whole region is covered by a web of small-scale filaments. In other words, large-scale perturbations, present in the model L100.100, have suppressed the growth of the density of filaments in void regions.

There exists a low-density smooth background, seen in the Fig. 8.5 in deep dark-blue colour. The density of this background, $D \approx 0.1$, is lower at the present epoch $z = 0$, i.e. the density of the smooth background decreases with time. Regions of very low density have much larger sizes in the model L100.100 than in the model L100.016.

Further evolution of wavelets can be followed using the density field along i-axis beams at fixed j, k coordinates of the model L100, shown in Fig. 8.7. The largest wavelet w5 of the strongly cut model L100.016 has for all redshifts approximately the same amplitude and a sinusoidal shape, suggesting that density perturbations on this scale are in the linear growth regime. The shape of the next wavelet w4 is very different from a sinusoid. In regions of maxima of the wavelet w5 the

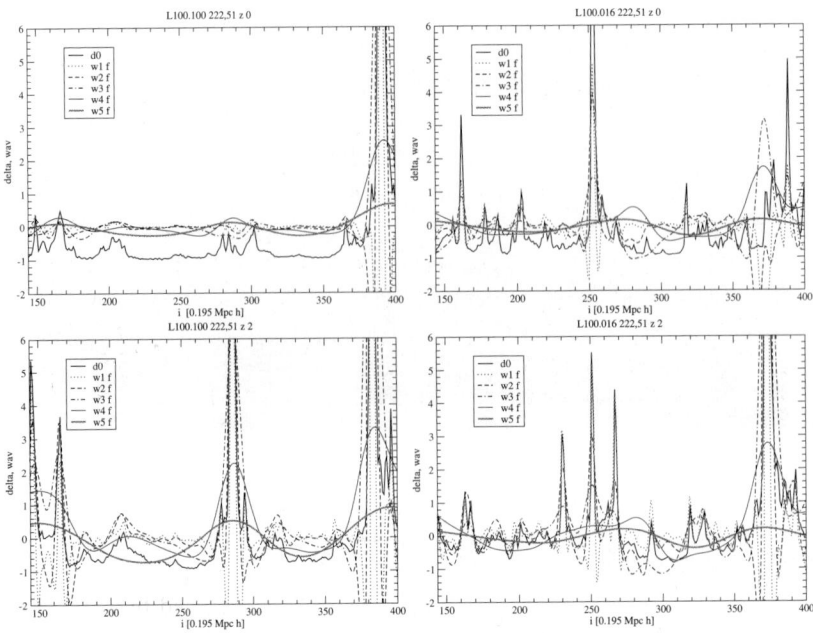

Fig. 8.7 The evolution of the local density and wavelets of the models L100.100 (left panels) and L100.016 (right panels) in beams along the i-coordinate at $j = 220$, $k = 51$. The same k-coordinate was used in plotting the density field in Fig. 8.5. Data are shown for epochs $z = 0$, 2 in the upper and lower panels, respectively. To see better details only the region $144 \leq i \leq 400$ of length $50\,h^{-1}$ Mpc is shown. The characteristic scale of the wavelet w5 is $12.5\,h^{-1}$ Mpc. Wavelets are divided by the growth factor $f \propto (1+z)^{-1}$ (Einasto et al., 2011b).

wavelet w4 has very strong maxima. An example is the region near $i \approx 370$. In this region all wavelets of lower scale also have strong maxima, and wavelets of all scales up to w4 are very well synchronised. The overall shape of the density profile is determined by the wavelet w3 that has a characteristic scale $3.1\,h^{-1}$ Mpc. Most peaks seen in the density profile are due to the maxima of this wavelet. In most of these peaks wavelets of smaller scale also have maxima, i.e. near the peaks small-scale wavelets are synchronised.

The evolution of the largest wavelet w5 of the full model L100.100 is almost linear up to the epoch $z \geq 1$. The shape of the wavelet w5 for different epochs is almost sinusoidal, and the heights of the maxima are approximately equal. The next wavelet w4 behaves as the first overtone of the wavelet w5 — near the minima of w5 there are maxima of w4, which have much lower heights than the maxima near the maxima of w5. This phenomenon is well seen also in the wavelet analysis of the Sloan Digital Sky Survey, see Fig. 8.1. Near the joint maxima of w5 and w4 there are very strong maxima of all wavelets of smaller order; this is very well seen at locations $i \approx 290$ and $i \approx 380$.

The density and wavelet distributions of the model L100.100 for the present epoch $z = 0$ are completely different from the distributions at higher redshifts. Here the dominant feature is the presence of a large under-dense region in the interval $120 < i < 380$. This large under-dense region is due to the influence of large-scale density perturbations not shown as wavelets in this Figure. These large-scale density waves have in this region their minima and have suppressed the amplitudes of all density waves of smaller scales, including w4 and w5. The amplitudes of wavelets w3 and lower orders are almost zero. Near the density maxima seen at higher redshifts at $i \approx 210$ and $i \approx 290$ there are very weak density peaks with maxima below the mean density level. These maxima are seen in the density field as weak filaments in Fig. 8.5.

The most remarkable feature of the density field of the model L100.100 at the present epoch is the presence of a large under-dense region of very low density $D \approx 0.1$, seen in Fig. 8.5 in deep-blue colour. At earlier epochs the density in this region was higher and there were numerous low-density peaks within it; for the present epoch most of these peaks are gone. This is caused by density perturbations of larger scales. The synchronisation of density waves of medium and large scales explains also the absence of even dwarf galaxies in voids, or the "void phenomenon", as discussed by Peebles (2001).

Einasto et al. (2011b) concludes the wavelet analysis as follows: *The wavelet analysis leads us to the conclusion that the properties of the large-scale cosmic web with filaments and voids depend on two connected properties of the evolution of density perturbations. The first property is the synchronisation of density waves of medium and large scales. Due to the synchronisation of density waves of different scales, positive amplitude regions of density waves add together to form rich systems of galaxies, and negative amplitude regions of density waves add together to decrease the mean overall density in voids. The amplification of density perturbations is another property of density evolution. Due to the addition of negative amplitudes of medium and large scale perturbations, there is no possibility for the growth of the initial small-scale positive density peaks in void regions. For this reason, small-scale protohaloes dissolve there. In the absence of medium and large-scale density perturbations, these peaks would contract to form haloes, which would also fill the void regions, i.e. there would be no void phenomenon as observed.*

The analysis by Suhhonenko et al. (2011) showed that density perturbations up to the scale $\sim 100\,h^{-1}$ Mpc determine the scale of the cosmic web in terms of void sizes. In contrast, waves of larger wavelengths do not influence the scale of but only amplify the web, leaving its scale unaffected, see Fig. 8.2 above.

There are probably two effects which affect the evolution of density waves of large scales. As mentioned above, all waves of scale ≥ 140 Mpc were outside the horizon during the early evolution of the Universe and could not grow. For this reason the supercluster–void network has just this characteristic scale. On the other hand, after recombination the amplitudes of density perturbations increased until a certain epoch, which corresponds to redshift ~ 0.7. Thereafter the value of the cosmological constant term exceeded the value of the density term, and the growth of the web was frozen, which hindered the formation of larger systems.

8.4 Dark energy

8.4.1 *The discovery of dark energy*

Observational evidence for the presence of a cosmological term in the mass/energy relation comes from the distant supernova experiments. Two teams, led by Riess et al. (1998, 2007) (High-Z Supernova Search Team) and Perlmutter et al. (1999) (Supernova Cosmology Project), initiated programs to detect distant type Ia supernovae in the early stage of their evolution, and to investigate with large telescopes their properties. These supernovae have an almost constant intrinsic brightness (depending slightly on their evolution). By comparing the luminosities and redshifts of nearby and distant supernovae it is possible to calculate how fast the Universe is expanding at different times. The supernova observations give strong support to the cosmological model with the Λ term.

A more detailed discussion of the discovery of dark energy is outside the scope of this book. Here we discuss only some consequences of the presence of dark energy in the evolution of the Universe.

The cosmological Λ term is presently interpreted as vacuum or dark energy. Dark energy has two important properties: its density ρ_v is constant, i.e. the density does not depend not on time nor on location; and it acts as a repulsive force, i.e. it accelerates the expansion of the Universe.

Studies of the Hubble flow in nearby space, using observations of type Ia supernovae with the Hubble Space Telescope (HST), were carried out by several groups. The major goal of the study was to determine the value of the Hubble constant. As a by-product also the smoothness of the Hubble flow was investigated. In this project supernovae were found up to the redshift (expansion speed) 20 000 km s^{-1}. This project (Sandage et al., 2006) confirmed earlier results that the Hubble flow is very quiet over a range of scales from our Local Supercluster to the most distant objects observed. This smoothness in spite of the inhomogeneous local mass distribution requires a special agent. Dark energy as the solution has been proposed by several authors (Chernin (2001); Baryshev et al. (2001) and others). Sandage emphasises that no viable alternative to dark energy is known at present, thus the quietness of the Hubble flow gives strong support for the existence of dark energy.

8.4.2 *The role of dark energy in the evolution of the Universe*

As noted above, dark energy has two important properties: its density depends neither on time nor on location; and it acts as a repulsive force or antigravity (for detailed discussions see Chernin (2003, 2008)).

The first property means that in an expanding universe in the earlier epoch the density of matter (ordinary + dark matter) exceeded the density of dark energy. As the universe expands the mean density of matter decreases and at a certain epoch the matter density and the absolute value of the dark energy effective gravitating density were equal. This happened at an epoch which corresponds to redshift $z \approx 0.7$. Before this epoch the gravity of matter decelerated the expansion; after this epoch the antigravity of dark energy accelerated the expansion. This is a global phenomenon — it happened for the whole Universe at once.

Dark energy influences also the local dynamics of astronomical bodies. The local effect of dark energy on the dynamics of bodies has been studied by Karachentsev, Chernin, Tully and collaborators. Using the Hubble Space Telescope and large ground-based telescopes Karachentsev determined accurate distances and redshifts of satellite galaxies in the Local group and several nearby groups of galaxies (Karachentsev et al., 2002, 2003, 2007, 2009; Tully et al., 2008). This study shows that near the group centre up to distance $R \sim 1.25\, h^{-1}$ Mpc satellite galaxies have both positive and negative velocities with respect to the group centre; at larger distance all relative velocities are positive and follow the Hubble flow.

This test demonstrates that dark energy influences both the local and the global dynamics of astronomical systems. For rich clusters of galaxies the zero gravity distance is about $10\, h^{-1}$ Mpc; for rich superclusters it is several tens of h^{-1} Mpc, which corresponds to the radius of cores of rich superclusters. The antigravity of dark energy also explains the absence of extremely large superclusters: even the richest superclusters have characteristic radii of about $50\, h^{-1}$ Mpc.

8.4.3 *Cosmological parameters*

Tartu astronomers participated in the determination of cosmological parameters in only a few cases. The first of these cases was the estimation of the total density of matter, including dark matter, by Einasto et al. (1974b). Actually this density estimate concerns only clustered matter in systems of galaxies. By definition, half of the matter was initially in low-density regions; this fraction decreases due to the infall of void matter to systems of galaxies. Presently the fraction of matter in voids is about 25%, depending slightly on the cosmological model (Einasto et al., 1994a). Taking this into account we get for the whole matter density about 0.27 in units of the critical density.

Together with Fernando Atrio-Barandela I made in the late 1990's a lot of calculations with the CMBFAST program (Seljak & Zaldarriaga, 1996) to calculate power spectra for various cosmological models, and to compare the results with observed power spectra of galaxies (Atrio-Barandela et al., 1997; Einasto et al., 1999a,b,c). Our results suggest that, if the neutrino contribution is negligible, then the matter density parameter is about 0.3; values as high as 0.4 and as low as 0.2 are definitely excluded. I had a chance to discuss these results with Mike Turner and Jerry Ostriker. We agreed that the mean value, 0.3, is favoured; it coincides with the density parameter of the 'concordant model' by Ostriker & Steinhardt (1995).

The second estimate is the determination of the Hubble constant using the gravitational lensing effect. Already Fritz Zwicky had suggested that a galaxy may act as a gravitational lens. Sjur Refsdal (1964) emphasised that this effect can be used to estimate the value of the Hubble constant. Suppose that we have at large distance an active object, such as a supernova or a quasar. If a galaxy is located near the line of sight toward the active object, the image of the distant object will be distorted or divided into two separated images, appearing on opposite sides of the galaxy. The path of the light from different images is different, and there is a difference in the arrival time of the light to the observer. As shown by Sjur, this time delay can be used to estimate the value of the Hubble constant, if the redshifts of both the distant object and the intervening galaxy are known.

Quasars are ideal objects to measure the time delay of two images, because their luminosity is not constant, but changes irregularly. By shifting light-curves of both images of the quasar it is possible to determine the time delay due to the gravitational lens. This idea was applied for the double quasar $0957 + 561$. Sjur with his colleagues in Hamburg (Bergedorf) Observatory got a preliminary value of the time delay of about 400 days, which corresponds to a Hubble constant of about 65 km s^{-1} Mpc^{-1}. Another value of about 540 days was found by Press et al. (1992a,b). This result made Sjur anxious, because the corresponding value of the Hubble constant is about 50 km s^{-1} Mpc^{-1}, outside the range allowed by other independent determinations in the early 1990's.

About this time one of the Sjur's collaborators, Tom Schramm, visited Tartu Observatory, and had a seminar talk on gravitational lensing. One Tartu astronomer, Jaan Pelt, has a very good knowledge of mathematics and statistics. So a discussion started: could he help to solve the problem of which of these time delay value is correct? Sjur invited Jaan Pelt to Hamburg Observatory. Jaan with the help of Sjur and his team started a careful comparison of statistical methods applied by Bill Press and other investigators. Jaan developed his own method to analyse data, and compared step-by-step Bill's analysis. Finally the joint Hamburg–Tartu team came to the conclusion that the correct value of the time delay should be close to 415 days. The authors made great efforts to prepare the paper (Pelt et al., 1994) in a suitable form, very clear but not aggressive against the Bill's papers.

This analysis caught the attention of the astronomical community. One young astronomer from Princeton, Tom Kundic, discovered a sharp drop in the light curve of component A of the double quasar $0957+561$ in late 1994 (Kundic et al., 1995). The authors predicted that a similar feature should be observed in the light curve of component B either 415 or 520 days later, depending on the correct value of the time delay.

The feature was detected — it was 415 days later than in image A (Kundic et al., 1997). This delay corresponds to a Hubble constant value 64 ± 14 km s^{-1} Mpc^{-1}, in good agreement with other independent estimates. The Hamburg–Tartu team had arrived at the more accurate value of the time delay! Jaan with his Hamburg colleagues reanalysed new data (Pelt et al., 1996, 1998). Sjur recognised that Jaan had helped to save the prestige of the gravitational lensing method, first suggested by him.

In the 1990's I visited Hamburg Observatory several times, and always Sjur helped me to find a place to stay overnight in the Observatory. I have not studied gravitational lensing myself, so Sjur explained to me in detail their results obtained together with Jaan Pelt.

Modern values of the Hubble constant and density parameters come from a combination of Hubble Space Telescope Key Program results, the CMB, and Sloan Digital Sky Survey observations. The mean values from this dataset are: $H_0 = 70.2 \pm 1.4$, $\Omega_b = 0.0458 \pm 0.0016$, $\Omega_{CDM} = 0.229 \pm 0.015$, $\Omega_\Lambda = 0.725 \pm 0.016$.

On March 21, 2013 new results obtained with the Planck satellite were announced. The Planck data suggest a small revision to the previously accepted set of cosmological parameters. According to Planck Collaboration et al. (2013) cosmological parameters have the following values: $H_0 = 67.4 \pm 1.4$, $\Omega_b = 0.0490 \pm 0.0004$, $\Omega_m = 0.314 \pm 0.020$, $\Omega_\Lambda = 0.686 \pm 0.020$. The reason for the discrepancy with previous results is not yet clear. The age of the Universe according to new Planck data is 13.813 ± 0.058 Gyr.

Using the Sloan Digital Sky Survey main galaxy sample, Tempel et al. (2011) derived for the mean luminosity density $0.01526 \times 10^{10} h L_\odot$ Mpc^{-3}. Mass-to-luminosity ratios in visible regions depend slightly on the morphological type of galaxies, and are in the range of 4–6 in solar units, including the contribution from dark matter. M/L of baryonic matter in visible galactic regions is about 3. From these data it follows that the mean density of baryonic matter in galaxies is about 0.5 % of the critical density.

8.4.4 New cosmology paradigm is ready: What next?

During the last 50 years the astronomy community has been witness to several paradigm changes. The most important paradigm changes are:

- Cosmic microwave background (CMB) radiation was detected which confirmed the paradigm of hot big bang cosmology.
- Most of the matter in the Universe is dark and consists of weakly interacting non-baryonic particles; the density of dark matter is about 0.25 of the critical cosmological density.
- The early evolution of the Universe includes a period of very rapid expansion or inflation which made the Universe homogeneous and its density equal to the critical density. The latest version of inflation theory suggests an initial chaotic stage within a multiverse.
- The Universe has structure in the form of a cosmic web. The seeds of the cosmic web were created at the very early stages of the evolution and give us information on the properties of Universe at this epoch.
- Most of the matter/energy content is the dark energy, about 0.7 of the critical density. Dark energy causes an accelerating expansion of the Universe.

Can we say that the modern cosmological paradigm is now complete? Probably not. The modern paradigm explains almost all observational facts, but there are still a few important problems open. As suggested by Peebles (2002), the situation now has some similarity with the situation in physics at the eve of the 20th century. A hundred years ago almost all experimental data were explained by classical 19th century physics. But there were some clouds on the horizon — the Michelson experiment which suggested the isotropy of the velocity of light, and difficulties in the theory of gases. To explain these clouds the whole of modern 20th century physics evolved: the theory of relativity, quantum physics, and much more.

As Peebles suggests, cosmology has now also several clouds on the horizon: What is the nature of dark matter and dark energy? Which physical processes created our Universe such as it is? Or in other words, what is the physics of the multiverse?

These questions concern both the particle physics and the cosmology of the very early Universe. In these domains scientists deal with energies which are outside the possibilities of present-day accelerators. So far the only information from these early events comes from astronomical data, in particular, indirectly from the structure of the cosmic web on large scales. Recent reviews on astrophysical probes of dark matter are given by Arneodo (2013) and Profumo (2013).

To measure the effects of dark energy, among other projects, the Dark Energy Survey (DES) has been initiated. The Survey aims to probe the dynamics of the expansion of the Universe and the growth of large scale structure. It is a collaboration of research institutes and universities from the United States, Brazil, the United Kingdom, Germany and Spain. The Survey will use the 4-meter Blanco Telescope at Cerro Tololo Inter-American Observatory in Chile. The main instrument is the DECam camera, which covers a field of view of diameter 2.2 degrees, and has 62 2048×4096 pixel CCDs, and g, r, i, z, y filters similar to the SDSS survey. Starting in September 2012 and continuing five years, the Survey covers a 5000 square degree field in the Southern sky, and has a depth of 24th magnitude in the i-band. It is expected that the Survey will get a sample of 300 million galaxies with photometric redshifts up to redshift $z \sim 1.4$. The Survey's principal aims are to measure BAO, to detect about 170,000 galaxy clusters, to investigate the large-scale distribution of galaxies at different redshifts, and to detect supernovas at high redshifts.

Such large surveys have some similarity with high-energy physics experiments, where very large teams of researchers, instrument designers and programmers are involved. Such large collectives have working cultures different from traditional astronomical cultures. White (2007) discussed in detail the differences between traditional astronomical and high-energy physics cultures. He emphasises that in doing such large projects one must not forget the major goals of astronomy: the understanding of the Universe as a complex system.

8.5 Remembering contacts with colleagues

8.5.1 *Encounters with astronomers from other centres*

My contacts with astronomers from other scientific centres started when I visited the Sternberg Institute in Moscow in 1948. These visits became more regular as Prof. Pavel Parenago from the Sternberg Institute was at the time the leader in galactic studies. With Sternberg Institute astronomers I maintained good contacts over the years. In my first visit to the Institute an older professor asked me to send his best greetings to Ernst Karlovich, this is the style in which Russian people call each other. He did not know that Ernst Öpik had left Estonia in 1944. I have some joint publications with my Sternberg colleagues. For my 80th birthday I got from Sternberg astronomers many books, one with the signatures of several tens of people — my old friends.

The next step was to visit the Leningrad University Astronomical Observatory and the Pulkovo Observatory. This resulted in good contacts with Prof. Kirill Ogorodnikov and his stellar dynamics group — Tatheus Agekian and younger colleagues. Our stellar astrophysics group collaborates very closely with the theoretical astrophysics group lead by Prof. V. V. Sobolev. Our previous collaborator Sergei Kutuzov, a student of Grigori Kuzmin, is presently working in the St. Petersburg Astronomical Observatory as Professor and head of the Department of Space Technologies and Applied Astrodynamics. With Sergei we developed the methods of models of galaxies using various data on galactic populations. He was born near Leningrad, but during WWII the town he lived in was occupied by the German army, and the family was evacuated to Tartu, so he got his education at Tartu University. The last time we met was in Pulkovo at a conference organised by St. Petersburg astronomers in honour of Grigori Kuzmin's 90th birthday.

Pulkovo Observatory is actually a younger sister of Tartu University Observatory. It was planned by Friedrich Georg Wilhelm Struve and, when finished, Struve moved with all his staff to Pulkovo. From Pulkovo the development of modern astronomy in the Russian Empire spread to other centres. With Prof. Aleksandr Mikhailov, director of the Pulkovo Observatory, I had many communications. Once during a visit to Pulkovo he invited me to his apartment in the Observatory building. We had a very interesting discussion on the history of astronomy, and he showed me his collection of historical cameras. For each camera he had a story: who built it and what role the camera had in the history of photography. In April 2011 Tartu University Observatory celebrated its 200th anniversary with a small conference. One of the main speakers was Prof. Viktor Abalakin, a former director of Pulkovo, who gave a very interesting overview on the history of both observatories.

My next visit was to the Abastumani Observatory in Georgia, where I spent my observational practicum in 1951. The supervisor of the practicum was Prof. Evgeny Kharadze, the director of the Observatory and the President of the Georgian Academy of Sciences. With him I became good friends. After the death of Pavel Parenago he was elected the Chairman of the Committee of Galactic Studies of the Astronomical Council of the USSR Academy of Sciences. After some years he suggested me as the Chairman of the Committee. Quite often we met either in Abastumani or in his office in the Academy in Tbilisi, discussing issues of Galactic studies.

With the Byurakan Observatory and Viktor Ambartsumian our contacts started even earlier. Aksel Kipper invited Ambartsumian to Tartu in 1948 to discuss plans of the building of a new observatory outside the town. At this time I was a student and happened to be present in these discussions. When the new observatory in Tõravere was officially opened in 1964, Ambartsumian was one of our main guests. For many years he was member of the Science Council of our Observatory. My own visits to Byurakan started in the early 1950's; the last visit was in September 2012 to receive the Viktor Ambartsumian Prize. We made jointly observations of galaxies in the late 1970's, and discussed the development of astronomy in the 1980's.

With foreign astronomers I had first contact in 1958, when the IAU General Assembly was held in Moscow. At this time I had no interesting results to discuss with other astronomers. But I met most influential astronomers whose work I had studied: Jan Henrik Oort, Fritz Zwicky, Harlow Shapley, Bengt Strömgren, Bertil Lindblad and many others. Very important was the participation of Grigori Kuzmin. He met in Moscow George Contopolous, a Greek astronomer, who had independently found the third integral of motions in stellar dynamics. They started an exchange of letters. Kuzmin sent him reprints of all his papers, and Contopolous acquainted Kuzmin's results to other astronomers in his publications and talks.

My first visit to a foreign country was in 1959 to Hungary. At this time I was the head of the Tartu University Satellite Tracking Station. The Astronomical Council coordinated the actions of all these stations, including stations located in 'Socialist Countries'. My trip was a few years after the 'Hungarian Revolution', thus quite soon we started to discuss these events with Hungarian astronomers. I made no secret of the attitude of the Estonian public of these events, and on the repression of the Revolution by the Soviet army. Hungarian astronomers told me a lot of details that I did not know before.

As an inspector of satellite stations I visited in 1962 the German Democratic Republic. Here I visited all the main astronomical institutions, including the Tautenburg Observatory, Jena University, Potsdam Institute of Astrophysics, and many other places of interest. Here astronomers told me the story of the 1953 uprising in East Germany, violently suppressed by Soviet tanks. I also saw the Berlin Wall, built just a year before my visit.

In 1967 the General Assembly of the IAU took place in Prague, and I was happy to be included in the Soviet delegation. Here I had a chance to discuss with Margaret Burbidge the structure of galaxies. I had interesting discussions with Luboŝ Perek, the Czech astronomer who had studied the modelling of galaxies. Soon he became the President of the IAU Commission 33 on the Structure and Dynamics of the Galaxy, and the Secretary General of IAU.

In Prague I met for the first time Alar Toomre, the MIT astronomer of Estonian origin. He is an admirer of Grigori Kuzmin's work, in fact he continued the study where Kuzmin has stopped, by using new possibilities opened up by fast-improving computers to simulate the evolution of galaxies. Together with his brother Juri he conducted the first computer simulations of galaxy mergers. They found that galaxies like the Antennae Galaxies (NGC 4038/4039) are actually colliding galaxies; the computer simulation reproduces very well the structure of the object. This process of the collision evolution is known as the Toomre sequence.

After the General Assembly the Soviet delegation had a tour through Czechoslovakia. We visited Karlovy Vary and other places of interest in the Western part of the country. In hotels I avoided speaking Russian; for me it was difficult to speak in the language of the occupying country. The administrator of one hotel asked me why I do not speak Russian as I am one of the Soviet delegates. What could I answer? Next year, after the Czech Spring and its suppression, I remembered this incident. Then I would be ready to answer his question. But I got from several Czech astronomers new-year cards. In the card the spectrum of a Nova was seen, taken in the night 20–21 August, the night when Soviet tanks entered Prague. The explanation was: *"Nova has reached a stadium of nebulous stage, the direction of the further evolution is unknown"*. Everything was clear!

During the IAU General Assembly in 1970 in Brighton I had a chance to talk to Luboŝ Perek again. I thanked him for the card, sent a few years ago, and explained why I was not able to thank him by mail earlier — all my mail was carefully controlled by 'competent organs'. Our astronomy community followed the 'Czech events' very carefully. Estonia had our own 'Spring' in the mid 1960's. After the suppression of the Czech Spring it was clear that we also had ahead years of suppression and stagnation. Recently one colleague of mine brought back to my memory that after the 'Czech events' I said *"Empires are not forever, perhaps next time we succeed"*.

My first report in an international conference was in 1969 in Basel at an IAU Symposium on the structure of the Galaxy. Next year there was the IAU General Assembly in Brighton. I also had a short report on galactic models in the Commission 33 Meeting. This time I was there together with Prof. Aksel Kipper, and we had a lot of time to discuss science, in particular our local problems in the Observatory. There were internal tensions in the Observatory, it was a fight between different groups. Kipper told me how to finish such fights: *"you have to separate both sides so that they cannot disturb each other, and do not accuse or justify any of the sides"*. Indeed, when we got back home Kipper acted according to these rules, and after some time the atmosphere in the Observatory was again normal.

In 1972 I got an invitation from George Contopolous to give an invited review lecture on galactic models at the First European Astronomy Meeting, to be held in Athens in September. I had calculated models for most Local Group galaxies and the giant elliptical galaxy M87 in the Virgo cluster, so I had a lot of stuff to speak on. So far the explanation of flat rotation curves of galaxies was difficult, since no stellar population could have properties needed to explain the rotation. I discussed the problem with Enn Saar, and he suggested abandoning the idea that only normal stellar populations exist in galaxies. As explained earlier, this assumption means that a population with unknown properties must be present in galaxies.

The next IAU General Assembly I participated was held in Grenoble, France, in 1976. The President of the Commission 33 Luboŝ Perek asked me to organise a session of the Commission to discuss dark matter, in particular its possible stellar origin. But I had a problem. I was member of a Soviet delegation, which this time was formed through Intourist, a Soviet organisation that manages foreign visits of Soviet citizens and foreign visitors in the USSR. The organisation was controlled by the KGB, and there were strict rules for participants of the delegation. The visit was made according to the standard itinerary of tours to France, which meant that most of time was allocated to Paris. To attend the IAU General Assembly, only a few days were reserved. The second rule for the Soviet delegates was — everybody must always be together with the rest of the delegation. The session where I had the talk on dark matter was outside the time-slot allocated for the Soviet delegation in Grenoble. Thus I contacted Luboŝ Perek and explained to him my problem. Perek understood the problem, and rearranged the timetable of Commission 33 Meetings, so that the session with my talk was shifted to the time-slot where I could participate in the session. Perhaps the most import result of this talk was that at least some astronomers started to understand that the corona is a new population with strange properties, not just an extended halo of faint stars. The other result of this visit was the meeting of the Scientific Organising Committee of the IAU Symposium on Large Scale Structure, to be held next year in Tallinn.

Surely the most important international astronomical conference in Estonia was the 1977 Symposium, discussed elsewhere. After the Symposium we got a number of invitations to visit observatories outside the Soviet Union. During the preparations for the Symposium we had frequent contacts with the Foreign Affairs Office of the USSR Academy of Sciences. We discovered that the head of the division of the Office responsible for handling the reception of foreigners was an Estonian, a very friendly man. He gave us advice on how to prepare applications for foreign visits in such a way that there was more chance of success. First of all, it was possible to send applications from the Estonian Academy directly to the Office, which meant that the Astronomical Council as a filter is eliminated. Second, we had to follow very accurately all the stupid rules for foreign visits. Then we have in reality much more freedom. One of important rules was: we must follow exactly the dates of the visit. Actually we needed visas not only from countries we planned to visit, but also permissions from Soviet 'competent organs' to visit a particular country for a fixed period. An additional rule was, soon after the visit we had to give a detailed report on the results of the visit, with recommendations on how to use the results of the visit to improve the Soviet science system.

The first person who used our new 'system' for visit applications was Enn Saar. He had received an invitation from Beatrice Tinsley, who was Director of the Astronomy Department at Yale University. I had with Beatrice already many years of mail exchange. She was the referee of one of my papers on the star formation rate in the Andromeda galaxy. I was very interested in details of her galaxy evolution calculations, and she sent me her PhD Thesis, which contained much more technical information than the published paper in Astrophysical Journal.

Enn had a very successful half-year visit to Yale. This was just the time when Bob Kirshner and his collaborators discovered the Bootes void. When Enn saw their results he told them that we had known and had described such big voids already several years ago (Jõeveer et al., 1977; Jõeveer & Einasto, 1978; Jõeveer et al., 1978). One of the large voids we found is located behind the Perseus–Pisces supercluster. To find a good title for the paper Yale authors even asked advice from an advertising company, who suggested "A million cubic megaparsec void in Bootes" — a really striking title. From now on voids were taken seriously by the astronomical community.

But Enn made a mistake. A conference on the structure of galaxies and the Universe was scheduled to place just after the period fixed for his visit, but Enn wanted to attend the conference. So he applied an extension of the USA visa, but forgot to ask an extension from Soviet authorities. After his return he was accused of violating important rules for foreign visits, and was banned from future visits. The ban was canceled only 7 years later.

I had also an accident in one of my visits. When I was a guest of the Astronomy Department of UCLA all my documents were stolen. I kept them in a wallet which was in my bag in my office. I went out from the office for only a minute. When I returned I discovered that my bag was open and the wallet and a pocket computer were gone. I purchased a new computer next day. I also searched together with George Abell all rubbish bins nearby in the hope that the thief might throw away documents he did not need, but futilely. Then I phoned the Estonian Academy and asked advice on what to do. I got an answer only a few days later. As I was told after my return, Academy officials initially suggested that I must return immediately, but then realised that it is more useful for me to continue the visit as planned. So they told me that the Soviet consulate in San Francisco is informed and shall give me a new passport. I had to go there and write an application with an explanation how my passport was stolen. After my return I had to explain the story to the Office of Foreign Affairs of the Academy in Moscow. And again in Tallinn to the Office of the Estonian Academy. I also got a ban for foreign visits. The President of the

Estonian Academy asked to cancel the ban next year, so it was not so severe as in the case of Enn's visit. But from now on I had to describe in detail the story whenever I wrote a new application for a foreign visit.

My visits to Cambridge in 1980 and ESO in 1981 and 1985 are described elsewhere. But I would like to add a few words on some later visits. In summer 1988 a workshop on cosmology was held in Krakow, and a fairly large group of Tartu astronomers attended. We all got conference badges. I cannot remember who was the first to cross out the word 'USSR' and replace it with 'Estonia', but then we all changed our country name. One of the participants was our good friend Gustav Andreas Tammann. Many years later in one of our meetings he told me that he was very surprised at our courage in doing this. In all subsequent conferences we attended we did the same.

In summer 1988 we had a visitor from USA: Joel Primack with his wife Nancy Abrams got permission to come to Tartu. The situation was a bit freer than in previous years, and they had the chance to stay in the Observatory for several days. We had a lot of time to discuss various topics from cosmology to the political situation in Estonia. His wife is a social scientist and a musician who likes singing. So we had in the Observatory a small concert where she among other songs performed a new song dedicated to our freedom movement. I drove them back to Tallinn, and during the trip we had time to exchange impressions on the visit. Nancy said that she had the possibility of meeting our social scientists. In her opinion our politicians were not aware how policy making worked, thus their activity looked like an amateur enterprise. Perhaps she was right. But, on the other hand, our situation was very unique, and the fall of an empire like the Soviet Union was completely different from the fall of other empires. Thus action in the style of trial-and-error was probably the only way to proceed.

Next year there was in Heidelberg an IAU Symposium on galaxy dynamics, and Enn and I had the chance to attend. One of the participants was Ivan King. During a break we had time to discuss cosmology and more general topics. Ivan praised the Soviet Union for the perestroika policy. Then I asked: *"What do you think, who is the actual initiator of perestroika?"* Ivan replied: *"Of course Gorbachev"*. Then I answered: *"I think that the actual initiator was Ronald Reagan. He pushed Soviet Union to the corner with his Star Wars program, what Russians could not follow. So an economic collapse followed, and the only way out was to change the policy of Cold War."* Ivan thought a bit and then agreed. Now this point of view is well recognised, but at the time not; it seems to me that I came to this conclusion independently.

In 1989 I had an invitation to visit the Institute of Astronomy in Cambridge. Together with me were several of my colleagues: Enn Saar, Ants Kaasik, and

Mikhail Babadzanjants from Leningrad University Observatory. At this time events in the Baltic countries were quite often in the headlines of newspapers, and often I was asked what was actually happening. Then I decided to answer these questions openly in a seminar talk. So I did, and slides of the talk *What Is Happening in Baltic Countries* are still in my archive. I said that this was not a problem of Baltic countries, but the whole Soviet Empire had started to collapse. I gave an overview of the thousand year history of the Russian Empire. The Russian Empire was the successor of the Roman and Mongolian Empires. The Mongolian Empire had as its goals to expand its territory "to the last ocean". The Russian Empire got first access to the Arctic Ocean near the White Sea. The next goals were to get access to the Baltic Sea and the Pacific Ocean. These attempts succeeded when Peter I won The Great Northern War, and Cossacks arrived in the Pacific. After the Russo–Turk wars Russia got access to the Black Sea. After World War II the Soviet Empire consisted of many layers like an onion: the central core was the Russian Federation, the next layer consisted of Soviet Republics, then came 'Socialist Countries' in East Europe, followed by 'friendly' countries like China, North Korea, Vietnam and Cuba. The last layer was formed by some African and Asian countries which were economically dependent on the USSR. Soviet leaders tried to control all these countries, but this was impossible, and now the disintegration of the Empire had started, similar to the disintegration of European Empires after WWI and II. The process was slow and painful since the Soviet Empire was a continental empire, which is easier to control than empires that consisted of a European core and overseas colonies.

To my surprise one of the listeners asked the question: *Why do you want to secede when we here build up a joint Europe?* To give a fair answer I should describe the terror regime of the Soviet System, with massive killings, deportations etc. But at this time I was not ready to do this. I just described succinctly what 'evil empire' actually means. In contrast, representatives of small nations, such as Danes, Dutch, Finns and Swedes, understood our problems much better.

In autumn 1989 the Australian Academy of Sciences arranged a joint Australia–Soviet seminar on cosmology. In the invitation Australian astronomers specially asked that observational cosmologists such as our Tartu team should participate in the seminar. So Enn and I were included in the Soviet delegation. We had the possibility of visiting all Australian astronomical facilities, both optical and radio observatories. At the end of the visit, Australian astronomers arranged an official reception, where the Diplomatic Corp was also invited. An intensive discussion started. Around Enn and me many participants gathered, they wanted to know what is happening in our countries. The most detailed questions were asked by an elderly gentleman. Finally he asked directly: *What you actually want?* Enn answered succinctly: *Freedom*. Later we were told that this gentleman was the USSR Ambassador in Australia.

Next day the head of the Soviet delegation, Igor Novikov, and I were invited to the Embassy of USSR. Here again the main questions were on what news there was from Moscow and the Baltic countries. Novikov spoke of the latest developments in Moscow. But most questions were addressed to me. The Embassy staff was very educated; they asked matter-of-fact questions, in particular questions about the culture, education and traditions in Estonia. I explained that one of our goals was to develop scientific terms in practically every field, since education in schools and in the university was in Estonian. Then one of the Embassy staff members, a Ukrainian, said that in Ukraine higher education was in Russian because the development of all terms in Ukrainian was too costly and time-consuming. My answer was, this was really a problem in Estonia too, but our own terminology is an essential feature of our culture, so we do this.

In the late 1980's when I was at the ESO, Sir Fred Hoyle had a talk on the origin of life on Earth. In his opinion the seeds for life can come from space where numerous organic molecules form in gaseous nebulae: the panspermia hypothesis. In his opinion *"The notion that not only the biopolymer but the operating program of a living cell could be arrived at by chance in a primordial organic soup here on the Earth is evidently nonsense of a high order"*. After the talk ESO Director Harry van der Laan invited the speaker and some astronomers to a dinner. I was happy to be one of astronomers to be invited. After some time Fred started to ask for news on the independence movement in Estonia. Some years later, when Estonia was already free, we met again in Cambridge. We looked back on our earlier discussion, and I explained all the difficulties we had during this process. But luckily everything happened quietly without bloodshed. I am not a supporter of his Steady State Universe and panspermia hypotheses, but he was an extremely interesting and inspiring astronomer. A few years later he together with his coauthors Geoff Burbidge and Jayant Narlikar revised their Steady State model taking into account new observational facts.

About ten years later I got an invitation from the Director of the Inter-University Centre for Astronomy and Astrophysics, Prof. Jayant Narlikar, to attend a small conference in Calcutta, and to visit thereafter the Centre in Pune University. So far I had had no chance to visit India, so I was happy to accept the invitation. The flight itself was interesting. Tallinn airport had at the time direct flights to Vienna, and from Vienna there was a direct flight to New Delhi, so I used this route. All local costs were covered by the Centre, so I was met at the New Delhi airport, a taxi brought me to a hotel, and very early in the next morning back to the airport. I had a first class ticket. The flight to Calcutta itself was very beautiful — to the left the whole range of Himalayas was seen on the horizon. Flights of bigger airplanes in India are scheduled for night, early morning or late evening hours. When I asked about this, it was explained to me that the runways are in daytime too hot for heavy planes to land or takeoff.

Indian cities were completely different from everything I had seen before. Streets were full of people, cars, bicycle and auto rickshaws. In every corner I saw tiny shops for food or small items; the noise was earsplitting. Cars looked like those in Europe some 50 years ago, and almost all were rather broken. I was warned that to eat on such street places can be hazardous, because Europeans are not accustomed to Indian micro-organisms. The food for us was served in the guesthouse — a large apartment was used for this purpose. The food itself was basically vegetarian and very tasty.

Among foreign participants invited to the small conference were Geoffrey Burbidge, Halton Arp and me. How I was included into this coterie I do not know. All of us had talks in the conference; Geoff was also invited to give a public lecture at the Birla Planetarium. One talk was Geoff's presentation of his picture on the structure and evolution of the Universe. My understanding of these phenomena is different, but it is always interesting to listen to arguments from the other side.

The trip from Calcutta to Pune was also interesting. The flight Calcutta–Mumbai was again very early in the morning, first class as in previous flight. Next was a trip by SUV to Pune, which is about a half day's drive away. It was interesting to see — everywhere in India we saw building works, new houses, roads, airports that were rather modern. The Astronomy Center was built in 1992 following the initiative of Jayant Narlikar. The Center is located in the Pune University park, at a very quiet location. Here we had plenty of time for discussions in a relaxed atmosphere. I discovered that Geoff Burbidge is an excellent story-teller. He told us how the organisation of science in USA works, how university committees function etc.

Of special interest to me were his recollections of galaxy rotation measurements he made together with his wife, Margaret Burbidge. The actual observer was Margaret; Geoff was assisting. The first thing to do was to find a powerful spectrograph. One unused spectrograph was found in a cellar of an observatory, and a lot of effort was needed to get it mounted at the prime focus of the 82-inch McDonald Observatory in Texas. They succeeded, and soon papers with galaxy rotation data started to appear in Astrophysical Journal. Once an accident occurred — one night Margaret fell from the observing platform and was injured. But her bones were not broken, and on the next night she was observing again. So the first large collection of galaxy rotation curves was obtained. Unfortunately, the equipment they used was not sensitive enough to see outlying sections of rotation curves, thus these observations

Fig. 8.8   Visiting North Holland with Adriaan Blaauw, December 2006 (author's photo).

did not yield hints of flat rotation curves, which a decade later were found in many galaxies by optical and radio observations.

8.5.2   *Collaboration with other centres*

During the Soviet period our cosmology team collaborated with several other astronomers. The most active collaboration was with the Zeldovich group in Moscow. Now the former Zeldovich team is spread over many different countries: Rashid Sunyaev is in Garching, Germany; Sergei Shandarin, Anatoly Klypin and Andrei Linde are in the USA; Igor Novikov and Andrei Doroshkevich in Copenhagen. Aleksei Starobinsky is still mostly in Moscow, and we have continued the collaboration with him. He is one of the best specialists in early Universe physics, and we have made use of his knowledge. One of our young collaborators, Gert Hütsi, did his PhD in Garching with Rashid Sunyaev; this collaboration is continuing.

During the Soviet period some collaboration also started with astronomers from Western countries. The most important of these had its beginning in the visit of Enn Saar and myself to Nordita in 1987. With Bernard Jones and Vicent Martinez we started to investigate the fractal nature of galaxy distribution. This led to several

Fig. 8.9 Jim Peebles at the void conference, Amsterdam, December 2006 (author's photo).

publications (Einasto & Saar, 1987; Jones et al., 1988). Presently Enn is a visiting Professor at Valencia University. With Vicent Martinez and other astronomers Enn started to explore in more detail the application of statistics to the study of the galaxy distribution. The most important results of this collaboration is realised in the monograph by Martínez & Saar (2002).

After a long period I visited again the Astrophysical Institute in Potsdam in 1990. This was the start of a continuing collaboration with Potsdamer astronomers, most importantly with Volker Müller and his group. Potsdamer astronomers have been coauthors of most our papers on the large-scale structure of the Universe. Their Institute has better computers, so we have often used N-body simulations made in Potsdam.

In 2000 a Finnish astronomer from Tuorla Observatory, Pekka Heinämäki, came to us as a postdoc. This was the start of a continuous collaboration with Finnish astronomers. Now we have every year joint Tartu–Tuorla cosmology workshops alternatively in Tuorla and in Tartu. In the last few years astronomers and physicists from other centres have also participated in these workshops, and the topics are now broader from microphysics of the early Universe to the structure and formation of galaxies.

Fig. 8.10 Bernard Jones at the void conference, Amsterdam, December 2006 (author's photo).

During his stay in Garching Gert Hütsi started collaboration with several groups studying the physics of the very early Universe. In Estonia one group in Tallinn, led by Martti Raidal, is also investigating similar problems connected with the nature of dark matter. Now we have started a collaboration with the Martti Raidal team, realised in the form of a joint Center of Excellence.

In 2006 I was invited to a three-week workshop in Aspen to discuss the void phenomenon and the evolution of galaxies. Volker Müller from Potsdam was also invited, so we rented a car together to drive from Denver airport to Aspen and have the freedom for some excursions in Colorado. This time I took with me my granddaughter Triin. On weekends we had time to look around in mountains around Aspen and Colorado river. The workshop was very interesting, all major problems of the structure of voids were discussed. One problem was: What is the explanation for the absence of even dwarf galaxies in voids? I had thought about this problem for many years already and was able to give a preliminary answer. This problem is discussed elsewhere in this book.

The discussion of the void problem was continued at a small conference in Amsterdam later in the year. A week before the Amsterdam conference I had the opportunity to visit the Kapteyn Astronomical Institut of Groningen University. This was my first visit to Netherlands after the 1994 IAU General Assembly. I was

Fig. 8.11 Participants of the Kuzmin90 conference in Pulkovo at lunch-time. In the foreground is Alar Toomre, while sitting on the far side is Simon White; August 2007 (author's photo).

very happy to have this opportunity, because astronomy in the Netherlands was always for us an example of what can be done in a small country. I had a talk on our work in the seminar. The most accurate and interesting questions were asked by Adriaan Blaauw. I have met Adriaan many times at various conferences, now we met in his home Observatory. After the seminar Adriaan invited me to a tour of North Holland. Next day he drove to the University campus with a brand-new Toyota Camry gasoline/electric hybrid car. We had a trip to the Northern coast of the country with high dams — a large part of the country is located below see level. Figure 8.8 is taken in a gas station where Adriaan added some gasoline to his car.

Next week the void conference took place in Amsterdam, in the main building of the Royal Netherlands Academy of Arts and Sciences. The meeting hall of the Academy had only 50 seats, so participation was very limited. This meeting was also very stimulating; some pictures taken during the breaks are shown in Figs. 8.9 and 8.10.

Another interesting conference was organised by St. Petersburg astronomers in Pulkovo Observatory in the summer of 2007 on the structure and dynamics of galaxies. With this conference the 90th birthday of Grigori Kuzmin was celebrated. The session devoted to the work and life of Grigori Kuzmin was lead by Alar Toomre. Alar's own work is a continuation of the theoretical work by Grigori Kuzmin on a new level, using the possibilities of modern computers — galaxy

Fig. 8.12 Participants of the Kuzmin90 conference at Pulkovo singing Russian folk songs, August 2007 (author's photo).

mergers, stability and other similar problems. It was summer time and lunches were served in the garden of the Observatory, as seen in Fig. 8.11. After the sessions in the evenings Russian astronomers gathered to sing old Russian folk songs; one such moment is shown in Fig. 8.12. When summer workshops on astrophysics were held in our Observatory in Tõravere in the Soviet period, participants also used to sit in the evenings around the fireplace and sing Russian songs. One of the enthusiastic and very good singers was Josif Shklovsky.

One evening in autumn 2008 I received a phone call from Remo Ruffini. With him we had in the 1980's very close collaboration, and several of our team members visited Rome, while he and his students visited our Observatory. Now he suggested I become an Adjunct Professor of ICRANet, starting 2009. I accepted the invitation, and since then I spend part of my time in the Pescara center of ICRANet. Also between Tartu Observatory and ICRANet an official agreement of scientific cooperation was signed. Several times I gave lectures to students of his Erasmus Mundus fellowship program.

8.6 Tartu Observatory and my life in the 2000's

8.6.1 *Transition years*

Throughout the new independence period right-wing parties had dominated Estonian policy. Free market economy was the major force during the whole transition period. This policy was rather successful, but its possibilities were exhausted by now. Foreign capital had floating in basically due to our low salaries. The basic problem is now — we have too few innovation-oriented enterprises. One reason for this is that during the transition period most young talented people studied soft sciences, such as management and law, because in these fields there were a lot of vacant positions with high salaries. The fraction of young people studying physics and engineering was decreasing.

These and other actual strategic problems were discussed by the President Academic Advisory Board. I was invited to this Board in 1995 by the first President of Estonia after the restoration of independence, Lennart Meri. He organised the work of the Board in a very interesting way. Every year one member of the Board was nominated as the head of the Board, who was responsible for the choice of topics for discussion, helping to find speakers etc. Because members of the Board had different background and interests, the choice of topics was also rather wide. One year I was in this role. My primary idea was to use the experience of Finland in the development of a modern and innovation-friendly society. Here Lennart Meri helped a lot — he had many connections in Finland, so we had as speakers leading specialists from Nokia and Finnish Bank.

I was invited several times to give talks in the Board, twice during the presidency of Lennart Meri, and twice during the present President Toomas Hendrik Ilves. I discussed several aspects of the innovation, science and education policy of Estonia. Unfortunately the role of the President of Estonia is rather restricted, thus suggestions made by the Advisory Board have basically only a moral value.

There is a tradition that the President has a reception on Independence Day, February 24. All members of the Estonian Parliament, Government, and diplomatic corpus are invited, as well as numerous active people in every field of the society. I had the honour to be invited several times. Once I was in the reception together with my sister Kersti and her husband Lepo Sumera, a famous Estonian composer, see Fig. 8.13. It is a tradition that at the reception after the talk by the President a concert will be held. This year a composition of Lepo was performed: "Amore et igne" ("With Love and Fire for narrator, mixed chorus and symphony orchestra"). Lepo told me that he has thought for years on how to write music on one of the most important periods of our history, the Christianisation of Estonia in the 13th century. This was the beginning of eight hundred years of dominance of Baltic Barons. Finally he found the form of the composition as a Passion, a modern version of the style used by Johann Sebastian Bach in "St. Matthew Passion" and "St. John Passion". His composition is based on the text of the "Heinrici Chronicon Livoniae", the only surviving document of this event. In this composition the afflictions of the whole nation are described, not those of one person — Jesus Christ.

Fig. 8.13 In the reception of the President with Kersti and Lepo, February 24, 1998 (author's photo).

8.6.2 *Center of Excellence*

In the 2000's our cosmology team got new students and postdocs. Most of them have now finished their theses and continue their work: Antti Tamm, Gert Hütsi, Ivan Suhhonenko, Elmo Tempel, Lauri Juhan Liivamägi. These young collaborators are interested in new problems. One of them, Gert Hütsi, got his PhD degree at the Max Planck Institute of Astrophysics in Garching under the supervision of Rashid Sunyaev. Presently he and Elmo Tempel are actively collaborating with the Martti Raidal group in Tallinn in the study of physical aspects of the dark matter problem.

We had applied several times in the competition to form a Center of Excellence. The last time we succeeded; the Center was formed together with the Martti Raidal group, and the main goal is to study in more detail the physics of dark matter, using both astronomical as well as experimental physics data. Raidal collaborated closely with CERN, thus we can now look at the problem from the astronomical as well from the physics point of view. A recent result of this collaboration was a

Fig. 8.14 The first picture of the Earth, taken with the onboard camera of ESTCube-1 (ESTCube team).

joint study of the gamma-ray observations made with the Fermi satellite (Hektor et al., 2013; Tempel et al., 2012b).

In 2007 Estonia became an associated member of European Space Agency. This has motivated our colleagues in Observatory and Tartu University to start a student satellite project[1]. The practical purpose is to build a small cubic satellite to check the viability of using Solar wind to accelerate satellite motion. The pedagogical goal is to give students real knowledge in the teamwork on international and interdisciplinary level, to learn to build with his/her hands a highly sophisticated and innovative device (the weight of ESTCube-1 is only 1.3 kg). The project involves over a hundred students from Tartu University, Tallinn Technical University, and universities in Finland, Latvia and Germany, as well specialists in various fields working in private enterprises. The satellite was launched on May 7, 2013 form the ESA site at French Guyana. Now radio amateurs from all continents send to Tartu Observatory e-mails that they have received signals from ESTCube-1. The first pictures of the Earth were already taken using the small onboard camera, see Fig. 8.14.

---

[1] http://www.estcube.eu

8.6.3 *Egeri*

My mother gave her farm Egeri to the state in 1940. Here our previous lessees and their children lived almost to the end of the Soviet period. However, in the final ten years of this period the last habitants went away, and Egeri was not inhabited. The houses decayed and copses grew. In the last years of the Soviet period a local agronomist purchased from the kolkhoz the ruins of the houses and started to rebuild the main house of the farm.

In the early 1990's our Parliament made the decision that all property taken by the Soviet regime can be returned to previous owners. With my brothers and sisters we were the devisees, and had the possibility of initiating the procedure to get the farm back. In 1990 I visited Egeri and discovered that the house was already occupied, so we did not start the procedure to reclaim the farm. However, in 1995 the new owner was ready to sell us the house he had restored. So I did, and we also proceeded to get the whole farm back. Now it is divided between my brothers and sisters. I own a part with the restored house, while the other part is a joint property together with my brothers and sisters.

With my brother Peeter, his son Laur and my son Indrek we started to restore one additional house. Now it is under a roof, and some rooms inside have been finished. In one of the last winters there was so much snow that the roof of the main house almost collapsed. The next summer was spent repairing the damage. Our experience shows that there is always something to do in the household.

We have had several times astronomy seminars there. Once the Estonian Astronomical Society had its annual weekly seminar there. Also we have had family gatherings in Egeri. In the summer I spend most of my time in Egeri; about half of my library is there. Internet connection is available, so I can also do my work there.

It is very quiet there. The next farm is about 2 kilometers away, and there is a large forest surrounding it. There are no city-lights nearby, so the sky is rather dark on a clear night, almost as it is in the mountains. In Egeri I feel at home: here are my roots, my connections to my earlier generations, to my homeland, to the Earth and to the Universe.

This book is accompanied by a webpage[2], which adds some important details to my story. First of all, the webpage contains copies of several papers of the dark matter story in their original form; the official published versions had in some cases omissions or were distorted to follow the referees' suggestions. The webpage contains several astronomical movies demonstrated in various conferences, both

---
[2] http://www.aai.ee/~einasto/DarkMatter

from the early period of the study of the cosmic web, as well as our recent movies based on Sloan Survey.

In addition, the webpage has my CV in pictures, and the Egeri Story. Both are accompanied by music, my CV by the music of my brother-in-law Lepo Sumera, played by his daughter Kadri-Ann. To express the spirit of the Egeri Story I used one of the best pieces of Bach's organ music in my music collection, Toccata and Fuga, BWV565, played by the Estonian organist Rolf Uusväli. My grandfather Jaan Lammas, who actually built up the Egeri farm together with his father Wiilip, was an enthusiastic organ player and liked Bach's music very much.

In the webpage there is also a movie, prepared by myself on the restoration of Egeri, and on a family gathering in 1999. The family gathering is interesting — the wife of my brother's son Mart has her roots in Setumaa, a region in South-East Estonia, where the traditions of the 'regilaul' are still alive. So during the family gathering she started to sing, as is traditional in family gatherings. Such songs were created and sung during the gathering. They are unique, used only for this particular case.

My brother Peeter lived several years in Egeri. He is a very talented photographer, and has captured a large number of moments which characterise our feelings and tempers in Egeri. I used a collection of these photos to prepare "Egeri contemplations". It is accompanied by Johann Sebastian Bach's music from "Das Wohltemperierte Klavier". I have in my music collection this Bach's masterpiece played by five different musicians, including Daniel Barenboim, Svjatoslav Richter and others, played on piano or harpsichord. In my opinion the spirit of the music which best suits my own feelings is reproduced by the Russian pianist Samuel Feinberg's recording from 1959. I have this rare mono recording on vinyl disks, and transformed it to mp3 files. I found in the web the following description of emotions of the Feinberg play: *The way I arrive at it, is that I imagine Feinberg, in Stalin's relentlessly grim Russia (dead though Stalin was in 1959), undertaking to record this music. There's something intensely human about pouring your soul out through music like this in the wake of Stalin — perhaps one of the very few allowable ways to finally let your pain and hope and faith in beauty find expression without vanishing into the censor's office or the Gulag yourself. But even without this historically dubious expatiation, I think that it is exceptionally apparent how much of a labor of love performing this music is for Feinberg. It seems as if the whole of the frailty of human existence is at stake in this recording — which does not mean that it is merely grim. There's as much light here as sadness, the whole world in fact* (from the review by Snow Leopard of Samuel Feinberg playing Bach "Das Wohltemperierte Klavier", www.amazon.com).

Chapter 9

Epilogue

In the spring of 1973 there was a conference on galaxies in Tbilisi, and one evening I walked with Rashid Sunyaev along the Rustaveli avenue discussing our galaxy models with dark halos. Rashid argued that nobody will take our results seriously until some American astronomer confirmed them. Indeed, our experience has confirmed his prediction several times. Our work on the presence of dark matter around galaxies was noticed after Ostriker et al. (1974) got similar results, and Vera Rubin et al. (1978) confirmed flat rotation curves of galaxies; the presence of the cosmic web in the Universe was accepted after the work by Geller & Huchra (1989).

I have noticed that actually Rashid's remark has a deeper meaning. People from leading scientific centres, especially Anglo-American ones, have a habit to noticing only the work done in established universities or centres, especially pioneering work. This practice reminds me the tendency in the Soviet period, where it was argued that all major scientific and technical innovations were done by Russian people — the steam machine was invented by Ivan Polzunov, radio by Alexander Popov etc.

It seems to me that different discoveries may have different characters. Most interesting are unexpected discoveries — one of the best example of such discoveries in cosmology is the discovery of pulsars by Jocelyn Bell and Antony Hewish (Hewish et al., 1968). Of a different type are planned discoveries — some phenomenon is expected and special observational techniques are developed to detect it. Perhaps the best example of such discoveries is the detection of CMB fluctuations by the COBE satellite by Smoot et al. (1992). But there was a surprise here too: the spectrum of the CMB radiation corresponds to a black-body spectrum with extremely high accuracy (Mather et al., 1994). These discoveries have extended enormously our knowledge on the Universe. The smallness of CMB fluctuations and its very accurate black-body spectrum were strong arguments supporting the inflation model of the early Universe, which together changed our paradigm on the evolution of the Universe.

Discoveries which change the foundation of our worldview have often a different character. The discoveries of dark matter and the cosmic web seem to belong to this category. People often ask: Who discovered dark matter? Who discovered the cosmic web? As is characteristic in a paradigm shift, there is no single discoverer; the new concepts were developed step-by-step by many scientists, see also Kuhn (1970) and Tremaine (1987).

The timeline of the study of dark matter is shown in Table 9.1. Actually there are two dark matter problems — the local dark matter close to the plane of our Galaxy, and the global dark matter surrounding galaxies and clusters of galaxies. However, this difference was understood only later, thus we show in the Table the whole story.

The first essential milestone of the dark matter story is the discovery of the possible presence of dark matter in the Galactic disk by Oort (1932), and in the Coma cluster of galaxies by Zwicky (1933); earlier work was mostly ignored. In the following years new data on the presence of dark matter both in the Galactic disk and in systems of galaxies slowly accumulated. Kuzmin (1952b, 1955) and his students in Tartu Observatory showed that the amount of DM in the Galactic disk is small; in contrast Hill (1960); Oort (1960) and some other astronomers found evidence that up to half of the matter in the Solar vicinity may be dark. More accurate data showed that the amount of local dark matter is small (Gilmore et al., 1989).

Data on masses of galaxies and systems of galaxies also accumulated, both from rotation data on the periphery of galaxies, and from statistical dynamical data of groups and clusters of galaxies. For some reason, these studies did not capture the attention of the astronomical community. However, awareness of the presence of a controversy between the masses of galaxies and galaxy systems slowly increased.

The next important milestone in the dark matter story was the understanding that there are two different DM problems, the local and the global DM, which have different distribution and origin (Einasto, 1972a, 1974a). The local DM belongs to the flat disk of the Galaxy, thus it must dissipative to disperse the energy left over during its formation. In contrast, Einasto et al. (1974b) found that the global DM must form a new previously unknown population — corona or halo, which is essentially dissipationless, and is probably of non-stellar origin. They found that the size and the mass of coronas exceed the size and mass of known stellar populations in galaxies about tenfold. Einasto et al. (1974b) and Ostriker et al. (1974) suggested that the total cosmological density of galaxies including their

Table 9.1 Dark matter timeline.

Year	Description
1915	First estimates of local DM: Öpik (1915), Kapteyn (1922), Jeans (1922)
1932	Galactic model and Local DM: Oort (1932)
1933	DM in Coma cluster: Zwicky (1933)
1939	Hints of large M/L on the periphery of galaxies: Babcock (1939), Oort (1940)
1952	Galactic models and local DM: Kuzmin (1952b,a, 1954, 1955, 1956a,b)
1957	Large M/L on the periphery of M31: van de Hulst et al. (1957), Roberts (1966)
1959	Mass of the Local Group: Kahn & Woltjer (1959)
1961	Cluster Stability Conference: Neyman et al. (1961)
1962	Galaxy dynamic evolution: Eggen et al. (1962)
1965	Galactic models with populations: Einasto (1965, 1969b)
1965	Discovery of CMB: Penzias & Wilson (1965)
1968	Galaxy physical evolution: Tinsley (1968), Einasto (1972b)
1972	Cluster X-ray data on mass of hot gas: Forman et al. (1972); Gursky et al. (1972)
1972	Local and global DM different, global DM non-stellar: Einasto (1972a, 1974a)
1974	Parameters of DM coronas/halos: Einasto et al. (1974b), Ostriker et al. (1974)
1975	Flat rotation of 14 galaxies: Roberts (1975)
1975	Tallinn DM Conference: Doroshkevich et al. (1975)
1975	DM contradicts classical cosmological paradigm: Materne & Tammann (1976)
1977	M/L of galactic bulges low: Faber et al. (1977)
1978	Extended flat rotation curves: Bosma (1978), (Rubin et al., 1978)
1978	Two-stage galaxy formation model: White & Rees (1978)
1979	Gravitational lensing data on cluster masses
1981	Non-baryonic DM discussion in Tallinn and Vatican conferences
1984	Cold DM accepted: Blumenthal et al. (1984)
1989	Absence of large amounts of local DM accepted: Gilmore et al. (1989)
1992	CMB fluctuations detected: Smoot et al. (1992), Mather et al. (1994)
1998	Acceleration of the Universe discovered: Riess et al. (1998), Perlmutter et al. (1999)
2012	CDM particle annihilation detected?: Weniger (2012), Tempel et al. (2012b)

coronae/halos is about 0.2 of the critical cosmological density, thus DM is the dominating population of the Universe.

The possible nature of galactic coronae was discussed in the conference on dark matter in Tallinn 1975. No known candidate (stars, cold or hot gas, neutrinos) fit all known observational data (Doroshkevich et al., 1975). General problems of dark matter were discussed during the Third European Astronomy Meeting in Tbilisi 1975. Here Materne & Tammann (1976) showed that the presence of dark matter contradicts fundamental data of the classical cosmological paradigm. The DM problem was acknowledged as a problem and a crisis.

In the following years experimenters found new evidence in favour of the presence of dark matter around galaxies, and in groups and clusters of galaxies. The work by Rubin and Bosma on galaxy rotation curves, X-ray studies of clusters, as well as investigation of gravitational lensing in clusters belong to this type of study.

The third important milestone in the dark matter story was the understanding that it is non-baryonic. The nature of DM and its role in the evolution of the Universe were discussed in 1981 in conferences in Tallinn and the Vatican. The dark matter concept was generally accepted when it was understood that its particles form a cold non-baryonic medium — Cold Dark Matter (Blumenthal et al., 1984). The close collaboration between cosmologists and particle physicists started — this was the birth of astro-particle physics.

The fourth essential milestone was the detection of CMB fluctuations by the COBE satellite (Smoot et al., 1992), its black-body spectrum (Mather et al., 1994), and the detection of the acceleration of the Universe due to dark energy (Riess et al., 1998; Perlmutter et al., 1999). These observations together with direct data from the distribution of galaxies allowed the density of all matter/energy components to be calculated with high accuracy; for dark matter these data yield $\Omega_{DM} = 0.229$. Also new data show that only a small fraction of dark matter is hot (neutrinos) or warm; mostly it should be cold.

So far physical experiments have not been able to determine the properties of dark matter particles. Some help comes from special astronomical observations of the results of the annihilation of DM particles. To such observations belong Fermi satellite Large Area Telescope observations of a double gamma-ray spectral line at 130 GeV (Weniger, 2012; Tempel et al., 2012b). However, confirmation by independent observations is needed to be sure that DM particle annihilation is actually observed.

The development of the understanding of the structure of the Universe was initially completely independent of the development of its matter content. The timeline of cosmic web study is presented in Table 9.2. Here I would like to emphasise the following milestones in the development.

The first milestone is the determination of distances to spiral nebulae, which demonstrated that there exist stellar systems outside the Milky Way. In other words, it was discovered that the Milky Way is not the whole Universe. This observation was quickly followed by the another very important discovery — the whole Universe is expanding and thus had a beginning.

There followed a gradual buildup of the classical paradigm of modern cosmology. This included the observations of redshifts of distant galaxies and the determination of the expansion speed — the Hubble constant. Also the distribution of galaxies was investigated, using galaxy counts on photographic plates. These efforts culminated with the Palomar Sky Survey. This survey was the basis behind the first relatively deep catalogues of galaxies and clusters of galaxies. The same

Table 9.2 Cosmic web timeline.

Year	Description
1922	Determination of distances of external galaxies: Öpik (1922a), Hubble (1925, 1926)
1929	Discovery of the expansion of the Universe: Hubble (1929a)
1935	Survey of Northern sky galaxies: Shapley (1935, 1937)
1949	Palomar Sky Survey
1953	Virgo supercluster: de Vaucouleurs (1953)
1958	Cluster catalogue: Abell (1958); Abell et al. (1989)
1967	Lick Galaxy Survey: Shane & Wirtanen (1967)
1968	Catalogue of galaxies and clusters: Zwicky et al. (1968)
1970	Galaxy clustering scenario: Peebles & Yu (1970)
1970	Pancake scenario of structure formation: Zeldovich (1970)
1972	Determination of the Hubble constant: Sandage (1972); Sandage & Tammann (1975, 1976)
1973	Statistical analysis of galaxy catalogues: Peebles (1973, 1974a)
1973	Simulation of pancake model: Doroshkevich & Shandarin (1973)
1976	Discovery of cosmic voids: Chincarini & Rood (1976); Gregory & Thompson (1978); Kirshner et al. (1981)
1977	Discovery of cosmic web: Jõeveer et al. (1977, 1978); Einasto et al. (1980a,b)
1977	Threshold galaxy formation: Jõeveer et al. (1977); Einasto et al. (1980a)
1980	Inflation theory: Starobinsky (1980); Guth (1981); Linde (1982)
1982	Quantitative comparison of models with observations: Zeldovich et al. (1982)
1983	Simulations of CDM models: Melott et al. (1983); Davis et al. (1985); White et al. (1987)
1984	Galaxy biasing: Kaiser (1984)
1986	Simulations of ΛCDM models: Einasto et al. (1986a); Gramann (1987, 1988); Efstathiou et al. (1990)
1986	Topology of cosmic web: Gott et al. (1986); Einasto et al. (1986a)
1986	Second CfA galaxy redshift survey: de Lapparent et al. (1986)
1994	Supercluster catalogues: Einasto et al. (1994b, 1997b, 2001); Liivamägi et al. (2012)
2000	Sloan Digital Sky Survey: York et al. (2000)
2001	Two-degree Field Galaxy redshift survey: Colless et al. (2001)
2005	Detection of baryon acoustic oscillations: Eisenstein et al. (2005); Hütsi (2006)

survey was also used to detect second order clusters of galaxies or superclusters. Also the mean density of matter in the Universe was determined using catalogues of galaxies in the nearby Universe, where distances of galaxies could be measured.

The third milestone in the understanding of the structure of the Universe is related to the introduction of new sensitive detectors, which allowed the determination of redshifts of faint galaxies. This was the fundamental data needed to study the 3-dimensional distribution of galaxies, and to discover the cosmic web with filaments of galaxies and clusters of galaxies, and voids between them.

The next important milestone was the development of the theory of the formation and evolution of the cosmic web. It was followed by the development of the inflation model of the early evolution of the Universe. It became evident that the seeds of the cosmic web were formed already in the very early Universe, perhaps during the inflation. The structure of the Universe evolves very slowly, thus the present structure contains imprints of the very early Universe. Since the dominant population of the Universe is dark matter, it is possible to get information on properties of DM particles from the structure of the cosmic web. This development emphasises that there exists a deep physical link between dark matter and the cosmic web. This caused the birth of the astro-particle physics. Since then the development of studies of dark matter and the cosmic web has occurred hand-in-hand. From the late 1970's to the early 1980's there was an intensive discussion between supporters of the classical and the new cosmological paradigms.

The rapid development of the new cosmological paradigm occurred in the 1970's and 1980's, at the peak of the Cold War between the Western and the Eastern Worlds. The East–West conflict influenced the development in several ways. First, this conflict made contact between scientists of both Worlds more difficult. Second, there was a clear tendency to ignore some developments in the other World.

Most experimental studies on dark matter and the cosmic web were made by Western astronomers. The interpretation of observational data and the buildup of the modern theory of formation and evolution of the Universe was made in parallel both in the West and East, as seen in both timelines. The new paradigm wins when its theoretical foundation is established. In the case of the dark matter this was done by Blumenthal et al. (1984) with the non-baryonic cold dark matter hypothesis. The connection between cosmology and particle physics was discussed in the Tallinn conference and in the Vatican Study Week, both in 1981. The new cosmological paradigm, which included all major new elements — dark matter, cosmic web and inflation — was discussed probably for first time in the lectures by Primack (1984).

Word on the development of the new cosmology paradigm spread more rapidly in the East: the first dark matter conference was held in Tallinn in 1975; the first official IAU dark matter conference was held only ten years later. The first popular discussions of the dark matter problem were given in "Priroda" and "Zemlya i Vselennaya" (the Russian counterparts of "Scientific American" and "Sky & Telescope") by Einasto et al. (1975c); Einasto (1975); Einasto & Jõeveer (1978) and in the respective journal in Estonian. In USA the first popular discussions of dark matter were given by Bok (1981) and Rubin (1983). The cosmic web was discussed in a popular book in Estonian by Einasto & Jõeveer (1979), in Russian by Einasto & Jaaniste (1982), and in English by Gregory & Thompson (1982) and Geller & Huchra (1989).

To conclude we can say that the story of dark matter and the cosmic web is not over yet — we still do not know of what non-baryonic particles dark matter is made of, and the nature of dark energy is completely unknown.

Bibliography

Aarseth, S. J. 1971a, *Direct Integration Methods of the N-Body Problem (Papers appear in the Proceedings of IAU Colloquium No. 10 Gravitational N-Body Problem (ed. by Myron Lecar), R. Reidel Publ. Co., Dordrecht-Holland)*, Ap&SS, 14, 118

Aarseth, S. J. 1971b, *Numerical Experiments on the N-Body Problem (Papers appear in the Proceedings of IAU Colloquium No. 10 Gravitational N-Body Problem (ed. by Myron Lecar), R. Reidel Publ. Co., Dordrecht-Holland)*, Ap&SS, 14, 20

Aarseth, S. J., Turner, E. L., & Gott, III, J. R. 1979, *N-body simulations of galaxy clustering. I - Initial conditions and galaxy collapse times*, ApJ, 228, 664

Abell, G. O. 1958, *The Distribution of Rich Clusters of Galaxies*, ApJS, 3, 211

Abell, G. O. 1977, *The Luminosity Function and Structure of the Coma Cluster*, ApJ, 213, 327

Abell, G. O., Corwin, Jr., H. G., & Olowin, R. P. 1989, *A catalog of rich clusters of galaxies*, ApJS, 70, 1

Alpher, R. A., Herman, R., & Gamow, G. A. 1948, *Thermonuclear Reactions in the Expanding Universe*, Physical Review, 74, 1198

Ambartsumian, V. A. 1958, *On the Problem of the Mechanism of the Origin of Stars in Stellar Associations*, Reviews of Modern Physics, 30, 944

Ambartsumian, V. A. 1961, *Instability phenomena in systems of galaxies*, AJ, 66, 536

Antonov, V. A. & Chernin, A. D. 1975, *The dynamics and cosmogony of galactic coronae*, Pis ma Astronomicheskii Zhurnal, 1, 18

Antonov, V. A., Osipkov, L. P., & Chernin, A. D. 1975a, *On the dynamics of galactic coronae.*, in Dynamics and Evolution of Stellar Systems, p. 54–57, 289

Antonov, V. A., Osipkov, L. P., & Chernin, A. D. 1975b, *Stellar motion in the unsteady gravitational field of an evolving galaxy*, Astrofizika, 11, 335

Aragon-Calvo, M., van de Weygaert, R., van der Hulst, T., Szalay, A., & Araya, P. 2006, *The Multiscale Morphology Filter*, in Bernard's Cosmic Stories: From Primordial Fluctuations to the Birth of Stars and Galaxies

Aragon-Calvo, M. A. 2012, *The MIP Ensemble Simulation: Local Ensemble Statistics in the Cosmic Web*, ArXiv:1210.7871

Aragón-Calvo, M. A., Jones, B. J. T., van de Weygaert, R., & van der Hulst, J. M. 2007a, *The multiscale morphology filter: identifying and extracting spatial patterns in the galaxy distribution*, A&A, 474, 315

Aragon-Calvo, M. A., Shandarin, S. F., & Szalay, A. 2010a, *Geometry of the Cosmic Web: Minkowski Functionals from the Delaunay Tessellation*, ArXiv:1006.4178

Aragon-Calvo, M. A., van de Weygaert, R., Araya-Melo, P. A., Platen, E., & Szalay, A. S. 2010b, *Unfolding the hierarchy of voids*, MNRAS, 404, L89

Aragon-Calvo, M. A., van de Weygaert, R., & Jones, B. J. T. 2010c, *Multiscale phenomenology of the cosmic web*, MNRAS, 408, 2163

Aragón-Calvo, M. A., van de Weygaert, R., Jones, B. J. T., & van der Hulst, J. M. 2007b, *Spin Alignment of Dark Matter Halos in Filaments and Walls*, ApJ, 655, L5

Arnalte-Mur, P., Labatie, A., Clerc, N., et al. 2012, *Wavelet analysis of baryon acoustic structures in the galaxy distribution*, A&A, 542, A34

Arneodo, F. 2013, *Dark Matter Searches*, ArXiv:1301.0441

Atrio-Barandela, F., Einasto, J., Gottlöber, S., Müller, V., & Starobinsky, A. 1997, *A built-in scale in the initial spectrum of density perturbations: Evidence from cluster and CMB data*, Soviet Journal of Experimental and Theoretical Physics Letters, 66, 397

Babcock, H. W. 1939, *The rotation of the Andromeda Nebula*, Lick Observatory Bulletin, 19, 41

Bahcall, J. N. 1984a, *K giants and the total amount of matter near the sun*, ApJ, 287, 926

Bahcall, J. N. 1984b, *Self-consistent determinations of the total amount of matter near the sun*, ApJ, 276, 169

Bahcall, J. N. 1984c, *The distribution of stars perpendicular to galactic disk*, ApJ, 276, 156

Bahcall, J. N. 1987, *Dark matter in the galactic disk*, in IAU Symposium, Vol. 117, Dark matter in the universe, ed. J. Kormendy & G. R. Knapp, 17

Bahcall, J. N. & Casertano, S. 1986, *Kinematics and density of the Galactic spheroid*, ApJ, 308, 347

Bahcall, J. N., Hut, P., & Tremaine, S. 1985, *Maximum mass of objects that constitute unseen disk material*, ApJ, 290, 15

Bahcall, J. N. & Soneira, R. M. 1980, *The universe at faint magnitudes. I - Models for the galaxy and the predicted star counts*, ApJS, 44, 73

Bahcall, J. N. & Soneira, R. M. 1984, *Comparisons of a standard galaxy model with stellar observations in five fields*, ApJS, 55, 67

Bahcall, J. N., Soneira, R. M., & Schmidt, M. 1983, *The galactic spheroid*, ApJ, 265, 730

Bahcall, N. A. & Soneira, R. M. 1983, *The spatial correlation function of rich clusters of galaxies*, ApJ, 270, 20

Bardeen, J. M., Bond, J. R., Kaiser, N., & Szalay, A. S. 1986, *The statistics of peaks of Gaussian random fields*, ApJ, 304, 15

Baryshev, Y. V., Chernin, A. D., & Teerikorpi, P. 2001, *The cold local Hubble flow as a signature of dark energy*, A&A, 378, 729

Bekenstein, J. D. 2004, *Relativistic gravitation theory for the modified Newtonian dynamics paradigm*, Phys. Rev. D, 70, 083509

Bennett, C. L., Banday, A. J., Gorski, K. M., et al. 1996, *Four-Year COBE DMR Cosmic Microwave Background Observations: Maps and Basic Results*, ApJ, 464, L1

Bennett, C. L., Halpern, M., Hinshaw, G., et al. 2003, *First-Year Wilkinson Microwave Anisotropy Probe (WMAP) Observations: Preliminary Maps and Basic Results*, ApJS, 148, 1

Binney, J. & Tremaine, S. 1987, Galactic dynamics (Princeton Univ. Press)

Bisnovatyi-Kogan, G. S. & Novikov, I. D. 1980, *Cosmology with a Nonzero Neutrino Rest Mass*, Soviet Astr.-AJ, 24, 516

Blanchard, A., Douspis, M., Rowan-Robinson, M., & Sarkar, S. 2003, *An alternative to the cosmological "concordance model"*, A&A, 412, 35

Blumenthal, G. R., Faber, S. M., Primack, J. R., & Rees, M. J. 1984, *Formation of galaxies and large-scale structure with cold dark matter*, Nature, 311, 517

Blumenthal, G. R., Pagels, H., & Primack, J. R. 1982, *Galaxy formation by dissipationless particles heavier than neutrinos*, Nature, 299, 37

Bobrova, N. A. & Ozernoi, L. M. 1975, *Expected distribution of 'hidden' mass in rich clusters of galaxies*, Pis ma Astronomicheskii Zhurnal, 1, 8

Bok, B. J. 1981, *The Milky Way Galaxy*, Scientific American, 244, 92

Bond, J. R., Kofman, L., & Pogosyan, D. 1996, *How filaments of galaxies are woven into the cosmic web*, Nature, 380, 603

Bond, J. R., Szalay, A. S., & Turner, M. S. 1982, *Formation of galaxies in a gravitino-dominated universe*, Physical Review Letters, 48, 1636

Bondi, H. & Gold, T. 1948, *The Steady-State Theory of the Expanding Universe*, MNRAS, 108, 252

Bosma, A. 1978, *The distribution and kinematics of neutral hydrogen in spiral galaxies of various morphological types*, PhD thesis, Groningen Univ.

Brandt, J. C. 1960, *On the Distribution of Mass in Galaxies. I. The Large-Scale Structure of Ordinary Spirals with Applications to M 31*, ApJ, 131, 293

Brandt, J. C. & Scheer, L. S. 1965, *A note on functions relating to galactic structure*, AJ, 70, 471

Bringmann, T., Huang, X., Ibarra, A., Vogl, S., & Weniger, C. 2012, *Fermi LAT search for internal bremsstrahlung signatures from dark matter annihilation*, J. Cosmology Astropart. Phys., 7, 54

Broadhurst, T. J., Ellis, R. S., Koo, D. C., & Szalay, A. S. 1990, *Large-scale distribution of galaxies at the Galactic poles*, Nature, 343, 726

Brosche, P., Einasto, J., & Rümmel, U. 1974, *Bibliography on the Structure of Galaxies*, Veroeffentlichungen des Astronomischen Rechen-Instituts Heidelberg, 26, 1

Brueck, H. A., Coyne, G. V., & Longair, M. S., eds. 1982, Astrophysical cosmology; Proceedings of the Study Week on Cosmology and Fundamental Physics, Vatican City State, September 28-October 2, 1981

Bullock, J. S., Kolatt, T. S., Sigad, Y., et al. 2001, *Profiles of dark haloes: evolution, scatter and environment*, MNRAS, 321, 559

Burbidge, E. M. & Burbidge, G. R. 1959, *Rotation and Internal Motions in NGC 5128*, ApJ, 129, 271

Burbidge, E. M., Burbidge, G. R., Fowler, W. A., & Hoyle, F. 1957, *Synthesis of the Elements in Stars*, Reviews of Modern Physics, 29, 547

Burbidge, E. M., Burbidge, G. R., & Prendergast, K. H. 1959, *The Rotation and Mass of NGC 2146*, ApJ, 130, 739

Burbidge, G. 1975, *On the masses and relative velocities of galaxies*, ApJ, 196, L7

Calzetti, D., Giavalisco, M., Ruffini, R., Einasto, J., & Saar, E. 1987, *The correlation function of galaxies in the direction of the Coma cluster*, Ap&SS, 137, 101

Cen, R. & Ostriker, J. P. 1992, *Galaxy formation and physical bias*, ApJ, 399, L113

Cen, R. & Ostriker, J. P. 2000, *Physical Bias of Galaxies from Large-Scale Hydrodynamic Simulations*, ApJ, 538, 83

Centrella, J. & Melott, A. L. 1983, *Three-dimensional simulation of large-scale structure in the universe*, Nature, 305, 196

Chemin, L., de Blok, W. J. G., & Mamon, G. A. 2011, *Improved Modeling of the Mass Distribution of Disk Galaxies by the Einasto Halo Model*, AJ, 142, 109

Chernin, A., Einasto, I., & Saar, E. 1976, *The role of diffuse matter in galactic coronas*, Ap&SS, 39, 53

Chernin, A. D. 1976, *Hidden mass and binding energy in systems of galaxies*, in Energy and Physics, 239

Chernin, A. D. 1981, *The rest mass of primordial neutrinos, and gravitational instability in the hot universe*, AZh, 58, 25

Chernin, A. D. 2001, *REVIEWS OF TOPICAL PROBLEMS: Cosmic vacuum*, Physics Uspekhi, 44, 1099

Chernin, A. D. 2003, *Cosmic vacuum and the 'flatness problem' in the concordant model*, New Astronomy, 8, 79

Chernin, A. D. 2008, *PHYSICS OF OUR DAYS: Dark energy and universal antigravitation*, Physics Uspekhi, 51, 253

Chincarini, G. & Rood, H. J. 1972, *Radial velocities of galaxies II*, AJ, 77, 448

Chincarini, G. & Rood, H. J. 1975, *Size of the Coma cluster*, Nature, 257, 294

Chincarini, G. & Rood, H. J. 1976, *The Coma supercluster — Analysis of Zwicky-Herzog cluster 16 in field 158*, ApJ, 206, 30

Clowe, D., Bradač, M., Gonzalez, A. H., et al. 2006, *A Direct Empirical Proof of the Existence of Dark Matter*, ApJ, 648, L109

Colberg, J. M., Pearce, F., Foster, C., et al. 2008, *The Aspen-Amsterdam void finder comparison project*, MNRAS, 387, 933

Coles, P. & Chiang, L. 2000, *Characterizing the nonlinear growth of large-scale structure in the Universe*, Nature, 406, 376

Colless, M., Dalton, G., Maddox, S., et al. 2001, *The 2dF Galaxy Redshift Survey: spectra and redshifts*, MNRAS, 328, 1039

Creze, M., Chereul, E., Bienayme, O., & Pichon, C. 1998, *The distribution of nearby stars in phase space mapped by Hipparcos. I-The potential well and local dynamical mass*, A&A, 329, 920

Croton, D. J., Springel, V., White, S. D. M., et al. 2006, *The many lives of active galactic nuclei: cooling flows, black holes and the luminosities and colours of galaxies*, MNRAS, 365, 11

Davis, M., Efstathiou, G., Frenk, C. S., & White, S. D. M. 1985, *The evolution of large-scale structure in a universe dominated by cold dark matter*, ApJ, 292, 371

de Blok, W. J. G. 2010, *The Core-Cusp Problem*, Advances in Astronomy, 2010

de Blok, W. J. G., Walter, F., Brinks, E., et al. 2008, *High-Resolution Rotation Curves and Galaxy Mass Models from THINGS*, AJ, 136, 2648

de Lapparent, V., Geller, M. J., & Huchra, J. P. 1986, *A slice of the universe*, ApJ, 302, L1

de Sitter, W. 1917, *Einstein's theory of gravitation and its astronomical consequences. Third paper*, MNRAS, 78, 3

de Vaucouleurs, G. 1953, *Evidence for a local super-galaxy*, AJ, 58, 30

de Vaucouleurs, G. 1970, *The Case for a Hierarchical Cosmology*, Science, 167, 1203

de Vaucouleurs, G. 1978, *The extragalactic distance scale. IV - Distances of nearest groups and field galaxies from secondary indicators*, ApJ, 224, 710

de Vaucouleurs, G. & de Vaucouleurs, A. 1964, Reference catalogue of bright galaxies, eds. G. de Vaucouleurs & A. de Vaucouleurs de Vaucouleurs, G., de Vaucouleurs, A., & Corwin, H. G. 1976, Second reference catalogue of bright galaxies (University of Texas Monographs in Astronomy, Austin: University of Texas Press)

Dekel, A. 1984, *Superpancakes and the cluster correlation function*, ApJ, 284, 445

Dekel, A. & Silk, J. 1986, *The origin of dwarf galaxies, cold dark matter, and biased galaxy formation*, ApJ, 303, 39

Dhar, B. K. & Williams, L. L. R. 2010, *Surface mass density of the Einasto family of dark matter haloes: are they Sersic-like?* MNRAS, 405, 340

Dhar, B. K. & Williams, L. L. R. 2011, *Surface Brightness and Intrinsic Luminosity of Ellipticals*, ArXiv:1112.3120

Di Matteo, T., Springel, V., & Hernquist, L. 2005, *Energy input from quasars regulates the growth and activity of black holes and their host galaxies*, Nature, 433, 604

Dicke, R. H., Peebles, P. J. E., Roll, P. G., & Wilkinson, D. T. 1965, *Cosmic Black-Body Radiation*, ApJ, 142, 414

Dimov, N. 1970, *Preliminary Note on the Astronomical Satellite Kosmos 215*, in IAU Symposium, Vol. 36, Ultraviolet Stellar Spectra and Related Ground-Based Observations, ed. R. Muller, L. Houziaux, & H. E. Butler, 138

Dolgov, A. D. 2002, *Neutrinos in cosmology*, Phys. Rep., 370, 333

Donato, F., Gentile, G., Salucci, P., et al. 2009, *A constant dark matter halo surface density in galaxies*, MNRAS, 397, 1169

Doroshkevich, A. G., Jõeveer, M., & Einasto, J. 1975, *Conference: Hidden Mass in the Universe*, AZh, 52, 1113

Doroshkevich, A. G., Kotok, E. V., Poliudov, A. N., et al. 1980, *Two-dimensional simulation of the gravitational system dynamics and formation of the large-scale structure of the universe*, MNRAS, 192, 321

Doroshkevich, A. G. & Novikov, I. D. 1964, *Mean Density of Radiation in the Metagalaxy and Certain Problems in Relativistic Cosmology*, Soviet Physics Doklady, 9, 111

Doroshkevich, A. G., Ryaben'kij, V. S., & Shandarin, S. F. 1973, *Non-linear theory of development of potential perturbations*, Astrofizika, 9, 257

Doroshkevich, A. G. & Shandarin, S. F. 1973, *Formation and evolution of inhomogeneities in non-linear theory of gravitational instability*, Astrofizika, 9, 549

Doroshkevich, A. G. & Shandarin, S. F. 1978, *Statistical problems in the theory of galaxy formation*, AZh, 55, 1144

Doroshkevich, A. G., Shandarin, S. F., & Saar, E. 1978, *Spatial structure of protoclusters and the formation of galaxies*, MNRAS, 184, 643

Doroshkevich, A. G., Sunyaev, R. A., & Zeldovich, I. B. 1974, *The formation of galaxies in Friedmannian universes*, in IAU Symposium, Vol. 63, Confrontation of Cosmological Theories with Observational Data, ed. M. S. Longair, 213

Dressler, A. 1980, *Galaxy morphology in rich clusters — Implications for the formation and evolution of galaxies*, ApJ, 236, 351

Eddington, A. S. 1914, Stellar Movement and the Structure of the Universe (London: McMillan and Co.)

Eddington, A. S. 1924, *On the relation between the masses and luminosities of the stars*, MNRAS, 84, 308

Eddington, A. S. 1926, The Internal Constitution of the Stars, ed. Eddington, A. S.

Eelsalu, H. 1959, *The gradient of the gravitational acceleration perpendicular to the Galactic plane near the Sun*, Tartu Astr. Obs. Publ., 33, 153

Efstathiou, G., Davis, M., White, S. D. M., & Frenk, C. S. 1985, *Numerical techniques for large cosmological N-body simulations*, ApJS, 57, 241

Efstathiou, G., Sutherland, W. J., & Maddox, S. J. 1990, *The cosmological constant and cold dark matter*, Nature, 348, 705

Eggen, O. J. 1950, *Photoelectric Studies. I-Color-Luminosity Array for Members of the Hyades Cluster*, ApJ, 111, 65

Eggen, O. J., Lynden-Bell, D., & Sandage, A. R. 1962, *Evidence from the motions of old stars that the Galaxy collapsed*, ApJ, 136, 748

Eggen, O. J. & Sandage, A. R. 1964, *New photoelectric observations of stars in the old galactic cluster M67*, ApJ, 140, 130

Einasto, J. 1952, *The kinematical division of the main sequence into two parts*, Tartu Astr. Obs. Publ., 32, 231

Einasto, J. 1955, *On the kinematical structure of the main sequence*, Tartu Astr. Obs. Publ., 32, 371

Einasto, J. 1961, *On the asymmetric shift of centroids of star groups*, Tartu Astr. Obs. Publ., 33, 371

Einasto, J. 1965, *On the construction of a composite model for the Galaxy and on the determination of the system of Galactic parameters*, Trudy Astrophys. Inst. Alma-Ata, 5, 87 (Tartu Astr. Obs. Teated 17)

Einasto, J. 1968a, *On constructing models of stellar systems. II-The descriptive functions and parameters*, Tartu Astr. Obs. Publ., 36, 357

Einasto, J. 1968b, *On constructing models of stellar systems. IV-The power-polynomial model*, Tartu Astr. Obs. Publ., 36, 396

Einasto, J. 1968c, *On constructing models of stellar systems. V-The binomial model*, Tartu Astr. Obs. Publ., 36, 414

Einasto, J. 1968d, *On constructing models of stellar systems. VI-On the methods of model constructing*, Tartu Astr. Obs. Publ., 36, 442

Einasto, J. 1969a, *On Galactic Descriptive Functions*, Astronomische Nachrichten, 291, 97

Einasto, J. 1969b, *The Andromeda galaxy M 31: I-A preliminary model*, Astrofizika, 5, 137

Einasto, J. 1970a, *On the structure and evolution of the Galaxy*, Tartu Astr. Obs. Teated, 26, 1

Einasto, J. 1970b, *Structural and kinematical properties of populations of the Andromeda galaxy*, Tartu Ast. Obs. Teated, 26, 23

Einasto, J. 1970c, *The Andromeda galaxy M31. II-Hydrodynamical model. Theory*, Astrofizika, 6, 149

Einasto, J. 1972a, *Galactic models and stellar orbits*, Tartu Astr. Obs. Teated, 40, 1

Einasto, J. 1972b, *Structure and Evolution of Regular Galaxies*, PhD thesis, Tartu University, Tartu

Einasto, J. 1972c, *The Rate of Star Formation*, Astrophys. Lett., 11, 195

Einasto, J. 1973, *The correlation between kinematical properties and ages of stellar populations*, Astronomicheskij Tsirkulyar, 790, 3

Einasto, J. 1974a, *Galactic Models and Stellar Orbits (Invited Lecture)*, in Stars and the Milky Way System, ed. L. N. Mavridis, 291

Einasto, J. 1974b, *The correlation between kinematical properties and ages of stellar populations*, in Kinematics and Ages of Stars Near the Sun, ed. L. Perek, 419

Einasto, J. 1975, *"Hidden" mass in galaxies*, Zemlia i Vselennaia, 3, 32

Einasto, J. 1978, *The Scatter in Mass-To-Luminosity Ratios*, in IAU Symposium, Vol. 79, Large Scale Structures in the Universe. ed. M. S. Longair & J. Einasto, 96

Einasto, J. 2001a, *Dark Matter and Large Scale Structure*, in Astronomical Society of the Pacific Conference Series, Vol. 252, Historical Development of Modern Cosmology, ed. V. J. Martínez, V. Trimble, & M. J. Pons-Bordería, 85

Einasto, J. 2001b, *Large scale structure*, New Astronomy Review, 45, 355

Einasto, J. 2005, *Dark Matter: Early Considerations*, in Frontiers of Cosmology, ed. A. Blanchard & M. Signore, 241

Einasto, J. 2009, *Dark Matter*, UNESCO EOLSS ENCYCLOPEDIA (ArXiv:0901.0632)

Einasto, J., Corwin, Jr., H. G., Huchra, J., Miller, R. H., & Tarenghi, M. 1983a, *Local cells of the universe — Two voids of diameter 200 MPC*, Highlights of Astronomy, 6, 757

Einasto, J. & Einasto, L. 1972a, *Descriptive functions of the Galaxy*, Tartu Astr. Obs. Teated, 36, 46

Einasto, J. & Einasto, L. 1972b, *Modified exponential models of stellar systems*, Tartu Astr. Obs. Teated, 36, 3

Einasto, J., Einasto, M., Gottloeber, S., et al. 1997a, *A 120 MPC Periodicity in the Three-Dimensional Distribution of Galaxy Superclusters*, Nature, 385, 139

Einasto, J., Einasto, M., & Gramann, M. 1989, *Structure and formation of superclusters. IX - Self-similarity of voids*, MNRAS, 238, 155

Einasto, J., Einasto, M., Gramann, M., & Saar, E. 1991, *Structure and formation of superclusters. XIII - The void probability function*, MNRAS, 248, 593

Einasto, J., Einasto, M., Hütsi, G., et al. 2003a, *Clusters and superclusters in the Las Campanas redshift survey*, A&A, 410, 425

Einasto, J., Einasto, M., Saar, E., Jones, B. J. T., & Martinez, V. J. 1988, *Superclustering: Theory Versus Observations*, in IAU Symposium, Vol. 130, Large Scale Structures of the Universe, ed. J. Audouze, M.-C. Pelletan, & S. Szalay, 245

Einasto, J., Einasto, M., Saar, E., et al. 2006, *Luminous superclusters: remnants from inflation?*, A&A, 459, L1

Einasto, J., Einasto, M., Saar, E., et al. 2007a, *Superclusters of galaxies from the 2dF redshift survey. II. Comparison with simulations*, A&A, 462, 397

Einasto, J., Einasto, M., Tago, E., et al. 1999a, *Steps toward the Power Spectrum of Matter. II-The Biasing Correction with σ_8 Normalization*, ApJ, 519, 456

Einasto, J., Einasto, M., Tago, E., et al. 2007b, *Superclusters of galaxies from the 2dF redshift survey. I-The catalogue*, A&A, 462, 811

Einasto, J., Einasto, M., Tago, E., et al. 1999b, *Steps toward the Power Spectrum of Matter. III-The Primordial Spectrum*, ApJ, 519, 469

Einasto, J., Einasto, M., Tago, E., et al. 1999c, *Steps toward the Power Spectrum of Matter. I-The Mean Spectrum of Galaxies*, ApJ, 519, 441

Einasto, J. & Gramann, M. 1993, *Transition scale to a homogeneous universe*, ApJ, 407, 443

Einasto, J., Gramann, M., Einasto, M., et al. 1986a, *Structure and Formation of Superclusters. Supercluster-Void Topology*, Tartu Astr. Obs. Preprint, 3

Einasto, J., Gramann, M., Saar, E., & Tago, E. 1993, *Power spectrum of the matter distribution in the universe on large scales*, MNRAS, 260, 705

Einasto, J. & Haud, U. 1989, *Galactic models with massive corona. I - Method. II - Galaxy*, A&A, 223, 89

Einasto, J., Haud, U., & Jõeveer, M. 1979a, *The Galactic Circular Velocity Near the Sun*, in IAU Symposium, Vol. 84, The Large-Scale Characteristics of the Galaxy, ed. W. B. Burton, 231

Einasto, J., Haud, U., Jõeveer, M., & Kaasik, A. 1976a, *The Magellanic Stream and the mass of our hypergalaxy*, MNRAS, 177, 357

Einasto, J., Haud, U., Jõeveer, M., Kaasik, A., & Traat, P. 1978, *Galactic model with a massive corona*, Astronomicheskij Tsirkulyar, 1023, 2

Einasto, J., Hütsi, G., Einasto, M., et al. 2003b, *Clusters and superclusters in the Sloan Digital Sky Survey*, A&A, 405, 425

Einasto, J., Hütsi, G., Saar, E., et al. 2011a, *Wavelet analysis of the cosmic web formation*, A&A, 531, A75

Einasto, J. & Jaaniste, J. 1982, *On the search for large-scale structure of the universe*, Priroda, 12, 80

Einasto, J., Jaaniste, J., Jõeveer, M., et al. 1974a, *Hypergalaxies*, Tartu Astr. Obs. Teated, 48, 3

Einasto, J. & Jõeveer, M. 1978, *Structure of the Galaxy*, Zemlia i Vselennaia, 6, 30

Einasto, J. & Jõeveer, M. 1979, *The structure of the Universe and the formation of galaxies*, in Science and Modern Times, ed. J. Kivi (Eesti Raamat), 57

Einasto, J., Jõeveer, M., & Kaasik, A. 1976b, *A new model of the Galaxy*, in Stars and Galaxies from Observational Points of View, ed. E. Kharadze, Third European Astronomy Meeting (Mecniereba, Tbilisi), 398

Einasto, J., Jõeveer, M., & Kaasik, A. 1976c, *The mass of the Galaxy*, Tartu Astr. Obs. Teated, 54, 3

Einasto, J., Jõeveer, M., Kaasik, A., Kalamees, P., & Vennik, J. 1977, *A list of hypergalaxies*, Tartu Astr. Obs. Teated, 49

Einasto, J., Jõeveer, M., Kaasik, A., & Vennik, J. 1976d, *The dynamics of aggregates of galaxies as related to their main galaxies*, A&A, 53, 35

Einasto, J., Jõeveer, M., Kaasik, A., & Vennik, J. 1976e, *The missing mass around galaxies*, in Stars and Galaxies from Observational Points of View, ed. E. K. Kharadze, 431

Einasto, J., Jõeveer, M., Kivila, A., & Tago, E. 1975a, *Superclusters of galaxies*, Astronomicheskij Tsirkulyar, 895, 2

Einasto, J., Jõeveer, M., & Saar, E. 1980a, *Structure of superclusters and supercluster formation*, MNRAS, 193, 353

Einasto, J., Jõeveer, M., & Saar, E. 1980b, *Superclusters and galaxy formation*, Nature, 283, 47

Einasto, J., Jõeveer, M., & Saar, E. 1987, *Dark matter — Observational aspects*, in IAU Symp. 117: Dark matter in the universe, eds. J. Kormendy & G. R. Knapp, 243

Einasto, J. & Kaasik, A. 1973, *The correlation between colour and mass-luminosity ratio of old galactic population*, Astronomicheskij Tsirkulyar, 790, 1

Einasto, J., Kaasik, A., Kalamees, P., & Vennik, J. 1975b, *The structure of groups of galaxies*, A&A, 40, 161

Einasto, J., Kaasik, A., & Saar, E. 1974b, *Dynamic Evidence on Massive coronas of galaxies*, Nature, 250, 309

Einasto, J., Klypin, A., & Shandarin, S. 1983b, *Structure of neighboring superclusters — A quantitative analysis*, in IAU Symposium, Vol. 104, Early Evolution of the Universe and its Present Structure, ed. G. O. Abell & G. Chincarini, 265

Einasto, J., Klypin, A. A., Saar, E., & Shandarin, S. F. 1984, *Structure of superclusters and supercluster formation. III-Quantitative study of the local supercluster*, MNRAS, 206, 529

Einasto, J. & Kutuzov, S. 1964, *On the system of Galactic parameters*, Tartu Astr. Obs. Teated, 11, 11

Einasto, J. & Laigo, R. 1973, *Thermal regime of telescope buildings*, Tartu Astr. Obs. Publ., 41, 211

Einasto, J. & Lynden-Bell, D. 1982, *On the mass of the Local Group and the motion of its barycentre*, MNRAS, 199, 67

Einasto, J. & Miller, R. H. 1983, *Neighboring Superclusters and Their Environs*, in IAU Symposium, Vol. 104, Early Evolution of the Universe and its Present Structure, eds. G. O. Abell & G. Chincarini, 405

Einasto, J. & Rümmel, U. 1970a, *Density Distribution and the Radial Velocity Field in the Spiral Arms of M31*, in IAU Symposium, Vol. 38, The Spiral Structure of our Galaxy, eds. W. Becker & G. I. Kontopoulos, 42

Einasto, J. & Rümmel, U. 1970b, *The Andromeda Galaxy M31. III-Hydrodynamical model*, Astrofizika, 6, 241

Einasto, J. & Rümmel, U. 1970c, *The Rotation Curve, Mass, Light and Velocity Distribution of M31*, in IAU Symposium, Vol. 38, The Spiral Structure of our Galaxy, eds. W. Becker & G. I. Kontopoulos, 51

Einasto, J. & Rümmel, U. 1972, *Descriptive functions of the Andromeda galaxy*, Tartu Astr. Obs. Teated, 36, 55

Einasto, J. & Saar, E. 1982, *Supercluster of Galaxies*, in Marcel Grossmann Meeting: General Relativity, 875

Einasto, J. & Saar, E. 1987, *Spatial distribution of galaxies — Biased galaxy formation, supercluster-void topology, and isolated galaxies*, in IAU Symposium, Vol. 124, Observational Cosmology, ed. A. Hewitt, G. Burbidge, & L. Z. Fang, 349

Einasto, J., Saar, E., Einasto, M., Freudling, W., & Gramann, M. 1994a, *The fraction of matter in voids*, ApJ, 429, 465

Einasto, J., Saar, E., Kaasik, A., & Chernin, A. D. 1974c, *Missing mass around galaxies — Morphological evidence*, Nature, 252, 111

Einasto, J., Saar, E., Kaasik, A., & Traat, P. 1974d, *Dynamical evidence for the presence of hidden mass in galaxies*, Astronomicheskij Tsirkulyar, 811, 3

Einasto, J., Saar, E., & Klypin, A. A. 1986b, *Structure of superclusters and supercluster formation. V - Spatial correlation and voids*, MNRAS, 219, 457

Einasto, J., Suhhonenko, I., Hütsi, G., et al. 2011b, *Towards understanding the structure of voids in the cosmic web*, A&A, 534, A128

Einasto, J., Tenjes, P., Barabanov, A. V., & Zasov, A. V. 1980c, *Central holes in disks of spiral galaxies*, Ap&SS, 67, 31

Einasto, J., Tenjes, P., & Traat, P. 1979b, *A Mass Distribution Model of the Andromeda Galaxy*, Astronomicheskij Tsirkulyar, 1032, 5

Einasto, J. E., Chernin, A. D., & Jõeveer, M. M. 1975c, *Hidden mass in galaxies*, Priroda, 5, 39

Einasto, M. 1991, *Structure and formation of superclusters. XIV - Correlation functions: Dependence on the intrinsic properties of galaxy samples*, MNRAS, 252, 261

Einasto, M. & Einasto, J. 1987, *Morphology of Isolated and Grouped Galaxies*, in IAU Symposium, Vol. 121, Observational Evidence of Activity in Galaxies, eds. E. E. Khachikian, K. J. Fricke, & J. Melnick, 101

Einasto, M. & Einasto, J. 1989, *Self similarity of voids in large-scale distribution of systems of galaxies*, Tartu Astr. Obs. Teated, 95, 21

Einasto, M., Einasto, J., Tago, E., Dalton, G. B., & Andernach, H. 1994b, *The Structure of the Universe Traced by Rich Clusters of Galaxies*, MNRAS, 269, 301

Einasto, M., Einasto, J., Tago, E., Müller, V., & Andernach, H. 2001, *Optical and X-Ray Clusters as Tracers of the Supercluster-Void Network. I-Superclusters of Abell and X-Ray Clusters*, AJ, 122, 2222

Einasto, M., Einasto, J., Tago, E., et al. 2007c, *Superclusters of galaxies in the 2dF redshift survey. III-The properties of galaxies in superclusters*, A&A, 464, 815

Einasto, M., Liivamägi, L. J., Saar, E., et al. 2011c, *SDSS DR7 superclusters. Principal component analysis*, A&A, 535, A36

Einasto, M., Liivamägi, L. J., Tago, E., et al. 2011d, *SDSS DR7 superclusters. Morphology*, A&A, 532, A5

Einasto, M., Liivamägi, L. J., Tempel, E., et al. 2011e, *The Sloan Great Wall. Morphology and Galaxy Content*, ApJ, 736, 51

Einasto, M., Saar, E., Liivamägi, L. J., et al. 2007d, *The richest superclusters. I. Morphology*, A&A, 476, 697

Einasto, M., Tago, E., Jaaniste, J., Einasto, J., & Andernach, H. 1997b, *The supercluster-void network I-The supercluster catalogue and large-scale distribution*, A&AS, 123, 119

Einasto, M., Tago, E., Saar, E., et al. 2010, *The Sloan great wall. Rich clusters*, A&A, 522, A92

Einasto, M., Vennik, J., Nurmi, P., et al. 2012, *Multimodality in galaxy clusters from SDSS DR8: substructure and velocity distribution*, A&A, 540, A123

Einstein, A. 1916, *Die Grundlage der allgemeinen Relativitätstheorie*, Annalen der Physik, 354, 769

Einstein, A. & de Sitter, W. 1932, *On the Relation between the Expansion and the Mean Density of the Universe*, Proceedings of the National Academy of Science, 18, 213

Eisenstein, D. J., Zehavi, I., Hogg, D. W., et al. 2005, *Detection of the Baryon Acoustic Peak in the Large-Scale Correlation Function of SDSS Luminous Red Galaxies*, ApJ, 633, 560

Eke, V. R., Baugh, C. M., Cole, S., et al. 2004, *Galaxy groups in the 2dFGRS: the group-finding algorithm and the 2PIGG catalogue*, MNRAS, 348, 866

Faber, S. M. 1982a, *Galaxy formation via hierarchical clustering and dissipation: the structure of disk systems*, in Astrophysical Cosmology Proceedings, eds. H. A. Brueck, G. V. Coyne, & M. S. Longair, 191

Faber, S. M. 1982b, *Galaxy formation via hierarchical clustering and dissipation: the structure of spheroids*, in Astrophysical Cosmology Proceedings, eds. H. A. Brueck, G. V. Coyne, & M. S. Longair, 219

Faber, S. M. 1984, *Galaxy Formation and Cosmology*, in Large-Scale Structure of the Universe, ed. G. Setti & L. Van Hove, 187

Faber, S. M., Balick, B., Gallagher, J. S., & Knapp, G. R. 1977, *The neutral hydrogen content, stellar rotation curve, and mass-to-light ratio of NGC 4594, the "Sombrero" galaxy*, ApJ, 214, 383

Faber, S. M. & Gallagher, J. S. 1979, *Masses and mass-to-light ratios of galaxies*, ARA&A, 17, 135

Faber, S. M. & Jackson, R. E. 1976, *Velocity dispersions and mass-to-light ratios for elliptical galaxies*, ApJ, 204, 668

Faber, S. M. & Lin, D. N. C. 1983, *Is there nonluminous matter in dwarf spheroidal galaxies*, ApJ, 266, L17

Fairall, A. P. 1998, Large-scale structure in the universe (Chichester: Wiley/Praxis)

Fesenko, B. I. 1976, *The missing mass and the apparent distribution of galaxies*, in Stars and Galaxies from Observational Points of View, ed. E. K. Kharadze, 486

Field, G. B. 1972, *Intergalactic Matter*, ARA&A, 10, 227

Forman, W., Kellogg, E., Gursky, H., Tananbaum, H., & Giacconi, R. 1972, *Observations of the Extended X-Ray Sources in the Perseus and Coma Clusters from UHURU*, ApJ, 178, 309

Freeman, K. C. & Munsuk, C. 1972, *Masses of old LMC globular clusters*, Proceedings of the Astronomical Society of Australia, 2, 151

Frenk, C. S., White, S. D. M., & Davis, M. 1983, *Nonlinear evolution of large-scale structure in the universe*, ApJ, 271, 417

Friedmann, A. 1922, *Über die Krümmung des Raumes*, Zeitschrift fur Physik, 10, 377

Friedmann, A. 1924, *Über die Möglichkeit einer Welt mit konstanter negativer Krümmung des Raumes*, Zeitschrift fur Physik, 21, 326

Fry, J. N. & Peebles, P. J. E. 1978, *Statistical analysis of catalogs of extragalactic objects. IX - The four-point galaxy correlation function*, ApJ, 221, 19

Gao, L., Navarro, J. F., Frenk, C. S., et al. 2012, *The Phoenix Project: the dark side of rich Galaxy clusters*, MNRAS, 425, 2169

Gao, L., Springel, V., & White, S. D. M. 2005a, *The age dependence of halo clustering*, MNRAS, 363, L66

Gao, L., White, S. D. M., Jenkins, A., Frenk, C. S., & Springel, V. 2005b, *Early structure in Lambda CDM*, MNRAS, 363, 379

Geller, M. J. 1987, *Large scale Structure in the Universe — Some Clues from Optical Data*, in Theory and Observational Limits in Cosmology, ed. W. R. Stoeger, 231

Geller, M. J., Diaferio, A., & Kurtz, M. J. 2011, *Mapping the Universe: The 2010 Russell Lecture*, AJ, 142, 133

Geller, M. J. & Huchra, J. P. 1989, *Mapping the universe*, Science, 246, 897

Gilmore, G., Wilkinson, M. I., Wyse, R. F. G., et al. 2007, *The Observed Properties of Dark Matter on Small Spatial Scales*, ApJ, 663, 948

Gilmore, G., Wyse, R. F. G., & Kuijken, K. 1989, *Kinematics, chemistry, and structure of the Galaxy*, ARA&A, 27, 555

Gott, III, J. R., Dickinson, M., & Melott, A. L. 1986, *The sponge-like topology of large-scale structure in the universe*, ApJ, 306, 341

Graham, A. W., Merritt, D., Moore, B., Diemand, J., & Terzić, B. 2006, *Empirical Models for Dark Matter Halos. II-Inner Profile Slopes, Dynamical Profiles, and ρ/σ^3*, AJ, 132, 2701

Gramann, M. 1987, *Formation of the structure in an axion Universe with a cosmological constant*, Tartu Astr. Obs. Publ., 52, 216

Gramann, M. 1988, *Structure and formation of superclusters. VIII - Evolution of structure in a model with cold dark matter and cosmological constant*, MNRAS, 234, 569

Gramann, M. 1990, *Structure and formation of superclusters. XI - Voids, connectivity and mean density in the CDM universes*, MNRAS, 244, 214

Gramann, M. & Einasto, J. 1992, *The power spectrum in nearby superclusters*, MNRAS, 254, 453

Gregory, S. A. & Thompson, L. A. 1978, *The Coma/A1367 supercluster and its environs*, ApJ, 222, 784

Gregory, S. A. & Thompson, L. A. 1982, *Superclusters and voids in the distribution of galaxies*, Scientific American, 246, 88

Groth, E. J. & Peebles, P. J. E. 1977, *Statistical analysis of catalogs of extragalactic objects. VII - Two- and three-point correlation functions for the high-resolution Shane-Wirtanen catalog of galaxies*, ApJ, 217, 385

Gunn, J. E. & Tinsley, B. M. 1975, *An accelerating Universe*, Nature, 257, 454

Gursky, H., Solinger, A., Kellogg, E. M., et al. 1972, *X-ray Emission from Rich Clusters of Galaxies*, ApJ, 173, L99

Guth, A. H. 1981, *Inflationary universe: A possible solution to the horizon and flatness problems*, Phys. Rev. D, 23, 347

Guzzo, L., Iovino, A., Chincarini, G., Giovanelli, R., & Haynes, M. P. 1991, *Scale-invariant clustering in the large-scale distribution of galaxies*, ApJ, 382, L5

Härm, R. & Schwarzschild, M. 1955, *Inhomogeneous Stellar Models. IV. Models with Continuously Varying Chemical Composition*, ApJ, 121, 445

Haud, U. & Einasto, J. 1989, *Galactic Models with Massive Corona — Part Two - Galaxy*, A&A, 223, 95

Hauser, M. G. & Peebles, P. J. E. 1973, *Statistical Analysis of Catalogs of Extragalactic Objects. 11. the Abell Catalog of Rich Clusters*, ApJ, 185, 757

Hektor, A., Raidal, M., & Tempel, E. 2013, *Evidence for Indirect Detection of Dark Matter from Galaxy Clusters in Fermi γ-Ray Data*, ApJ, 762, L22

Hen, L. & Schwarzschild, M. 1949, *Red-Giant Models with Chemical Inhomogeneities*, MNRAS, 109, 631

Hewish, A., Bell, S. J., Pilkington, J. D. H., Scott, P. F., & Collins, R. A. 1968, *Observation of a Rapidly Pulsating Radio Source*, Nature, 217, 709

Hill, E. R. 1960, *The component of the galactic gravitational field perpendicular to the galactic plane, K_z*, Bull. Astron. Inst. Netherlands, 15, 1

Hockney, R. W. & Eastwood, J. W. 1981, Computer Simulation Using Particles, eds. Hockney, R. W. & Eastwood, J. W.

Holberg, J., Bowyer, S., & Lampton, M. 1973, *Upper Limits on an Ionized Intracluster Medium in the Coma Cluster*, ApJ, 180, L55

Holmberg, E. 1937, *A study of double and multiple galaxies*, Annals of the Observatory of Lund, 6, 1

Holmberg, J. & Flynn, C. 2000, *The local density of matter mapped by Hipparcos*, MNRAS, 313, 209
Hoyle, F. 1948, *A New Model for the Expanding Universe*, MNRAS, 108, 372
Hoyle, F. & Schwarzschild, M. 1955, *On the Evolution of Type II Stars*, ApJS, 2, 1
Hubble, E. 1929a, *A Relation between Distance and Radial Velocity among Extra-Galactic Nebulae*, Proceedings of the National Academy of Science, 15, 168
Hubble, E. P. 1925, *NGC 6822, a remote stellar system*, ApJ, 62, 409
Hubble, E. P. 1926, *A spiral nebula as a stellar system: Messier 33*, ApJ, 63, 236
Hubble, E. P. 1929b, *A spiral nebula as a stellar system, Messier 31*, ApJ, 69, 103
Huchra, J. P. & Geller, M. J. 1982, *Groups of galaxies. I - Nearby groups*, ApJ, 257, 423
Humphrey, P. J. & Buote, D. A. 2010, *The slope of the mass profile and the tilt of the Fundamental Plane in early-type galaxies*, MNRAS, 403, 2143
Humphrey, P. J., Buote, D. A., Gastaldello, F., et al. 2006, *A Chandra View of Dark Matter in Early-Type Galaxies*, ApJ, 646, 899
Hütsi, G. 2006, *Acoustic oscillations in the SDSS DR4 luminous red galaxy sample power spectrum*, A&A, 449, 891
Hütsi, G., Chluba, J., Hektor, A., & Raidal, M. 2011, *WMAP7 and future CMB constraints on annihilating dark matter: implications for GeV-scale WIMPs*, A&A, 535, A26
Hütsi, G., Hektor, A., & Raidal, M. 2009, *Constraints on leptonically annihilating dark matter from reionization and extragalactic gamma background*, A&A, 505, 999
Hütsi, G., Hektor, A., & Raidal, M. 2010, *Implications of the Fermi-LAT diffuse gamma-ray measurements on annihilating or decaying dark matter*, J. Cosmology Astropart. Phys., 7, 8
Idlis, G. 1956, *Gravitational potential of the Galaxy and some problems of the evolution of the Galaxy*, AZh, 33, 20
Idlis, G. 1957, *Gravitational potential of the Galaxy and some problems of the evolution of the Galaxy*, Izv. Astrofiz. Inst. Kazakh Acad. Sciences, 5–6, 3
Illingworth, G. & Freeman, K. C. 1974, *The Mass of the Globular Cluster NGC 6388*, ApJ, 188, L83
Jõeveer, M. 1968a, *On the kinematics of the B stars*, Tartu Astr. Obs. Publ., 36, 84
Jõeveer, M. 1968b, *The galactovertical motions and luminosities of K stars*, Tartu Astr. Obs. Publ., 36, 70
Jõeveer, M. 1972, *An attempt to estimate the Galactic mass density in the vicinity of the Sun*, Tartu Astr. Obs. Teated, 37, 3
Jõeveer, M. 1974, *Ages of delta Cephei stars and the Galactic mass density near the Sun*, Tartu Astr. Obs. Teated, 46, 35
Jõeveer, M. & Einasto, J. 1978, *Has the universe the cell structure*, in IAU Symposium, Vol. 79, Large Scale Structures in the Universe, eds. M. S. Longair & J. Einasto, 241
Jõeveer, M., Einasto, J., & Tago, E. 1977, *The cell structure of the Universe*, Tartu Astr. Obs. Preprint, 3
Jõeveer, M., Einasto, J., & Tago, E. 1978, *Spatial distribution of galaxies and of clusters of galaxies in the southern galactic hemisphere*, MNRAS, 185, 357
Jaaniste, J. & Saar, E. 1975, *On the stellar component of galactic coronae*, Tartu Astr. Obs. Publ., 43, 216
Jeans, J. H. 1922, *The motions of stars in a Kapteyn universe*, MNRAS, 82, 122

Jones, B. J. T. 2009, *The Sea of Wavelets*, in Lecture Notes in Physics, Berlin Springer Verlag, Vol. 665, Data Analysis in Cosmology, ed. V. J. Martinez, E. Saar, E. M. Gonzales, & M. J. Pons-Borderia, 3

Jones, B. J. T., Martinez, V. J., Saar, E., & Einasto, J. 1988, *Multifractal description of the large-scale structure of the universe*, ApJ, 332, L1

Kahn, F. D. & Woltjer, L. 1959, *Intergalactic Matter and the Galaxy*, ApJ, 130, 705

Kaiser, N. 1984, *On the spatial correlations of Abell clusters*, ApJ, 284, L9

Kalnajs, A. J. 1987, *Halos and disk stability*, in IAU Symposium, Vol. 117, Dark matter in the universe, ed. J. Kormendy & G. R. Knapp, 289

Kapteyn, J. C. 1922, *First Attempt at a Theory of the Arrangement and Motion of the Sidereal System*, ApJ, 55, 302

Kapteyn, J. C. & van Rhijn, P. J. 1920, *On the Distribution of the Stars in Space Especially in the High Galactic Latitudes*, ApJ, 52, 23

Karachentsev, I. D., Kashibadze, O. G., Makarov, D. I., & Tully, R. B. 2009, *The Hubble flow around the Local Group*, MNRAS, 393, 1265

Karachentsev, I. D., Makarov, D. I., Sharina, M. E., et al. 2003, *Local galaxy flows within 5 Mpc*, A&A, 398, 479

Karachentsev, I. D., Sharina, M. E., Makarov, D. I., et al. 2002, *The very local Hubble flow*, A&A, 389, 812

Karachentsev, I. D., Tully, R. B., Dolphin, A., et al. 2007, *The Hubble Flow around the Centaurus A/M83 Galaxy Complex*, AJ, 133, 504

Kellogg, E., Murray, S., Giacconi, R., Tananbaum, T., & Gursky, H. 1973, *Clusters of Galaxies with a Wide Range of X-Ray Luminosities*, ApJ, 185, L13

Kerr, F. J. & Lynden-Bell, D. 1986, *Review of galactic constants*, MNRAS, 221, 1023

Kiang, T. 1967, *On the clustering of rich clusters of galaxies*, MNRAS, 135, 1

Kim, J., Park, C., Rossi, G., Lee, S. M., & Gott, III, J. R. 2011, *The New Horizon Run Cosmological N-Body Simulations*, Journal of Korean Astronomical Society, 44, 217

King, I. R. 1977, *Introduction: Galaxies and Their Populations — The View on a Cloudy Day*, in Evolution of Galaxies and Stellar Populations, eds. B. M. Tinsley & R. B. G. Larson, D. Campbell, 1

Kirshner, R. P., Oemler, Jr., A., Schechter, P. L., & Shectman, S. A. 1981, *A million cubic megaparsec void in Bootes*, ApJ, 248, L57

Klypin, A., Kravtsov, A. V., Valenzuela, O., & Prada, F. 1999, *Where Are the Missing Galactic Satellites?*, ApJ, 522, 82

Klypin, A. A., Einasto, J., Einasto, M., & Saar, E. 1989, *Structure and formation of superclusters. X - Fractal properties of superclusters*, MNRAS, 237, 929

Klypin, A. A. & Kopylov, A. I. 1983, *The Spatial Covariance Function for Rich Clusters of Galaxies*, Soviet Astronomy Letters, 9, 41

Klypin, A. A. & Shandarin, S. F. 1983, *Three-dimensional numerical model of the formation of large-scale structure in the Universe*, MNRAS, 204, 891

Kofman, L. A., Einasto, J., & Linde, A. D. 1987, *Cosmic bubbles as remnants from inflation*, Nature, 326, 48

Kofman, L. A. & Shandarin, S. F. 1988, *Theory of adhesion for the large-scale structure of the universe*, Nature, 334, 129

Kofman, L. A. & Starobinsky, A. A. 1985, *Effect of the Cosmological Constant on Largescale Anisotropies in the Microwave Background*, Soviet Astronomy Letters, 11, 271

Kofman, L. A., Starobinsky, A. A., Shandarin, S. F., & Einasto, J. E. 1986, *Second Tartu Cosmology Seminar — Missing Mass Largescale Structure Microwave Background*, Soviet Astronomy, 30, 729

Komatsu, E., Smith, K. M., Dunkley, J., et al. 2011, *Seven-year Wilkinson Microwave Anisotropy Probe (WMAP) Observations: Cosmological Interpretation*, ApJS, 192, 18

Komberg, B. V. & Novikov, I. D. 1975, *Nature of the coronae of spiral galaxies*, Pis ma Astronomicheskii Zhurnal, 1, 3

Kormendy, J. & Bender, R. 2011, *Supermassive black holes do not correlate with dark matter haloes of galaxies*, Nature, 469, 377

Kormendy, J., Bender, R., & Cornell, M. E. 2011, *Supermassive black holes do not correlate with galaxy disks or pseudobulges*, Nature, 469, 374

Kormendy, J. & Freeman, K. C. 2004, *Scaling Laws for Dark Matter Halos in Late-Type and Dwarf Spheroidal Galaxies*, in IAU Symposium, Vol. 220, Dark Matter in Galaxies, ed. S. Ryder, D. Pisano, M. Walker, & K. Freeman, 377

Kuhn, T. S. 1970, The structure of scientific revolutions (Chicago: University of Chicago Press, 2nd ed., enlarged)

Kuijken, K. & Gilmore, G. 1989a, *The Mass Distribution in the Galactic Disc — II - Determination of the Surface Mass Density of the Galactic Disc Near the Sun*, MNRAS, 239, 605

Kuijken, K. & Gilmore, G. 1989b, *The Mass Distribution in the Galactic Disc — Part III - the Local Volume Mass Density*, MNRAS, 239, 651

Kuijken, K. & Gilmore, G. 1989c, *The mass distribution in the galactic disc. I - A technique to determine the integral surface mass density of the disc near the sun*, MNRAS, 239, 571

Kundic, T., Colley, W. N., Gott, III, J. R., et al. 1995, *An Event in the Light Curve of 0957+561A and Prediction of the 1996 Image B Light Curve*, ApJ, 455, L5

Kundic, T., Turner, E. L., Colley, W. N., et al. 1997, *A Robust Determination of the Time Delay in 0957+561A, B and a Measurement of the Global Value of Hubble's Constant*, ApJ, 482, 75

Kutuzov, S. A. 1965, *The problem of estimating a galactic parameter system*, Trudy Astrophys. Inst. Acad. Nauk. Kazakh SSR, 5, 78

Kutuzov, S. A. 1968, *On constructing models of stellar systems. III. On the description of the mass distribution in the models*, Tartu Astr. Obs. Publ., 36, 379

Kutuzov, S. A. & Einasto, J. 1968, *On constructing models of stellar systems. I-On the classification of the models*, Tartu Astr. Obs. Publ., 36, 341

Kuzmin, G. 1952a, *On the distribution of mass in the Galaxy*, Tartu Astr. Obs. Publ., 32, 211

Kuzmin, G. 1952b, *Proper motions of galactic-equatorial A and K stars perpendicular to the Galactic plane, and the dynamical density in Galactic plane*, Tartu Astr. Obs. Publ., 32, 5

Kuzmin, G. 1953, *The third integral of motions of stars and the dynamics of the stationar Galaxy. I*, Tartu Astr. Obs. Publ., 32, 332

Kuzmin, G. 1954, *On the gravitational potential of the Galaxy and third integral of motion of stars*, Tartu Astr. Obs. Teated, 1, 3 (Proc. Estonian Academy of Sciences, 2, 368, 1953)

Kuzmin, G. 1955, *On the value of the dynamical parameter C and the density of matter in the Solar neighbourhood*, Tartu Astr. Obs. Publ., 33, 3

Kuzmin, G. 1956a, *Model of the stationary Galaxy allowing three-axial distribution of velocities*, AZh, 33, 27

Kuzmin, G. 1956b, *Some Problems Concerning the Dynamics of the Galaxy*, Proceedings of the Estonian Academy of Sciences. Series of technical and physical sciences, 5, 91

Lea, S. M., Silk, J., Kellogg, E., & Murray, S. 1973, *Thermal-Bremsstrahlung Interpretation of Cluster X-Ray Sources*, ApJ, 184, L105

Lemaître, G. 1927, *Un Univers homogène de masse constante et de rayon croissant rendant compte de la vitesse radiale des nébuleuses extra-galactiques*, Annales de la Societe Scietifique de Bruxelles, 47, 49

Lightman, A. & Braver, R. 1992, *Origins; the Lives and Worlds of Modern Cosmologists* (Harvard Univ. Press)

Liigant, M. & Einasto, J. 1960, *The Theory of Automatic Satellite Tracking Telescopes*, AZh, 37, 1087

Liivamägi, L. J., Tempel, E., & Saar, E. 2012, *SDSS DR7 superclusters. The catalogues*, A&A, 539, A80

Lin, D. N. C. & Faber, S. M. 1983, *Some implications of nonluminous matter in dwarf spheroidal galaxies*, ApJ, 266, L21

Lindblad, B. 1927, *On the state of motion in the galactic system*, MNRAS, 87, 553

Lindblad, B. 1933, *Die Milchstraße*, Handbuch der Astrophysik, 5, 937

Linde, A. 2002a, *Inflation, Quantum Cosmology and the Anthropic Principle*, ArXiv:hep-th/0211048

Linde, A. 2002b, *Inflationary Theory versus Ekpyrotic/Cyclic Scenario*, arXiv:astro-ph/0205259v1

Linde, A. D. 1982, *A new inflationary universe scenario: A possible solution of the horizon, flatness, homogeneity, isotropy and primordial monopole problems*, Physics Letters B, 108, 389

Linde, A. D. 1983, *Chaotic inflation*, Physics Letters B, 129, 177

Lindner, U., Einasto, J., Einasto, M., et al. 1995, *The structure of supervoids. I. Void hierarchy in the Northern Local Supervoid*, A&A, 301, 329

Lindner, U., Einasto, J., Einasto, M., & Fricke, K. J. 1997, *Void Hierarchy in the Northern Local Void: Faint structures in low density regions of the nearby Universe*, ArXiv: astro-phy/9711046

Lindner, U., Einasto, M., Einasto, J., et al. 1996, *The distribution of galaxies in voids*, A&A, 314, 1

Longair, M. S. 1978, *Personal View — The Large Scale Structure of the Universe*, in IAU Symposium, Vol. 79, Large Scale Structures in the Universe, ed. M. S. Longair & J. Einasto, 451

Longair, M. S. 2006, *The cosmic century: a history of astrophysics and cosmology* (Cambridge University Press)

Lovell, M. R., Eke, V., Frenk, C. S., et al. 2012, *The haloes of bright satellite galaxies in a warm dark matter universe*, MNRAS, 420, 2318

Lubimov, V. A., Novikov, E. G., Nozik, V. Z., Tretyakov, E. F., & Kosik, V. S. 1980, *An estimate of the ν_e mass from the β-spectrum of tritium in the valine molecule*, Physics Letters B, 94, 266

Ludlow, A. D., Navarro, J. F., Springel, V., et al. 2010, *Secondary infall and the pseudo-phase-space density profiles of cold dark matter haloes*, MNRAS, 406, 137

Ludlow, A. D., Navarro, J. F., White, S. D. M., et al. 2011, *The density and pseudo-phase-space density profiles of cold dark matter haloes*, MNRAS, 415, 3895

Lundmark, K. 1924, *The determination of the curvature of space-time in de Sitter's world*, MNRAS, 84, 747

Lundmark, K. 1925, *Nebulæ, The motions and the distances of spiral*, MNRAS, 85, 865

Mandelbrot, B. B. 1982, The Fractal Geometry of Nature, ed. Mandelbrot, B. B.

Martinez, V. J. & Jones, B. J. T. 1990, *Why the universe is not a fractal*, MNRAS, 242, 517

Martinez, V. J., Jones, B. J. T., Dominguez-Tenreiro, R., & van de Weygaert, R. 1990, *Clustering paradigms and multifractal measures*, ApJ, 357, 50

Martínez, V. J. & Saar, E. 2002, Statistics of the Galaxy Distribution, eds. V. J. Martínez & E. Saar (Chapman & Hall/CRC)

Materne, J. & Tammann, G. A. 1974a, *The NGC 7331 Group, a Stable Group of Galaxies Projected on Stephan's Quartet*, A&A, 35, 441

Materne, J. & Tammann, G. A. 1974b, *Virial Tests for Fourteen Nearby Groups of Galaxies*, A&A, 37, 383

Materne, J. & Tammann, G. A. 1976, *On the stability of groups of galaxies and the question of hidden matter*, in Stars and Galaxies from Observational Points of View, ed. E. K. Kharadze, 455

Mather, J. C., Cheng, E. S., Cottingham, D. A., et al. 1994, *Measurement of the cosmic microwave background spectrum by the COBE FIRAS instrument*, ApJ, 420, 439

Mather, J. C., Cheng, E. S., Eplee, Jr., R. E., et al. 1990, *A preliminary measurement of the cosmic microwave background spectrum by the Cosmic Background Explorer (COBE) satellite*, ApJ, 354, L37

Melott, A. L. 1983, *Two-dimensional simulation of the gravitational superclustering of collisionless particles*, MNRAS, 202, 595

Melott, A. L., Einasto, J., Saar, E., et al. 1983, *Cluster analysis of the nonlinear evolution of large-scale structure in an axion/gravitino/photino-dominated universe*, Physical Review Letters, 51, 935

Merritt, D., Graham, A. W., Moore, B., Diemand, J., & Terzić, B. 2006, *Empirical Models for Dark Matter Halos. I-Nonparametric Construction of Density Profiles and Comparison with Parametric Models*, AJ, 132, 2685

Merritt, D., Navarro, J. F., Ludlow, A., & Jenkins, A. 2005, *A Universal Density Profile for Dark and Luminous Matter?* ApJ, 624, L85

Milgrom, M. & Bekenstein, J. 1987, *The modified Newtonian dynamics as an alternative to hidden matter*, in IAU Symposium, Vol. 117, Dark matter in the universe, ed. J. Kormendy & G. R. Knapp, 319

Miller, R. H. 1978, *Free collapse of a rotating sphere of stars*, ApJ, 223, 122

Navarro, J. F., Frenk, C. S., & White, S. D. M. 1997, *A Universal Density Profile from Hierarchical Clustering*, ApJ, 490, 493

Navarro, J. F., Hayashi, E., Power, C., et al. 2004, *The inner structure of Lambda CDM haloes - III. Universality and asymptotic slopes*, MNRAS, 349, 1039

Navarro, J. F., Ludlow, A., Springel, V., et al. 2010, *The diversity and similarity of simulated cold dark matter haloes*, MNRAS, 402, 21

Neyman, J., Page, T., & Scott, E. 1961, *CONFERENCE on the Instability of Systems of Galaxies (Santa Barbara, California, August 10–12, 1961): Summary of the conference*, AJ, 66, 633

Nilson, P. 1973, Uppsala general catalogue of galaxies, Vol. 6 (Uppsala Astr. Obs. Ann.)

Oh, S.-H., de Blok, W. J. G., Brinks, E., Walter, F., & Kennicutt, Jr., R. C. 2011, *Dark and Luminous Matter in THINGS Dwarf Galaxies*, AJ, 141, 193

Oh, S.-H., de Blok, W. J. G., Walter, F., Brinks, E., & Kennicutt, Jr., R. C. 2008, *High-Resolution Dark Matter Density Profiles of THINGS Dwarf Galaxies: Correcting for Noncircular Motions*, AJ, 136, 2761

Oke, J. B. & Schwarzschild, M. 1952, *Inhomogeneous Stellar Models. I. Models with a Convective Core and a Discontinuity in the Chemical Composition*, ApJ, 116, 317

Oort, J. H. 1927, *Observational evidence confirming Lindblad's hypothesis of a rotation of the galactic system*, Bull. Astron. Inst. Netherlands, 3, 275

Oort, J. H. 1928, *Dynamics of the galactic system in the vicinity of the Sun*, Bull. Astron. Inst. Netherlands, 4, 269

Oort, J. H. 1932, *The force exerted by the stellar system in the direction perpendicular to the galactic plane and some related problems*, Bull. Astron. Inst. Netherlands, 6, 249

Oort, J. H. 1940, *Some Problems Concerning the Structure and Dynamics of the Galactic System and the Elliptical Nebulae NGC 3115 and 4494*, ApJ, 91, 273

Oort, J. H. 1958, *Dynamics and Evolution of the Galaxy, in so far as Relevant to the Problem of the Populations*, Ricerche Astronomiche, Vol. 5, Specola Vaticana, Proceedings of a Conference at Vatican Observatory, Castel Gandolfo, May 20–28, 1957, Amsterdam: North-Holland, and New York: Interscience, 1958, edited by D.J.K. O'Connell., p. 415, 5, 415

Oort, J. H. 1960, *Note on the determination of K_z and on the mass density near the Sun*, Bull. Astron. Inst. Netherlands, 15, 45

Oort, J. H. 1983, *Superclusters*, ARA&A, 21, 373

Öpik, E. 1915, *Selective absorption of light in space, and the dynamics of the Universe*, Bull. de la Soc. Astr. de Russie, 21, 150

Öpik, E. 1922a, *An estimate of the distance of the Andromeda Nebula*, ApJ, 55, 406

Öpik, E. 1922b, *Notes on Stellar Statistics and Stellar Evolution*, Tartu Astr. Obs. Publ., 25, 1

Öpik, E. 1923, *On the Luminosity-Curve of Components of Double Stars*, Tartu Astr. Obs. Publ., 25, 1

Öpik, E. 1924, *Statistical Studies of Double Stars: On the Distribution of Relative Luminosities and Distances of Double Stars* in the *Harvard Revised Photometry North of Declination -31 deg*, Tartu Astr. Obs. Publ., 25, 1

Öpik, E. 1933, *Meteorites and the age of the universe*, Popular Astronomy, 41, 71

Öpik, E. 1938, *Stellar structure, source of energy, and evolution*, Tartu Astr. Obs. Publ., 30, 1

Öpik, E. J. 1977, *About Dogma in Science, and Other Recollections of an Astronomer*, ARA&A, 15, 1

Ostriker, J. P. & Cowie, L. L. 1981, *Galaxy formation in an intergalactic medium dominated by explosions*, ApJ, 243, L127

Ostriker, J. P. & Peebles, P. J. E. 1973, *A Numerical Study of the Stability of Flattened Galaxies: or, can Cold Galaxies Survive?*, ApJ, 186, 467

Ostriker, J. P., Peebles, P. J. E., & Yahil, A. 1974, *The size and mass of galaxies, and the mass of the universe*, ApJ, 193, L1

Ostriker, J. P. & Steinhardt, P. J. 1995, *The Observational Case for a Low Density Universe with a Non-Zero Cosmological Constant*, Nature, 377, 600

Ozernoi, L. M. 1974, *Where is the 'hidden' mass localized*, AZh, 51, 1108

Page, T. 1952, *Radial Velocities and Masses of Double Galaxies*, ApJ, 116, 63

Page, T. 1959, *Masses of the double galaxies*, AJ, 64, 53

Page, T. 1960, *Average Masses and Mass-Luminosity Ratios of the Double Galaxies*, ApJ, 132, 910

Parenago, P. 1948, *Structure of the Galaxy*, Uspekhi Astr. Nauk, 4, 69

Parenago, P. 1950, *On the Gravitational Potential of the Galaxy*, AZh, 27, 329

Parenago, P. 1952, *On the Gravitational Potential of the Galaxy II*, AZh, 29, 245

Parijskij, Y. N. 1978, *Search for Primordial Perturbations of the Universe: Observations with RATAN-600 Radio Telescope*, in IAU Symposium, Vol. 79, Large Scale Structures in the Universe, ed. M. S. Longair & J. Einasto, 315

Park, C., Choi, Y.-Y., Kim, J., et al. 2012, *The Challenge of the Largest Structures in the Universe to Cosmology*, ApJ, 759, L7

Park, C. & Gott, III, J. R. 1991, *Dynamical evolution of topology of large-scale structure*, ApJ, 378, 457

Park, C., Gott, III, J. R., & da Costa, L. N. 1992a, *Large-scale structure in the Southern Sky Redshift Survey*, ApJ, 392, L51

Park, C., Gott, III, J. R., Melott, A. L., & Karachentsev, I. D. 1992b, *The topology of large-scale structure. VI - Slices of the universe*, ApJ, 387, 1

Parry, O. H., Eke, V. R., Frenk, C. S., & Okamoto, T. 2012, *The baryons in the Milky Way satellites*, MNRAS, 419, 3304

Peacock, J. A. 1991, *The power spectrum of galaxy clustering*, MNRAS, 253, 1P

Peebles, P. J. E. 1970, *Structure of the Coma Cluster of Galaxies*, AJ, 75, 13

Peebles, P. J. E. 1971a, Physical cosmology (Princeton Series in Physics, Princeton, N.J.: Princeton University Press, 1971)

Peebles, P. J. E. 1971b, *Rotation of Galaxies and the Gravitational Instability Picture*, A&A, 11, 377

Peebles, P. J. E. 1973, *Statistical Analysis of Catalogs of Extragalactic Objects. I-Theory*, ApJ, 185, 413

Peebles, P. J. E. 1974a, *Statistical Analysis of Catalogs of Extragalactic Objects. IV-Cross-Correlation of the Abell and Shane-Wirtanen Catalogs*, ApJS, 28, 37

Peebles, P. J. E. 1974b, *The Gravitational-Instability Picture and the Nature of the Distribution of Galaxies*, ApJ, 189, L51+

Peebles, P. J. E. 1982, *Primeval adiabatic perturbations — Effect of massive neutrinos*, ApJ, 258, 415

Peebles, P. J. E. 2001, *The Void Phenomenon*, ApJ, 557, 495

Peebles, P. J. E. 2002, *Nineteenth and Twentieth Century Clouds Over the Twenty-First Century Virtual Observatory*, ArXiv:astro-ph/0209403

Peebles, P. J. E. 2012, *Seeing Cosmology Grow*, ARA&A, 50, 1

Peebles, P. J. E. & Groth, E. J. 1975, *Statistical analysis of catalogs of extragalactic objects. V - Three-point correlation function for the galaxy distribution in the Zwicky catalog*, ApJ, 196, 1

Peebles, P. J. E. & Hauser, M. G. 1974, *Statistical Analysis of Catalogs of Extragalactic Objects. III-The Shane-Wirtanen and Zwicky Catalogs*, ApJS, 28, 19

Peebles, P. J. E., Page, Jr., L. A., & Partridge, R. B. 2009, Finding the Big Bang

Peebles, P. J. E. & Yu, J. T. 1970, *Primeval Adiabatic Perturbation in an Expanding Universe*, ApJ, 162, 815

Pelt, J., Hoff, W., Kayser, R., Refsdal, S., & Schramm, T. 1994, *Time delay controversy on QSO 0957+561 not yet decided*, A&A, 286, 775

Pelt, J., Kayser, R., Refsdal, S., & Schramm, T. 1996, *The light curve and the time delay of QSO 0957+561*, A&A, 305, 97

Pelt, J., Schild, R., Refsdal, S., & Stabell, R. 1998, *Microlensing on different timescales in the lightcurves of QSO 0957+561 A,B*, A&A, 336, 829

Penzias, A. A. & Wilson, R. W. 1965, *A Measurement of Excess Antenna Temperature at 4080 Mc/s*, ApJ, 142, 419

Perek, L. 1948, *A model of the Galaxy*, Contr. Astr. Inst. Masaryk Univ., 1

Perek, L. 1962, *Models of galaxies*, Advances in Astron. Astrophys., 1, 165

Perlmutter, S., Aldering, G., Goldhaber, G., et al. 1999, *Measurements of Omega and Lambda from 42 High-Redshift Supernovae*, ApJ, 517, 565

Persic, M. & Salucci, P. 1991, *The universal galaxy rotation curve*, ApJ, 368, 60

Persic, M., Salucci, P., & Stel, F. 1996, *The universal rotation curve of spiral galaxies — I. The dark matter connection*, MNRAS, 281, 27

Pietronero, L. 1987, *The fractal structure of the universe: Correlations of galaxies and clusters and the average mass density*, Physica A Statistical Mechanics and its Applications, 144, 257

Planck Collaboration, Ade, P. A. R., Aghanim, N., et al. 2013, *Planck 2013 results. XVI. Cosmological parameters*, ArXiv:1303.5076

Platen, E., van de Weygaert, R., & Jones, B. J. T. 2007, *A cosmic watershed: the WVF void detection technique*, MNRAS, 380, 551

Polisensky, E. & Ricotti, M. 2011, *Constraints on the dark matter particle mass from the number of Milky Way satellites*, Phys. Rev. D, 83, 043506

Poveda, A. & Allen, C. 1975, *The Mass and Tidal Radius of Omega Centauri*, ApJ, 197, 155

Praton, E. A., Melott, A. L., & McKee, M. Q. 1997, *The Bull's-Eye Effect: Are Galaxy Walls Observationally Enhanced?* ApJ, 479, L15

Press, W. H., Rybicki, G. B., & Hewitt, J. N. 1992a, *The time delay of gravitational lens 0957 + 561. I - Methodology and analysis of optical photometric data. II - Analysis of radio data and combined optical-radio analysis*, ApJ, 385, 404

Press, W. H., Rybicki, G. B., & Hewitt, J. N. 1992b, *The Time Delay of Gravitational Lens 0957+561. II. Analysis of Radio Data and Combined Optical-Radio Analysis*, ApJ, 385, 416

Press, W. H. & Schechter, P. 1974, *Formation of Galaxies and Clusters of Galaxies by Self-Similar Gravitational Condensation*, ApJ, 187, 425

Primack, J. R. 1984, *Dark matter, galaxies, and large scale structure in the universe*, NASA STI/Recon Technical Report N, 85, 27786

Primack, J. R. & Blumenthal, G. R. 1984, *What is the dark matter? — Implications for galaxy formation and particle physics*, in NATO ASIC Proc. 117: Formation and Evolution of Galaxies and Large Structures in the Universe, ed. J. Audouze & J. Tran Thanh Van, 163

Profumo, S. 2013, *TASI 2012 Lectures on Astrophysical Probes of Dark Matter*, ArXiv: astro-ph.HZ1001.4086

Reed, D. S., Bower, R., Frenk, C. S., et al. 2005, *The first generation of star-forming haloes*, MNRAS, 363, 393

Rees, M. J. 2000, Just six numbers: the deep forces that shape the universe, ed. Rees, M. (Basic Books)

Rees, M. J. 1977, *Cosmology and Galaxy Formation*, in Evolution of Galaxies and Stellar Populations, ed. B. M. Tinsley & R. B. Larson, 339

Refsdal, S. 1964, *On the possibility of determining Hubble's parameter and the masses of galaxies from the gravitational lens effect*, MNRAS, 128, 307

Retana-Montenegro, E., van Hese, E., Gentile, G., Baes, M., & Frutos-Alfaro, F. 2012, *Analytical properties of Einasto dark matter haloes*, A&A, 540, A70

Riess, A. G., Filippenko, A. V., Challis, P., et al. 1998, *Observational Evidence from Supernovae for an Accelerating Universe and a Cosmological Constant*, AJ, 116, 1009

Riess, A. G., Strolger, L.-G., Casertano, S., et al. 2007, *New Hubble Space Telescope Discoveries of Type Ia Supernovae at $z > 1$: Narrowing Constraints on the Early Behavior of Dark Energy*, ApJ, 659, 98

Roberts, M. S. 1966, *A High-Resolution 21-cm Hydrogen-Line Survey of the Andromeda Nebula*, ApJ, 144, 639

Roberts, M. S. 1975, *The Rotation Curve of Galaxies*, in IAU Symposium, Vol. 69, Dynamics of the Stellar Systems, ed. A. Hayli, 331

Roberts, M. S. & Rots, A. H. 1973, *Comparison of Rotation Curves of Different Galaxy Types*, A&A, 26, 483

Rohlfs, K. & Kreitschmann, J. 1980, *A two component mass model for M81 /NGC 3031*, A&A, 87, 175

Rohlfs, K. & Kreitschmann, J. 1988, *Dynamical mass modelling of the Galaxy*, A&A, 201, 51

Rootsmäe, T. 1961, *The problem of stellar evolution in connection with the regularities in their kinematics*, Tartu Astr. Obs. Publ., 33, 322

Rubin, V. C. 1983, *The rotation of spiral galaxies*, Science, 220, 1339

Rubin, V. C., Burstein, D., Ford, Jr., W. K., & Thonnard, N. 1985, *Rotation velocities of 16 SA galaxies and a comparison of Sa, Sb, and SC rotation properties*, ApJ, 289, 81

Rubin, V. C., Ford, Jr., W. K., & Rubin, J. S. 1973, *A Curious Distribution of Radial Velocities of SC i Galaxies with $14.0 < M < 15.0$*, ApJ, 183, L111

Rubin, V. C. & Ford, W. K. J. 1970, *Rotation of the Andromeda Nebula from a Spectroscopic Survey of Emission Regions*, ApJ, 159, 379

Rubin, V. C., Ford, W. K. J., & . Thonnard, N. 1980, *Rotational properties of 21 SC galaxies with a large range of luminosities and radii, from NGC 4605 R = 4kpc to UGC 2885 R = 122 kpc*, ApJ, 238, 471

Rubin, V. C., Roberts, M. S., Graham, J. A., Ford, Jr., W. K., & Thonnard, N. 1976a, *Motion of the Galaxy and the local group determined from the velocity anisotropy of distant SC I galaxies. I - The data*, AJ, 81, 687

Rubin, V. C., Thonnard, N., & Ford, Jr., W. K. 1978, *Extended rotation curves of high-luminosity spiral galaxies. IV - Systematic dynamical properties, SA through SC*, ApJ, 225, L107

Rubin, V. C., Thonnard, N., Ford, Jr., W. K., & Roberts, M. S. 1976b, *Motion of the Galaxy and the local group determined from the velocity anisotropy of distant SC I galaxies. II - The analysis for the motion*, AJ, 81, 719

Ryden, B. S. & Gramann, M. 1991, *Phase shifts in gravitationally evolving density fields*, ApJ, 383, L33

Sahni, V., Sathyaprakash, B. S., & Shandarin, S. F. 1998, *Shapefinders: A New Shape Diagnostic for Large-Scale Structure*, ApJ, 495, L5

Sales, L. V., Navarro, J. F., Theuns, T., et al. 2012, *The origin of discs and spheroids in simulated galaxies*, MNRAS, 423, 1544

Salpeter, E. E. 1955, *The Luminosity Function and Stellar Evolution*, ApJ, 121, 161

Salucci, P., Lapi, A., Tonini, C., et al. 2007, *The universal rotation curve of spiral galaxies - II-The dark matter distribution out to the virial radius*, MNRAS, 378, 41

Salucci, P., Wilkinson, M. I., Walker, M. G., et al. 2012, *Dwarf spheroidal galaxy kinematics and spiral galaxy scaling laws*, MNRAS, 420, 2034

Salvador-Solé, E., Viñas, J., Manrique, A., & Serra, S. 2012, *Theoretical dark matter halo density profile*, MNRAS, 423, 2190

Sandage, A. 1961, *The Ability of the 200-INCH Telescope to Discriminate Between Selected World Models*, ApJ, 133, 355

Sandage, A. 1972, *The redshift-distance relation. II. The Hubble diagram and its scatter for first-ranked cluster galaxies: A formal value for q_0*, ApJ, 178, 1

Sandage, A. 1978, *Optical redshifts for 719 bright galaxies*, AJ, 83, 904

Sandage, A. & Tammann, G. A. 1975, *Steps toward the Hubble constant. V - The Hubble constant from nearby galaxies and the regularity of the local velocity field*, ApJ, 196, 313

Sandage, A. & Tammann, G. A. 1976, *Steps toward the Hubble constant. VII - Distances to NGC 2403, M101, and the Virgo cluster using 21 centimeter line widths compared with optical methods: The global value of H sub 0*, ApJ, 210, 7

Sandage, A. & Tammann, G. A. 1981, *Revised Shapley-Ames Catalog of Bright Galaxies*, in Carnegie Inst. of Washington, Publ. 635

Sandage, A., Tammann, G. A., Saha, A., et al. 2006, *The Hubble Constant: A Summary of the Hubble Space Telescope Program for the Luminosity Calibration of Type Ia Supernovae by Means of Cepheids*, ApJ, 653, 843

Sandage, A. R. & Schwarzschild, M. 1952, *Inhomogeneous Stellar Models. II. Models with Exhausted Cores in Gravitational Contraction*, ApJ, 116, 463

Sanders, R. H. 2010, The Dark Matter Problem: A Historical Perspective (Cambridge Univ. Press)

Schechter, P. 1976, *An analytic expression for the luminosity function for galaxies*, ApJ, 203, 297

Schmidt, M. 1956, *A model of the distribution of mass in the Galactic System*, Bull. Astron. Inst. Netherlands, 13, 15

Schmidt, M. 1959, *The Rate of Star Formation*, ApJ, 129, 243

Schwarzschild, M. 1958, Structure and evolution of the stars, ed. Schwarzschild, M.

Schwarzschild, M., Rabinowitz, I., & Härm, R. 1953, *Inhomogeneous Stellar Models. III. Models with Partially Degenerate Isothermal Cores*, ApJ, 118, 326

Seljak, U. & Zaldarriaga, M. 1996, *A Line-of-Sight Integration Approach to Cosmic Microwave Background Anisotropies*, ApJ, 469, 437

Sérsic, J. L. 1963, *Influence of the atmospheric and instrumental dispersion on the brightness distribution in a galaxy*, Boletin de la Asociacion Argentina de Astronomia La Plata Argentina, 6, 41

Shandarin, S. F., Sheth, J. V., & Sahni, V. 2004, *Morphology of the supercluster-void network in ΛCDM cosmology*, MNRAS, 353, 162

Shandarin, S. F. & Zeldovich, I. B. 1983, *Topology of the large-scale structure of the universe*, Comments on Astrophysics, 10, 33

Shane, C. & Wirtanen, C. 1967, *The distribution of galaxies*, Publ. Lick Obs., 22

Shapley, H. 1935, *A catalogue of 7,889 external galaxies in Horologium and surrounding regions*, Annals of Harvard College Observatory, 88, 105

Shapley, H. 1937, *The Distribution of Eighty-Nine Thousand Galaxies Over the South Galactic Cap*, Harvard College Observatory Circular, 423, 1

Shapley, H. 1940, *Galactic and Extragalactic Studies, VI. Summary of a Photometric Survey of 35,500 Galaxies in High Southern Latitudes*, Proceedings of the National Academy of Science, 26, 166

Shaver, P. A. 1991, *Radio Surveys and Large Scale Structure*, Australian Journal of Physics, 44, 759

Sheth, R. K. & van de Weygaert, R. 2004, *A hierarchy of voids: much ado about nothing*, MNRAS, 350, 517

Silk, J. 1974, *Does the galaxy possess a gaseous halo*, Comments on Astrophysics and Space Physics, 6, 1

Silk, J. 1982, *Fundamental tests of galaxy formation theory*, in Astrophysical Cosmology Proceedings, ed. H. A. Brueck, G. V. Coyne, & M. S. Longair, 427

Sizikov, V. S. 1968, *A model of the distribution of mass in M31. I*, Astrofizika, 4, 633

Sizikov, V. S. 1969, *A model of the distribution of mass in M31. II*, Astrofizika, 5, 317

Smith, S. 1936, *The Mass of the Virgo Cluster*, ApJ, 83, 23

Smoot, G. F., Bennett, C. L., Kogut, A., et al. 1992, *Structure in the COBE differential microwave radiometer first-year maps*, ApJ, 396, L1

Soneira, R. M. & Peebles, P. J. E. 1978, *A computer model universe — Simulation of the nature of the galaxy distribution in the Lick catalog*, AJ, 83, 845

Spinrad, H. 1966, *Normal Galaxies in the Post-Baade ERA*, PASP, 78, 367

Spinrad, H., Greenstein, J. L., Taylor, B. J., & King, I. R. 1970, *On the Supermetallicity of the Main-Sequence Stars in M67 and NGC 188*, ApJ, 162, 891

Spinrad, H., Gunn, J. E., Taylor, B. J., McClure, R. D., & Young, J. W. 1971, *Color Changes and Absorption-Line Variations Over the Inner Disk of M31 and the Central Regions of M32 and NGC 4472*, ApJ, 164, 11

Spinrad, H. & Taylor, B. J. 1971, *Scanner Abundance Studies. III-The Super-Metal Cluster NGC 6791*, ApJ, 163, 303

Spinrad, H., Taylor, B. J., & van den Bergh, S. 1969, *The M7 giants in the nuclear bulge of the Galaxy*, AJ, 74, 525

Springel, V., Frenk, C. S., & White, S. D. M. 2006, *The large-scale structure of the Universe*, Nature, 440, 1137

Springel, V., Wang, J., Vogelsberger, M., et al. 2008a, *The Aquarius Project: the subhaloes of galactic haloes*, MNRAS, 391, 1685

Springel, V., White, S. D. M., Frenk, C. S., et al. 2008b, *Prospects for detecting supersymmetric dark matter in the Galactic halo*, Nature, 456, 73

Springel, V., White, S. D. M., Jenkins, A., et al. 2005, *Simulations of the formation, evolution and clustering of galaxies and quasars*, Nature, 435, 629

Starobinsky, A. A. 1980, *A new type of isotropic cosmological models without singularity*, Physics Letters B, 91, 99

Starobinsky, A. A. 1982, *Dynamics of phase transition in the new inflationary universe scenario and generation of perturbations*, Physics Letters B, 117, 175

Starobinsky, A. A. 1985, *Multicomponent de Sitter (inflationary) stages and the generation of perturbations*, ZhETF Pis ma Redaktsiiu, 42, 124

Stoica, R. S., Martínez, V. J., & Saar, E. 2010, *Filaments in observed and mock galaxy catalogues*, A&A, 510, A38

Strom, R. G. 2012, *How was atomic HI ($\lambda = 21$ cm line) in space discovered?*, in International Journal of Modern Physics: Conference Series, Vol. 1 (World Scientific), 1

Strömberg, G. 1924, *The Asymmetry in Stellar Motions and the Existence of a Velocity-Restriction in Space*, ApJ, 59, 228

Su, M. & Finkbeiner, D. P. 2012a, *Double Gamma-ray Lines from Unassociated Fermi-LAT Sources*, ArXiv:1207.7060

Su, M. & Finkbeiner, D. P. 2012b, *Strong Evidence for Gamma-ray Line Emission from the Inner Galaxy*, ArXiv:1206.1616

Suhhonenko, I., Einasto, J., Liivamägi, L. J., et al. 2011, *The cosmic web for density perturbations of various scales*, A&A, 531, A149+

Sunyaev, R. A. & Zeldovich, Y. B. 1970, *Small-scale fluctuations of relic radiation*, Ap&SS, 6, 358

Szalay, A. S. & Marx, G. 1976, *Neutrino rest mass from cosmology*, A&A, 49, 437

Tago, E., Einasto, J., & Saar, E. 1984, *Structure of superclusters and superclusters formation. IV Spatial distribution of clusters of galaxies in the Coma supercluster and its large-scale environment*, MNRAS, 206, 559

Tago, E., Einasto, J., & Saar, E. 1986, *A prominent string of galaxies in Bootes — Evidence for a Lagrangian singularity?*, MNRAS, 218, 177

Tago, E., Einasto, J., Saar, E., et al. 2006, *Clusters and groups of galaxies in the 2dF galaxy redshift survey: A new catalogue*, Astronomische Nachrichten, 327, 365

Tago, E., Einasto, J., Saar, E., et al. 2008, *Groups of galaxies in the SDSS Data Release 5. A group-finder and a catalogue*, A&A, 479, 927

Tamm, A., Tempel, E., & Tenjes, P. 2007, *Visible and dark matter in M31 — I. Properties of stellar components*, ArXiv:0707.4375

Tamm, A., Tempel, E., Tenjes, P., Tihhonova, O., & Tuvikene, T. 2012, *Stellar mass map and dark matter distribution in M 31*, A&A, 546, A4

Tarenghi, M., Tifft, W. G., Chincarini, G., Rood, H. J., & Thompson, L. A. 1978, *The Structure of the Hercules Supercluster*, in IAU Symposium, Vol. 79, Large Scale Structures in the Universe, eds. M. S. Longair & J. Einasto, 263

Tempel, E., Einasto, J., Einasto, M., Saar, E., & Tago, E. 2009, *Anatomy of luminosity functions: the 2dFGRS example*, A&A, 495, 37

Tempel, E., Hektor, A., & Raidal, M. 2012a, *Addendum: Fermi 130 GeV gamma-ray excess and dark matter annihilation in sub-haloes and in the Galactic centre Addendum: Fermi 130 GeV gamma-ray excess and dark matter annihilation in sub-haloes and in the Galactic centre*, J. Cosmology Astropart. Phys., 11, 0

Tempel, E., Hektor, A., & Raidal, M. 2012b, *Fermi 130 GeV gamma-ray excess and dark matter annihilation in sub-haloes and in the Galactic centre*, J. Cosmology Astropart. Phys., 9, 32

Tempel, E., Saar, E., Liivamägi, L. J., et al. 2011, *Galaxy morphology, luminosity, and environment in the SDSS DR7*, A&A, 529, A53

Tempel, E., Stoica, R. S., & Saar, E. 2013, *Evidence for spin alignment of spiral and elliptical/S0 galaxies in filaments*, MNRAS, 428, 1827

Tempel, E., Tago, E., & Liivamägi, L. J. 2012c, *Groups and clusters of galaxies in the SDSS DR8. Value-added catalogues*, A&A, 540, A106

Tempel, E., Tamm, A., Kipper, R., & Tenjes, P. 2012d, *Uncertainties in SDSS galaxy parameter determination: 3D photometrical modelling of test galaxies and restoration of their structural parameters*, ArXiv:1205.6319

Tempel, E., Tamm, A., & Tenjes, P. 2007, *Visible and dark matter in M 31 - II. A dynamical model and dark matter density distribution*, ArXiv:0707.4374

Tenjes, P., Einasto, J., & Haud, U. 1991, *Galactic models with massive coronae. III - Giant elliptical galaxy M 87*, A&A, 248, 395

Tenjes, P., Haud, U., & Einasto, J. 1994, *Galactic models with massive coronae IV-The Andromeda galaxy, M 31*, A&A, 286, 753

Tenjes, P., Haud, U., & Einasto, J. 1998, *Galactic models with massive coronae. V-The spiral SAB galaxy M 81*, A&A, 335, 449

Thompson, L. A. & Gregory, S. A. 2011, *An Historical View: The Discovery of Voids in the Galaxy Distribution*, ArXiv:1109.1268

Tifft, W. G. & Gregory, S. A. 1978, *Observations of the Large Scale Distribution of Galaxies*, in IAU Symposium, Vol. 79, Large Scale Structures in the Universe, eds. M. S. Longair & J. Einasto, 267

Tiit, E. & Einasto, J. 1964, *Factor Analysis of Red Dwarfs*, Tartu Astr. Obs. Publ., 34, 156

Tinsley, B. M. 1968, *Evolution of the Stars and Gas in Galaxies*, ApJ, 151, 547

Tinsley, B. M. & Spinrad, H. 1971, *Evolution of the M31 Disk Population*, Ap&SS, 12, 118

Toomre, A. 1964, *On the gravitational stability of a disk of stars*, ApJ, 139, 1217

Toomre, A. 1977, *Mergers and Some Consequences*, in Evolution of Galaxies and Stellar Populations, ed. B. M. Tinsley & R. B. G. Larson, D. Campbell, 401

Toomre, A. & Toomre, J. 1972, *Galactic Bridges and Tails*, ApJ, 178, 623

Trachternach, C., de Blok, W. J. G., Walter, F., Brinks, E., & Kennicutt, Jr., R. C. 2008, *Dynamical Centers and Noncircular Motions in THINGS Galaxies: Implications for Dark Matter Halos*, AJ, 136, 2720

Tremaine, S. 1987, *A Historical Perspective on Dark Matter*, in IAU Symposium, Vol. 117, Dark matter in the universe, ed. J. Kormendy & G. R. Knapp, 547

Trimble, V. 1987, *Existence and nature of dark matter in the universe*, ARA&A, 25, 425

Trimble, V. 1988a, *Dark matter in the universe: where, what, and why?* Contemporary Physics, 29, 373

Trimble, V. 1988b, Existence and Nature of Dark Matter in the Universe, ed. E. W. Kolb & M. S. Turner, 67

Trimble, V. 1988c, *The search for dark matter*, Astronomy, 16, 18

Trimble, V. 1990, History of dark matter in the universe (1922-1974), ed. B. Bertotti, R. Balbinot, & S. Bergia, 355

Trimble, V. 1995, *The World Line of Dark Matter: Its Existence and Nature Through Time*, in Sources of Dark Matter in the Universe, ed. D. B. Cline, 9

Trimble, V. 2008, *BOOK REVIEW: Universe or Multiverse?* Classical and Quantum Gravity, 25, 229001

Trimble, V. 2010, *History of Dark Matter in Galaxies*, in Planets, Stars and Stellar Systems, Vol. 5, Planets, Stars and Stellar Systems, ed. G. Gilmore (Springer)

Tully, R. B. & Fisher, J. R. 1978, *A Tour of the Local Supercluster*, in IAU Symposium, Vol. 79, Large Scale Structures in the Universe, eds. M. S. Longair & J. Einasto, 214

Tully, R. B. & Fisher, J. R. 1987, Atlas of Nearby Galaxies

Tully, R. B., Shaya, E. J., Karachentsev, I. D., et al. 2008, *Our Peculiar Motion Away from the Local Void*, ApJ, 676, 184

Turner, M. S., Steigman, G., & Krauss, L. M. 1984, *Flatness of the universe - Reconciling theoretical prejudices with observational data*, Physical Review Letters, 52, 2090

van de Hulst, H. C., Muller, C. A., & Oort, J. H. 1954, *The spiral structure of the outer part of the Galactic System derived from the hydrogen emission at 21 cm wavelength*, Bull. Astron. Inst. Netherlands, 12, 117

van de Hulst, H. C., Raimond, E., & van Woerden, H. 1957, *Rotation and density distribution of the Andromeda nebula derived from observations of the 21-cm line*, Bull. Astron. Inst. Netherlands, 14, 1

van de Weygaert, R. 2002, *Froth across the Universe Dynamics and Stochastic Geometry of the Cosmic Foam*, ArXiv:astro-ph/0206427

van de Weygaert, R., Aragon-Calvo, M. A., Jones, B. J. T., & Platen, E. 2009, *Geometry and Morphology of the Cosmic Web: Analyzing Spatial Patterns in the Universe*, ArXiv:0912.3448

van de Weygaert, R., Kreckel, K., Platen, E., et al. 2011, The Void Galaxy Survey, eds. I. Ferreras & A. Pasquali, 17

van de Weygaert, R. & Platen, E. 2009, *Cosmic Voids: structure, dynamics and galaxies*, ArXiv:0912.2997

van de Weygaert, R., Platen, E., Tigrak, E., et al. 2010, *The Cosmically Depressed: Life, Sociology and Identity of Voids*, in Astronomical Society of the Pacific Conference Series, Vol. 421, Galaxies in Isolation: Exploring Nature Versus Nurture, eds. L. Verdes-Montenegro, A. Del Olmo, & J. Sulentic, 99

van de Weygaert, R. & van Kampen, E. 1993, *Voids in Gravitational Instability Scenarios - Part One - Global Density and Velocity Fields in an Einstein - de-Sitter Universe*, MNRAS, 263, 481

van den Bergh, S. 1961, *The stability of clusters of galaxies*, AJ, 66, 566

van den Bergh, S. 1962, *The Stability of Clusters of Galaxies*, Zeitschrift fur Astrophysik, 55, 21

van den Bergh, S. 1972, *A New Method for Estimating the Hubble Constant*, A&A, 20, 469

van den Bergh, S. 1973, *The age of the universe*, in Stellar Ages, 40

van den Bergh, S. 1999, *The Early History of Dark Matter*, PASP, 111, 657

van den Bergh, S. 2001, *A Short History of the Missing Mass and Dark Energy Paradigms*, in Astronomical Society of the Pacific Conference Series, Vol. 252, Historical Development of Modern Cosmology, eds. V. J. Martínez, V. Trimble, & M. J. Pons-Bordería, 75

Vennik, J. 1984, *A list of nearby groups of galaxies*, Tartu Astr. Obs. Teated, 73, 1

Vennik, J. & Kaasik, A. 1982, *Radial velocities of galaxies in neighborhoods of groups of galaxies. I*, Astrofizika, 18, 523

Vennik, J., Kaasik, A., & Amirkhanian, A. 1982, *Radial velocities of galaxies in neighborhoods of groups of galaxies. II*, Astrofizika, 18, 533

Vogeley, M. S., Geller, M. J., Park, C., & Huchra, J. P. 1994a, *Voids and constraints on nonlinear clustering of galaxies*, AJ, 108, 745

Vogeley, M. S., Hoyle, F., Rojas, R. R., & Goldberg, D. M. 2004, *Mapping the cosmic web with the Sloan Digital Sky Survey*, in IAU Colloq. 195: Outskirts of Galaxy Clusters: Intense Life in the Suburbs, ed. A. Diaferio, 5

Vogeley, M. S., Park, C., Geller, M. J., & Huchra, J. P. 1992, *Large-scale clustering of galaxies in the CfA Redshift Survey*, ApJ, 391, L5

Vogeley, M. S., Park, C., Geller, M. J., Huchra, J. P., & Gott, III, J. R. 1994b, *Topological analysis of the CfA redshift survey*, ApJ, 420, 525

Walter, F., Brinks, E., de Blok, W. J. G., et al. 2008, *THINGS: The H I Nearby Galaxy Survey*, AJ, 136, 2563

Weniger, C. 2012, *A tentative gamma-ray line from Dark Matter annihilation at the Fermi Large Area Telescope*, J. Cosmology Astropart. Phys., 8, 7

White, M., Blanton, M., Bolton, A., et al. 2011, *The Clustering of Massive Galaxies at z 0.5 from the First Semester of BOSS Data*, ApJ, 728, 126

White, S. D. M. 1979, *The hierarchy of correlation functions and its relation to other measures of galaxy clustering*, MNRAS, 186, 145

White, S. D. M. 2007, *Fundamentalist physics: why Dark Energy is bad for astronomy*, Reports on Progress in Physics, 70, 883

White, S. D. M., Frenk, C. S., & Davis, M. 1983, *Clustering in a neutrino-dominated universe*, ApJ, 274, L1

White, S. D. M., Frenk, C. S., Davis, M., & Efstathiou, G. 1987, *Clusters, filaments, and voids in a universe dominated by cold dark matter*, ApJ, 313, 505

White, S. D. M. & Rees, M. J. 1978, *Core condensation in heavy halos - A two-stage theory for galaxy formation and clustering*, MNRAS, 183, 341

Wirtz, C. 1922, *Einiges zur Statistik der Radialbewegungen von Spiralnebeln und Kugelsternhaufen*, Astronomische Nachrichten, 215, 349

Wirtz, C. 1924, *De Sitters Kosmologie und die Radialbewegungen der Spiralnebel*, Astronomische Nachrichten, 222, 21

Wolf, J., Martinez, G. D., Bullock, J. S., et al. 2010, *Accurate masses for dispersion-supported galaxies*, MNRAS, 406, 1220

Wyse, A. B. & Mayall, N. U. 1942, *Distribution of Mass in the Spiral Nebulae Messier 31 and Messier 33*, ApJ, 95, 24

York, D. G., Adelman, J., Anderson, Jr., J. E., et al. 2000, *The Sloan Digital Sky Survey: Technical Summary*, AJ, 120, 1579

Zeldovich, Y. B. 1970, *Gravitational instability: An approximate theory for large density perturbations*, A&A, 5, 84

Zeldovich, Y. B. 1975, *Deuterium of cosmological origin and the mean density of the universe*, Soviet Astronomy Letters, 1, 5

Zeldovich, Y. B. 1978, *The theory of the large scale structure of the universe*, in IAU Symposium, Vol. 79, Large Scale Structures in the Universe, eds. M. S. Longair & J. Einasto, 409

Zeldovich, Y. B., Einasto, J., & Shandarin, S. F. 1982, *Giant voids in the universe*, Nature, 300, 407

Zwicky, F. 1933, *Die Rotverschiebung von extragalaktischen Nebeln*, Helvetica Physica Acta, 6, 110

Zwicky, F. 1937, *On the Masses of Nebulae and of Clusters of Nebulae*, ApJ, 86, 217

Zwicky, F., Herzog, E., & Wild, P. 1968, Catalogue of galaxies and of clusters of galaxies (Pasadena: California Institute of Technology (CIT), 1961–1968)

Zwicky, F. & Zwicky, M. A. 1971, Catalogue of selected compact galaxies and of post-eruptive galaxies

General Index

1.5-m telescope, 72, 74
(2-dimensional) distribution of galaxies, 9, 124, 137, 176, 195
13th Marcel Grossmann Meeting, 173

A194, 131
A194 supercluster, 130
A262, 131
A2634, 132
A2666, 132
A347, 131
A397, 132
A400, 132
A426, 131, 134
Abastumani Observatory, 34, 95, 286
Abell clusters, 122, 130–136, 188, 189, 200, 201, 203–204, 208–214, 224, 228, 230–232, 235–236, 240, 242, 272
Academy of Sciences, 169
acceleration of the Universe, 308
alternatives to dark matter, 164
Andromeda galaxy, see M31
annihilation of DM particles, 308
apogalactic distances, 59, 60
Apple II, 114
Aquarius, 160
Aquarius Project, 105
Astronomical Council of the USSR Academy of Sciences, 38–40, 70, 71, 289
Astronomical Institute of the Basel University, 96

Astronomy Department of UCLA, 204, 290
Astronomy Department of Yale University, 290
astroparticle physics, 156
August Coup, 23, 248, 249
axions, 153, 157

Baltic Barons, 18, 300
Baltic Germans, 19, 20, 27
baryonic dark matter, 147–151
baryonic matter, 92, 101, 155
Belorussia, 22
Big Bang, 14, 70, 151–153
    Big Bang theory, 5
Big Bang nucleosynthesis, 15, 96, 147, 151
Big Crunch, 7
black-body spectrum, 308
Bootes void, 136
bridges, 139
bulges, 54, 62, 65
"bullet" cluster 1E 0657-558, 165
Byurakan Astrophysical Observatory, 70, 77, 140–142

California Institue of Technology, 124
catalogue of galaxies and clusters of galaxies, 8, 122
catalogue of rich clusters of galaxies, 8
Catalogue of Selected Compact Galaxies, 123

Catholicos of All Armenians, 142
Caucasus Winter Schools, 89, 120, 148, 152
CDM model, 158
cellular structure of the Universe, 107, 136
Center of Excellence, 161, 301
CERN, 161
Chandra, 101
circular velocity, 51, 53, 54, 58, 99
classical cosmological paradigm, 14, 96, 308
COBE, 154
cold dark matter (CDM), 102, 157–159, 161, 308
Cold War, 139, 310
Coma cluster, 85, 120, 122, 126, 136
Coma supercluster, 120, 128, 129, 138
Commission 33, 59, 60, 81, 97
Commodore VIC-20, 115
conference on dark matter in Tallinn, 155, 307
Congress of People's Deputies, 167, 170, 171–173
connection formulae, 54
convection, 12
core radius, 57
corona, 83, 84, 306
cosmic microwave background (CMB), 152, 155, 157
    CMB fluctuations, 305, 308
    CMB radiation, 153
cosmic web, 138, 139
Crimean Astrophysical Observatory, 77, 78
critical cosmological density, 155, 164

dark corona, 82, 89
dark energy, 165
dark matter, 46, 50, 83, 90–93, 95, 165
    distribution of, 101, 102
    local, 84, 306
DEC VAX computer, 115
declaration of sovereignty, 171
density of matter, 8, 50, 54, 56, 57, 96
description functions, 54, 56, 57, 68
de Sitter universe, 5

deuterium abundance, 157
deuterium nucleosynthesis, 95
discovery of pulsars, 305
disks, 54, 62, 65
dissolution of the USSR, 171, 172
distance of the Sun from the Galactic centre, 53
distribution of galaxies, 8, 124–127, 134–139, 176, 195, 203–206
Division of Chemistry, Biology and Geology, 168
DM annihilations, 162, 163
double elliptical galaxies, 87
dSph galaxies, 104
Dutch, 98
dwarf spheroidal galaxies, 159
dynamical density, 47, 49

Egeri, 24, 26, 145, 303–304
Einasto index, 57
Einasto profile, 57, 103, 104, 110, 114, 160
Einstein X-ray orbiting observatory, 100
Einstein–de Sitter model, 7
elliptical galaxies, 65
escape velocity, 53, 54
ESTCube, 302
Estonia, 19, 20, 173
Estonian Academy of Sciences, 37, 116, 168–170, 289
Estonian Biocenter, 28, 168
Estonian Communist Party, 167
Estonian Congress, 170
Estonian flag, 166
Estonian Greens Movement, 169
Estonian IME program, 169
Estonian Popular Front, 173
Estonian Society of Prehistoric Art, 144
Estonian Supreme Soviet, 170, 171
European Astronomical Society, 96
European Southern Observatory, 140, 180, 185, 228, 252, 293
evolution models, 63
evolution of galaxies, 63, 84, 107, 110
expansion of the Universe, 15

fall of the Berlin wall, 172
Faza, 78
Fermi Gamma-ray Space Telescope, 162
Fermi satellite Large Area Telescope, 162, 308
Fermilab, 161
filaments of galaxies, 139, 224–227, 265
filling factor of the Universe, 135
fine structure of the Universe, 157, 271, 275–279
Finland, 20
Finnish TV, 166
Finno-Ugric languages, 15
First European Astronomy Meeting, 81, 86, 109
flags, 166
flat disk, 62
fluctuations of the CMB, 153
formation of galaxies, 106
Fornax, 67

galactic constants, 53, 58
galactic coronas, 81
galactic models, 51, 54, 81, 82
galactic outer radius, 52
Galaxies Intergalactic Medium Calculation (GIMIC), 108
Galaxy, 47, 57, 67, 68, 87, 88, 98, 109, 148, 160
galaxy formation, 106–108
gamma rays, 162
gamma-ray spectra, 163, 308
gaseous coronae, 151
generalised exponential model, 56
German, 20, 21
giant elliptical galaxies, 104
glasnost, 23
global dark matter, 84, 97, 306
globular clusters, 63, 64, 66, 98, 104, 159
gravitational lensing, 101, 282–283
gravitational potential, 54, 59
gravitinos, 157
Great Northern War, 17, 20, 27, 37
groups of galaxies, 88, 207–211
  clusters of galaxies, 85, 87, 90, 101
guilds, 18

halo, 54, 62, 65
Hanseatic Days, 165
Hanseatic League, 17
harmonic mean radius, 57
Harvard Center for Astrophysics, 8, 100, 162, 203–206
Hercules superclusters, 120, 122, 126, 136
Hertzsprung–Russell diagram, 10, 11
HESS, 161
hierarchical clustering scenario of structure formation, 9, 107, 120, 176–179, 185, 221, 258, 261, 265
hot dark matter (HDM), 158, 159, 161
Hubble constant, 7, 14, 96, 282–283, 308
Hubble Space Telescope, 100, 105, 114, 280, 281, 283
Hubble time, 6

IAU General Assembly, 59, 82, 87, 97, 99, 114, 118, 123, 129, 159
IAU Symposium on Dynamics of Stellar Systems, 99
IAU Symposium on External Galaxies and Quasi-Stellar Objects, 109
IAU Symposium on galaxies, 114
IAU Symposium on the Spiral Structure of Galaxies, 109
IBM Personal Computer, 115
Ice Age, 16
impact factor, 74
Independence Day, 168
initial mass function (IMF), 64
Institute of Astronomy of Cambridge University, 115, 263, 264, 291
Institute of Astrophysics and Atmospheric Physics, 116
Institute of Chemical and Biological Physics, 161, 168
Institute of Cybernetics, 113
Institute of Physics, 116, 168
Institute of Physics and Astronomy, 116
International Astronomical Union (IAU), 58–60, 114, 118, 287–289
International Geophysical Year 1957, 78
Interregional Group of Deputies (IRGD), 173
iPhone, 115

Kaali lake, 16
Kapteyn Astronomical Institute of Groningen University, 297
Kapustin Yar, 77
Katyn, 22
Keplerian law, 99
KGB, 30, 166, 168, 289
Kosmos 215, 77
Kuzmin constant, 48, 58

ΛCDM model, 102, 103, 160, 262–265
Large Area Telescope (LAT), 162
Large Scale Structure of the Universe, 129, 136, 156
Las Campanas Redshift Survey, 214
Last Glacial Maximum refugia, 16
Latvia, 20, 166, 173
Leningrad University, 69
Lick counts, 9, 137
limiting radii, 60
limiting velocity, 59, 60
Lithuania, 20, 166, 173
Livonia, 19
Livonian Brothers, 27
Local Group, 85, 88
Local Supercluster, 128–130, 134
low-density Universe, 96
luminosity density field, 214–216, 223, 237, 266–268, 272
luminosity function (LF), 64, 209, 210, 219–220, 222, 231–232
Lutheran reformation, 18

M31, 4, 47, 51, 61–64, 66, 67, 70, 83, 85, 87, 88, 98, 101, 109, 148, 160
  nucleus of, 62
M32, 63, 65, 67
M33, 4
M67, 62
M7, 62
M81, 94, 109
M87, 65, 67, 82, 83, 94
MacBook, 115
MACHO, 101
Magellanic Clouds, 101
Magellanic Stream, 94
Maidanak, 141

Markarian galaxies, 123, 126
mass distribution function, 60, 271–272
mass paradox in clusters of galaxies, 15
mass-to-luminosity ratio, 4, 61, 63, 65, 66, 68, 83, 84, 87, 88, 97–99, 104, 105, 110, 147–149, 159, 283
massive coronae, 89
matter density, 58
Max-Planck-Institut für Physik, 162
Max-Planck-Institut für Astrophysik, Garching, 264, 301
maximum disk, 66, 110
merging, 110
metal content, 64
metal-poor populations, 98
Michurin–Lysenko type biology, 168
microscopic structure, 156
Mikron, 78
Milky Way, 3, 45, 101
Millennium simulation, 108
Millennium-II simulations, 103
Milne's model, 5, 7
missing satellite problem, 160
model of the Galaxy, 52, 53, 67, 83
models of stellar interiors, 12
Modified Newtonian Dynamics (MOND), 164
Molotov–Ribbentrop Pact, 20, 22, 172, 173
morphological properties of companion galaxies, 94, 151
Moscow University, 34

N-body simulations, 102, 176–186, 192, 195, 197–199, 206, 215, 218, 226, 230, 232–233, 241, 242, 257, 259–265, 269–270, 273, 276
N4169 group of galaxies, 120
National Institute of Chemical Physics and Biophysics, 161
National Optical Astronomy Observatory, 140
National Radio Astronomy Observatory, 61, 98
National Singing Festival, 25
Nazi Germany, 22, 172
near clusters, 120

neutrino-dominated dark matter, 153, 154–155, 158, 159, 161
neutrinos, 91, 154, 157
New Year parties, 72
Newtonian gravity, 164
NFW profile, 102, 104, 110
NGC 1835, 63
NGC 188, 62
NGC 2210, 63
NGC 3115, 97, 147
NGC 4472, 63
NGC 6388, 63
NGC 6791, 62
NGC 6822, 4
NGC 6946, 94
noctilucent clouds, 78
non-baryonic dark matter, 152, 154, 155
Nordita, 115
Nordic Optical Telescope, 143
Northern Crusades, 17
nucleosynthesis, 12, 152
nucleosynthesis constraints, 92, 151, 155, 164

Old City Days, 165, 166
Omega Centauri, 63
Oort's constants, 47, 58
Oort's limiting velocity, 53
open clusters, 64
Ozernoi scenario, 125

pairs of galaxies, 89
Palomar Observatory Sky Survey, 124, 126
PAMELA, 161
Peace of Tartu, 20
Peebles scenario, 125
perestroika, 23
Perseus–Pisces supercluster, 120, 122, 127–129, 132, 135
philosophical seminars, 72
Phoenix Project, 105
photinos, 157
photometer Faza, 79
physical criteria, 55
physical nature of dark matter, 91, 95, 147–158

Planck epoch, 156
Planck satellite, 154
Poland, 20, 22, 172
Popular Front of Estonia, 167, 170, 173
Post Office Radio Division, 98
Princeton University, 100
principal descriptive function, 55
projected density, 54, 57, 106
Pulkovo Observatory, 70, 286, 298, 299

radio galaxies, 126, 127
RATAN-600, 154
Reference catalogue of galaxies, 122
regilaul, 144
Revised Shapley–Ames catalog of bright galaxies, 124
ROSAT, 101
Rotation curves, 97
Russia, 166
Russian Empire, 18, 37
Russian Soviet Federative Republic, 172

Salyut 6, 78
Salyut 7, 78
Sangaste, 97
satellite station of the Tartu University, 41, 42, 72
satellites of our Galaxy, 160
scientific revolution, 93
Sculptor, 67
Second Reference catalogue of galaxies, 124
Shapley Supercluster, 8
singing revolution, 165
Sloan Digital Sky Survey (SDSS), 111, 143, 206, 210, 215–217, 221–223, 231–233, 244–246, 266–268, 283, 309
SDSS galaxies, 111
Smersh, 30
smoothness of the Hubble flow, 96, 147, 280, 281
Solar eclipse observations, 112
Sombrero galaxy, 65
Song Festival Arena, 166, 167
Soros Foundation, 115
Soviet Academy of Sciences, 116, 118
Soviet Army, 20

Soviet deportation, 22
Soviet occupation/annexation, 21, 250
Soviet Union, 20, 22, 23, 131, 134, 141, 143, 166–169, 171–174, 191
Space Station Mir, 79
spatial properties of stellar populations, 10
stability of a flat galactic disk, 110
star clusters, 63, 64
star formation rate, 69
Steady State theory, 6, 153
stellar evolution, 11–14, 64
stellar populations, 10, 64
Sternberg Astronomical Institute, 34, 38, 59, 76, 78, 255
Strömberg diagram, 9, 148, 149
Strömberg equation, 57
Study Week on Cosmology and Fundamental Physics, 108, 155
subsystem of globular clusters, 10
subsystems of stars, 10
superclusters, 8, 101, 120, 122, 125–136, 139, 211–217, 228–235, 241–243, 265–268
supervoids, 136, 180, 183, 201, 227, 228, 230, 235–238, 240, 241, 242, 275
Shapley's survey of galaxies, 8
Swedish Empire, 17
system of galactic constants, 57

Tallinn conference on dark matter, 150
Tallinn IAU Symposium, 127, 150, 154
Tallinn Technical University, 37
Tandy TRS-80, 114
Tartu Observatory, 31, 34, 37, 39, 42, 43, 77, 78, 112, 114, 116, 124, 131, 140, 158, 167, 168, 170, 173
Tartu Real Gymnasium, 28, 31
Tartu Teachers Seminar, 24
Tartu Treffner Gymnasium, 31
Tartu University, 20, 34, 37, 40, 41, 96, 112–114, 145, 165, 169, 173
Tensor-vector-scalar gravity (TeVeS), 164
Teutonic Order, 17
The HI Nearby Galaxy Survey (THINGS), 103
Third European Astronomy Meeting, 69, 95, 109, 307

Türgi farm, 24, 29, 42
Two-degree Field Galaxy Redshift Survey, 206–207, 209, 211, 214–215, 219, 220, 222, 231–233, 309
Two Micron All-Sky Survey (2MASS), 206

UHURU, 99, 100, 151
Ukraine, 166
Universal Rotation Curve, 106
Uppsala General Catalogue, 135
Uroborus, 156
USSR Academy of Sciences, 141, 289
Uzbek SSR, 141

Vatican Observatory, 172
Vatican Study Week, 158
velocity dispersion, 69
velvet revolution in Prague, 172
Viljandi Culture Academy, 165
Virgo cluster, 86, 105
Virgo supercluster, 8, 120, 126
voids, 120, 128–132, 235–243
Vorontsov-Veljaminov interacting galaxies, 123

warm dark matter (WDM), 159, 160
weak gravitational lensing, 101, 165
Westerbork Synthesis Radio Telescope, 99
Western Allies, 22
Winter War, 20
WMAP, 154
World War I, 18, 19, 96, 143, 174
World War II, 20–22, 31, 47, 98, 174

X-ray clusters, 101
X-ray gas, 165
X-rays, 100
XMM-Newton, 101

ZCAT, 124, 180, 181, 183, 188, 203
Zeldovich pancake, 107, 120, 125, 126, 140
Zwicky clusters, 124, 126, 127, 131–133, 135
Zwicky list of compact galaxies, 126

Name Index

Aarseth, S. J., 257
Abalakin, V., 286
Abell, G. O., 8, 115, 118, 203, 204, 210, 212, 226, 228, 290, 309
Abrams, N., 291
Agekian, T., 286
Alexander I, Tsar, 18
Allen, C., 63
Alpher, R. A., 151, 152
Ambartsumian, V. A., 13, 38, 70, 72, 73, 86, 122, 141, 142, 218, 286
Andernach, H., 212, 213
Antonov, V. A., 91
Aragon-Calvo, M. A., 227, 237, 238, 241
Arnalte-Mur, P., 245, 246, 275
Arneodo, F., 285
Arp, H., 294
Atrio-Barandela, F., 282

Baade, W., 7
Babadzanjants, M., 292
Babcock, H. W., 47, 51, 88, 97, 307
Bahcall, J. N., 50, 197
Barclay de Tolly, M.A., 18
Bardeen, J. M., 194
Barenboim, D., 36
Baryshev, Y. V., 281
Bekenstein, J., 164
Bell, J., 305
Bender, R., 104
Bennett, G. F., 154
Binney, J., 1, 157

Bisnovatyi-Kogan, G. S., 155
Blaauw, A., 295, 298
Blanchard, A., 244
Blumenthal, G. R., 157, 158, 307, 308, 310
Bobrova, N. A., 91
Bok, B., 59, 204, 310
Bond, J. R., 157, 265
Bosma, A., 99, 159, 307
Brandt, J. C. , 61
Bringmann, T., 162
Broadhurst, T. J., 201, 203, 242
Bronshtein, M., 173
Brosche, P., 217
Brueck, H. A., 155
Bullock, J. S., 102
Buote, D. A., 104
Burbidge, G., 61, 88, 93, 98, 293, 294
Burbidge, M., 14, 61, 88, 98, 287, 294

Calzetti, D., 173, 190
Cen, R., 198
Centrella, J., 184, 260
Chemin, L., 57, 103
Chernin, A., 89, 91, 94, 155, 281
Chiang, L., 267
Chincarini, G., 120, 138, 309
Clowe, D., 165
Colberg, J. M., 237
Coles, P., 267
Colless, M., 309
Contopoulos, G., 81, 251, 287, 288
Corwin, H., 212, 228

345

Cowie, L. L., 206, 226
Coyne, G., 172
Creze, M., 50
Croton, D. J., 231, 264
Curtis, H., 4

Davis, M., 199, 262, 309
de Blok, W. J. G., 103, 104
de Lapparent, V., 205, 226, 230, 309
de Sitter, W., 5
de Vaucouleurs, A., 122
de Vaucouleurs, G., 8, 9, 118, 122, 124, 134, 203, 204, 226, 228, 309
de Voogt, A. M., 98
de Zeeuw, T., 74, 117
Dekel, A., 197, 219
Dhar, B. K., 57, 105
Di Matteo, T., 264
Dicke, R. H., 153
Dimov, N., 77
Dolgov, A. D., 161
Donato, F., 105, 106
Doroshkevich, A. G., 95, 107, 124, 125, 128, 152, 184, 259, 307, 309
Dressler, A., 218

Eastwood, J. W., 258
Eddington, A. S., 3, 10, 11
Eelsalu, H., 49, 118, 143
Efstathiou, G., 201, 260, 262–264
Eggen, O. J., 10, 59, 106, 121, 307
Einasto, M., 96, 144, 191, 213, 216, 220, 223, 230, 234–240, 242, 253, 263, 267, 275
Einasto (Tiit), L., 35–37, 113
Einstein, A., 5
Eisenstein, D. J., 244, 275, 309
Eke, V. R., 208, 209
Ewen, H., 98

Faber, S. M., vii, 65, 87, 99, 108, 150, 157–159, 217, 219, 307
Fairfall, A. P., 139
Fall, M., 192
Fesenko, B. I., 96
Field, G. B., 151

Finkbeiner, D. P., 162
Fisher, J. R., 134, 230
Flynn, C., 50
Ford, W. K. J., 61, 62, 88, 98
Forman, W., 100, 307
Freeman, K. C., 63, 104
Frenk, C. S., 158, 260, 262, 263
Friedmann, A., 5

Gallagher, J. S., v, 87, 100, 150
Gamow, G., 34
Gao, L., 105, 264
Geller, M. J., 204, 206–208, 226, 305, 310
Giacconi, R., 101, 203
Gilmore, G., 50, 106, 306, 307
Glebovskaya, V., 173
Gold, T., 6
Gorbachev, M., 23, 167, 171, 172, 291
Gott, III, J. R., 185, 188, 309
Graham, A. W., 103
Gramann, M., 96, 186, 187, 192, 199–201, 206, 253, 262–264, 273, 309
Grechko, G., 79
Green, A., 171
Gregory, S. A., 120, 136, 138, 139, 205, 310
Grinberg, M., 36
Grossberg, M., 23
Gunn, J. E., 262
Gursky, H., 100, 307
Gustav Adolf, 20, 37
Guth, A. H., 254, 309
Guzzo, L., 191

Hänni, L., 247–250
Hallimäe, R., 28
Harm, R., 14
Harrison, E. R., 259
Haud, U., 109
Hawking, S., 255
Heinämäki, P., 296
Hektor, A., 162, 163, 302
Herschel, J., 8
Hershel, W., 8
Hewish, A., 305
Hill, E. R., 49, 306

Name Index

Hitler, A., 20
Hockney, R. W., 258
Holberg, J., 100
Holmberg, J., 50, 87
Hoyle, F., 5, 14, 293
Hubble, E. P., 4, 5, 7, 309
Huchra, J. P., 7, 100, 183, 188, 203, 204, 206, 208, 226, 228, 305, 310
Humphrey, P. J., 104
Hütsi, G., 162, 244, 245, 275, 295, 297, 301, 309

Idlis, G., 52, 53
Illingworth, G., 63
Ilves, T. H., 300

Jürgenson, K., 168
Jõeveer, M., 49, 89, 107, 118, 120, 126–128, 130, 132–135, 137–139, 181, 183, 185, 201, 206, 212, 224, 226, 228, 238, 275, 290, 309, 310
Jannsen, J. V., 19
Jaaniste, J., 91, 93, 95, 117, 121, 126, 150, 310
Jackson, R. E., 65, 99, 217, 219
Jakobson, C. R., 19
Jeans, J. H., 46, 52, 83, 307
Jones, B. J. T., 190, 191, 267, 295, 297

Kõiv, M., 75
Kaasik, A., 63, 64, 89, 117, 140, 217, 291
Kahn, F. D., 85, 87, 96, 148, 151, 307
Kaiser, N., 194, 309
Kaljuste, T., 132
Kallis, A., 76
Kalnajs, A. J., 110
Kapteyn, J. C., 4, 46, 88, 307
Karachentsev, I. D., 281
Kellogg, E., 100
Keres, H., 37, 38, 43, 87
Kerr, F. J., 61
Khalatnikov, I., 173
Kharadze, E., 86, 95, 96, 286
Kiang, T., 9
Kim, J., 232
King, I., 97, 291

Kipper, A., 31, 37–39, 41, 43, 72, 78, 79, 87, 112, 113, 116–118, 286, 288
Kirshner, R. P., 136, 242, 290, 309
Klypin, A., 158, 160, 176, 180, 184, 191, 197, 259, 260, 263, 295
Kofman, L., 153, 250, 253, 257, 261, 262, 274
Komatsu, E., 154
Komberg, B. V., 90, 94, 95, 151
Koppel, M., 23
Kopylov, A. I., 197
Kormendy, J., 104
Kreitschmann, J., 109
Kruus, H., 22
Kuhn, T. S., 1, 93, 306
Kuijken, K., 50
Kukarkin, B., 10
Kundic, T., 283
Kuperjanov, A., 168
Kuperjanov, J., 19
Kutuzov, S., 55, 56, 58, 286
Kuzmin, G., 1, 32–35, 37–40, 43, 47–50, 52, 53, 55, 58, 59, 69, 70, 72, 73, 116, 117, 286, 287, 298, 306, 307

Laigo, R., 72
Landau, L., 261
Lauristin, M., 169, 173, 249, 250
Lee, S. M., 100
Leedjärv, L., 74, 252
Lemaître, G., 5
Lifshitz, E., 261
Lightman, A., 140
Liigant, M., 70
Liivamägi, L. J., 215, 217, 230, 301, 309
Lin, D. N. C., 159
Lindblad, B., 9, 10, 51, 287
Linde, A. D., 253, 255–257, 261, 295, 309
Lindner, U., 136, 235, 238, 275
Lippmaa, E., 169, 174, 248
Longair, M., 129, 133, 136
Lovell, M. R., 160
Lubimov, V. A., 155
Ludlow, A. D., 103
Lundmark, K., 5, 7
Lynden-Bell, D., 61, 88, 255

Männil, A., 114
Müller, V., 269, 296, 297
Mandelbrot, B. B., 189, 190
Markarian, B., 127
Martinez, V., 190, 191, 209, 215, 233, 267, 295, 296
Marx, K., 155
Masing, U., 73
Massevich, A., 34, 71
Materne, J., 69, 88, 96, 265, 307
Mather, J. C., 153, 305, 307, 308
Mayall, N. U., 51
Melott, A. L., 158, 184, 186, 188, 194, 260–262, 309
Meri, L., 16, 300
Merritt, D., 102
Mikhailov, A., 70, 72, 286
Miller, D., 180, 181, 183, 185, 225, 228, 230, 257
Milgrom, M., 164
Molotov, V., 20
Munsuk, C., 63
Mustel, E., 70, 72

Naan, G., 72
Narlikar, J., 293
Navarro, J. F., 102, 160
Neyman, J., 87, 307
Nikonov, V., 86
Novikov, I. D., 90, 94, 95, 151, 152, 155, 293

Ogorodnikov, K. F., 58, 72, 286
Oh, S.-H., 103
Oja, T., 123
Oke, J.B., 14
Oort, J. H., 9, 10, 46–51, 53, 55, 59, 88, 97, 147, 181, 182, 287, 306, 307
Öpik, E., 1, 4–6, 11–13, 28, 32–34, 45, 47, 48, 50, 51, 62, 81, 82, 88, 138, 188, 285, 307, 309
Ostriker, J. P., 92, 110, 150, 152, 198, 199, 206, 226, 262, 282, 305–307
Ozernoi, L. M., 91, 95, 120

Page, T., 87

Palm, V., 173
Parenago, P., 1, 10, 34, 38, 39, 49, 50, 53, 285
Parijskij, Y. N., 154
Park, C., 188, 201, 232
Parry, O. H., 160
Peacock, J. A., 201
Peebles, P. J. E., vii, 9, 92, 107, 110, 120, 137, 157, 176, 185, 230, 244, 257, 279, 284, 296, 309
Pelt, J., 283
Penzias, A. A., 152, 307
Perek, L., 51, 54, 287, 289
Perlmutter, S., 165, 280, 307, 308
Persic, M., 105
Peter I (the Great), 17
Pietronero, L., 190, 191
Pikelner, S., 150
Platen, E., 237, 238
Pogosyan, D., 250, 253
Polisensky, E., 160
Pontecorvo, B., 71
Poveda, A., 63
Praton, E. A., 230
Press, W. H., 198, 282
Primack, J. R., 156–158, 291, 310
Profumo, S., 285
Purcell, E., 98

Raidal, M., 161, 297, 301
Reed, D. S., 67
Rees, L., 22, 307
Rees, M. J., 108, 155, 156, 256
Refsdal, S., 282
Retana-Montenegro, E., 57, 103
Richter, S., 36
Ricotti, M., 160
Riess, A. G., 165, 280, 307, 308
Riives, V., 37, 43
Ristlaan, R., 76
Roberts, M. S., 61, 62, 85, 88, 98
Rohlfs, K., 109
Romano, S., 172
Rood, H. J., 120, 138, 309
Rootsmäe, T., 1, 10, 28, 29, 31, 33, 34, 37, 43, 47, 69, 70, 121

Rots, A. H., 98
Rubin, F., 120
Rubin, V. C., 61, 62, 88, 98, 99, 105, 119, 127, 159, 252, 305, 307, 310
Ruffini, R., 172, 189, 247, 299
Rümmel, U., 57, 61, 62, 67, 69, 109
Russell, H. N., 11
Rüütel, A., 170, 171, 249
Ryden, B., 273

Sérsic, J. L., 57
Saar, E., 78, 82, 89, 91, 93, 95, 117, 121, 126, 150, 158, 168, 184, 185, 189, 192, 194, 198, 209, 215, 233, 242, 263, 266, 267, 269, 289–292, 295, 296
Sahni, V., 233
Sakharov, A., 171, 173
Sales, L. V., 108
Salpeter, E. E., 64
Salucci, P., 106
Salvador-Solé, E., 103
Sandage, A., 7, 10, 96, 124, 203, 280, 309
Sanders, R. H., v, 87
Sapar, A., 75
Sarv, M., 144, 145, 165
Schechter, P., 198, 210, 222
Scheer, L. S., 61
Schmidt, M., 49, 50, 53, 55, 58, 64, 69
Schramm, T., 282
Schwarzschild, M., 14, 34
Seljak, U., 282
Sevastyanov, V., 79
Shandarin, S. F., 107, 124, 125, 158, 176, 180, 184, 185, 233, 257, 259, 260, 274, 295, 309
Shane, C., 9, 309
Shapley, H., 4, 8, 287, 309
Shaver, P., 230
Sheth, R. K., 241
Shevardnadze, E., 171
Shklovsky, J. S., 90, 247, 261, 299
Silk, J., 140, 151, 155, 197, 219
Sizikov, V. S., 70
Smith, S., 86
Smoot, G. F., 154, 305, 307, 308
Sobolev, V. V., 286

Soneira, R. M., 50, 137, 185, 197
Spinrad, H., 62, 107
Springel, V., 103, 108, 160, 161, 215, 264
Stalin, J., 20
Starobinsky, A. A., 253, 254, 262, 295, 309
Steinhardt, P. J., 282
Stoica, R. S., 227
Strömgren, B., 287
Strom, R. G., 97
Stromberg, G., 9
Struve, F. G. W., 20, 37, 70, 286
Su, M., 162, 163
Suburg, L., 26
Suhhonenko, I., 216, 269–271, 279, 301
Sulla, O., 28
Sunyaev, R., 244, 295, 301, 305
Szalay, A. S., 155

Tago, E., 117, 126, 136, 176, 183, 209, 210, 212, 213, 222, 225, 227
Tamm, A., 57, 110, 301
Tammann, G. A., 7, 69, 88, 96, 124, 203, 265, 291, 307, 309
Tarenghi, M., 136, 138, 228
Taylor, B. J., 62
Tchaikovsky, P. I., 76
Tempel, E., 110, 111, 162, 163, 210, 211, 215, 219–222, 227, 283, 301, 307, 308
Tenjes, P., 66, 85, 109, 117
Thompson, L. A., 120, 138, 139, 205, 310
Tifft, W. G., 136, 138
Tiit (Humal), E., 40, 42, 217
Tiit, V., 35, 40, 41, 77, 170
Tinsley, B. M., 64, 67, 107, 150, 262, 290, 307
Toomre, A., 13, 21, 107, 110, 225, 287, 298
Toomre, J., 107, 110, 225
Traat, P., 117, 168
Trachternach, C., 103
Tremaine, S., 1, 137, 157, 306
Trimble, V., v, 87, 256
Tully, R. B., 134, 181, 230, 281
Turner, M. S., 262, 282

Unt, V., 79

Väljas, V., 167, 170, 171
van de Hulst, H. C., 61, 97, 307
van de Weygaert, R., 227, 237, 238, 241, 275
van den Bergh, S., 7, 87
van der Laan, H., 252, 293
van Kampen, E., 238, 275
van Rhijin, P. J., 4, 88
van Woerden, H., 133
Veltmann, Ü., 118
Vennik, J., 117, 140, 207
Vilbaste, E., 248
Vilbaste, J., 248
Villems, R., 28, 168
Villmann, C., 78, 79
Vogeley, M. S., 188, 201, 230, 237
von Baer, K. E., 18, 20
von Bellingshausen, F. G., 18
von Ribbentrop, J., 20
Vorontsov-Velyaminov, B., 95

Wajda, A., 22
Walter, F., 103
Weniger, C., 162, 307, 308
West, R., 95
White, S. D. M., 108, 184, 194, 199, 207, 236, 260–263, 285, 298, 307, 309

Williams, L. L. R., 57, 105
Wilson, R. W., 307
Wirtanen, C., 9, 309
Wirtz, C., 5
Wolf, J., 104
Woltjer, L., 85, 96, 148, 151, 307
Wyse, A. B., 51

Yeltsin, B., 173, 248, 249
York, D. G., 309
Yu, J. T., 244, 258, 309

Zaldarriaga, M., 282
Zasov, A. V., 95
Zeldovich, Y. B., 70, 71, 90, 91, 95, 96, 107, 120, 121, 124, 127, 128 136, 140, 152, 154, 155, 157, 175–180, 182, 184, 185, 188, 190–192, 198, 206, 212, 225, 226, 229, 244, 247, 255, 258–260, 262, 265, 295, 309
Zwicky, F., 8, 15, 85, 86, 90, 96, 120, 122–124, 135, 147, 282, 287, 306, 307, 309